The language issue in the teaching of mathematics in South Africa

Published by African Sun Media under the SUN PReSS imprint

This publication was subjected to an independent double-blind peer evaluation by the publisher.

Based on the thesis titled *Multiple levels and aspects of language competency in English and intermediate phase mathematics teachers: An analysis of case of the Eastern Cape Province.*
Promoter: Professor M. L. A. Le Cordeur.
Thesis published: December 2017.

First edition 2020

ISBN 978-1-928480-96-9
ISBN 978-1-928480-97-6 (e-book)
https://doi.org/10.18820/9781928480976

Set in Chaparral Pro 11/13

Cover design, typesetting and production by African Sun Media

SUN PReSS is an imprint of African Sun Media. Scholarly, professional and reference works are published under this imprint in print and electronic formats.

This publication can be ordered from:

orders@africansunmedia.co.za
Takealot: bit.ly/2monsfl
Google Books: bit.ly/2k1Uilm
africansunmedia.store.it.si (e-books)
Amazon Kindle: amzn.to/2ktL.pkL

Visit africansunmedia.co.za for more information.

THE LANGUAGE ISSUE IN THE TEACHING OF MATHEMATICS IN SOUTH AFRICA

Intermediate Phase research from one province

Lindiwe Tshuma

SYNOPSIS

This book is a platform for further dialogue towards better serving learners in under-resourced multilingual settings. In this book, material is specifically intended for **specialist Intermediate Phase Mathematics lecturers** tasked to educate mathematics teachers who must deliver content in English or through translanguaging /code switching/code mixing. The book has a clear and straightforward writing style that pre- and in-service teachers will find extremely useful. It includes practical cases that should be helpful in illustrating applicable learning points as support for academics. The book draws on a broad international perspective and at the same time refers to examples from one of the South African Provinces: The Eastern Cape. The following points are key features of the book:

- *Levels of Analysis*: One of the characteristics of the text is the organisation of the text around the four-tiered conceptual model that distinguishes between the global, African sub-Saharan, South African and the Eastern Cape perspectives. After all, South Africa is still part of the global village and all the challenges it poses.
- *Writing Style*: Considerable effort was made to present the content in a simple, clear, and straightforward way. Extensive use of examples illustrates complex concept topics.
- *Comprehensive Literature Coverage:* A compelling feature of this book is its use of comprehensive and up-to-date coverage of international, regional and local information captured from academic books, journals and periodicals on mathematics education. Specifically highlighting information on the Eastern Cape Province illustrates the Southern African context and the challenges faced by mathematics teachers in under-resourced multilingual settings. While extensive literature highlights learners' contexts and challenges, the book fundamentally highlights teachers as key stakeholders in any education system.
- *Research and Information:* In order to illustrate and explain concepts and topics, topical comprehensive case studies that depict what is happening on the ground, examples from real classrooms, are included in this book. This brings the themes, topics, and concepts closer to pre-service teachers while raising interesting debates on these matters to in-service teachers, teacher educators, curriculum developers and researchers alike. A value-add for academics, researchers and curriculum planners is that it poses a challenge to plug the gaps in current research.
- *Pedagogy:* From an instructional perspective, this book features sections on pedagogical implications, using real-time examples that highlight real cases of what is currently happening in multilingual under-resourced mathematics

classrooms, against the backdrop of the country's continued hefty investments and numerous strategies to improve the quality of education and be on par with other countries worldwide.

I trust that you will enjoy an innovative volume designed to highlight language and mathematics as pivotal parts of one whole, for the purposes of bringing a new twist and propelling the on-going debate on language in mathematics education.

ABOUT THE AUTHOR

Dr Lindiwe Tshuma is currently a Teaching and Learning Advisor at University of the Witwatersrand Faculty of Humanities and a Research Fellow at University of Stellenbosch Curriculum Studies Department. Previously, she was a lecturer, a Primary Mathematics Specialist at the Schools Enrichment Centre of the African Institute for Mathematical Sciences. Before becoming a Teacher Educator, she was a Curriculum Development Specialist responsible for developing Mathematics and Science teaching and learning resources for Dinaledi Schools in Gauteng Province and taught Mathematics and English in various under-resourced public schools in South Africa for several years. Her interest in educational linguistics emerged from her experience as an Intermediate Phase (IP) mathematics and science teacher within the deaf section of a school for the deaf and blind in Thaba Nchu, Free State Province. The experience of presenting mathematics and science content through languages foreign to both the teacher and learners made her realise the importance of harnessing together subject content and the language of instruction, rather than treating them as separate entities. Her research thus focuses on teaching and learning in multilingual settings, including the signposting of language issues in mathematics education for the purposes of better serving the learners.

ACKNOWLEDGEMENTS

I owe a debt of gratitude to a number of institutions and people who have contributed immensely to the book.

A great thank you to all the case study participants in this book: 55 Intermediate Phase mathematics teachers from the Eastern Cape Province and 10 Intermediate Phase mathematics lecturers from various university departments of education in South Africa. The participants in the case studies are not merely objects of study, but are very reliable and crucial knowledge partners.

A big thank you to the following institutions that contributed to the success of the project: University of Stellenbosch, African Institute of Mathematical Sciences Schools Enrichment Centre (AIMSSEC), Joint Education Services (JET) and the Eastern Cape Department of Education (ECDoE).

Most sincere gratitude to the University of the Witwatersrand Faculty of Humanities for partly funding this book.

Finally, abundant appreciation goes to African Sun Media Publishing for their expertise and assistance throughout this project – their support was crucial.

BRIEF CONTENTS

TABLE OF CONTENTS

LIST OF ABBREVIATIONS AND ACRONYMS

AIMS	African Institute for Mathematical Sciences
AIMSSEC	African Institute for Mathematical Sciences Schools Enrichment Centre
AMESA	Association of Mathematics Educators of South Africa
ANA	Annual National Assessment
BICS	Basic Interpersonal Communication Skills
CALP	Cognitive Academic Language Proficiency
CAPS	Curriculum and Assessment Policy Statements
CPTD	Continuing Professional Teacher Development
CSC	Centre for Statistical Consultation
CUP	Common Underlying Proficiency
DBE	Department of Basic Education
DHET	Department of Higher Education and Training
DoE	Department of Education
ECDoE	Eastern Cape Department of Education
ELLs	English Language Learners
FET	Further Education and Training
FP	Foundation Phase
INSET	In-service Education and Training
IP	Intermediate Phase
ITE	Initial Teacher Education
JET	Joint Education Trust
LiEP	Language in Education Policy
LoLT	Language of Learning and Teaching
MRTEQ	Minimum Requirements for Teacher Education Qualifications
NEEDU	National Education Evaluation and Development Unit
NCS	National Curriculum Statement
NCTM	National Council of Teachers of Mathematics
REQV	Relative Education Qualification Value
SACE	South African Council for Educators
SACMEQ	Southern and Eastern African Consortium for Monitoring Educational Quality
SAQA	South African Qualifications Authority
SASL	South African Sign Language
SGB	School Governing Board
SP	Senior Phase
STEM	Science, Technology, Engineering and Mathematics
SUP	Separate Underlying Proficiency
TLDCIP	Teaching and Learning Development Capacity Improvement Programme
UNESCO	United Nations Educational, Scientific and Cultural Organisation

LIST OF ACRONYMS: Teacher Education Qualifications in South Africa

ABET	Adult Basic Education and Training
ACE[1]	Advanced Certificate in Education
ACT	Advanced Certificate in Teaching
ADE	Advanced Diploma in Education
BEd	Bachelor's Degree in Education
MEd	Master's Degree in Education
NPDE	National Professional Diploma in Education
PGCE	Postgraduate Certificate in Education
PTD	Primary Teacher's Diploma
STD	Secondary Teacher's Diploma

LIST OF ACRONYMS: Universities in South Africa

CPUT	Cape Peninsula University of Technology
CUT	Central University of Technology
DUT	Durban University of Technology
MEDUNSA	Medical University of Southern Africa
MUT	Mangosuthu University of Technology
NMMU	Nelson Mandela Metropolitan University
NWU	North-West University
RU	Rhodes University
SPU	Sol Plaatje University
SU	Stellenbosch University
TUT	Tshwane University of Technology
UCT	University of Cape Town
UFH	University of Fort Hare
UFS	University of the Free State
UJ	University of Johannesburg
UKZN	University of KwaZulu-Natal
UL	University of Limpopo
UMP	University of Mpumalanga
UNISA	University of South Africa
Univen	University of Venda
UP	University of Pretoria
UWC	University of the Western Cape
UZ	University of Zululand
VUT	Vaal University of Technology
Wits	University of the Witwatersrand
WSU	Walter Sisulu University

1 The Advanced Certificate in Education (ACE) was phased out in 2014.

FOREWORD

What is the relationship between teacher language competency and mathematics instruction?

Teacher language competency is not well-researched partly because as custodians of knowledge, teachers are generally expected to be competent in both the language of instruction and the content to be taught. With Mathematics being the high-risk subject that it is, it is worth investigating the relationship(s) that exist between mathematics instruction and teacher language competency in the language of instruction.

Language plays a crucial role in content delivery. In a mathematics classroom, language is used to construct meaning, to express understanding and to develop mathematical thinking skills. However, there is the danger that this role may be interpreted superficially without taking into consideration some underlying linguistic factors that may interfere with the understanding of mathematics. In the Intermediate Phase (IP) years, these factors include, among others, the teacher's mastery of the Language of Learning and Teaching (LoLT), the switch from mother-tongue instruction at Foundation Phase (FP) to the use of English as the LoLT (at IP level and beyond), and the teacher's use of code switching and translanguaging to facilitate meaningful mathematics instruction. It therefore follows that for improved mathematics instruction teachers need first to acknowledge and recognise the importance of language in mathematics learning and teaching, as well as being proficient in the LoLT. Only then can the quality of mathematics instruction improve.

To influence mathematics instruction in a positive way, specific initiatives need to focus on the change in the language of instruction at IP level. Initiatives need to target teachers of mathematics with a broader view to developing mathematics teachers who are better able to handle the linguistic barriers currently challenging both teachers and learners in the IP years.

Interrogating how (***if at all***) language competency in the LoLT relates to content delivery by IP mathematics teachers seeks to:

i. Determine the extent to which IP mathematics teachers in the ECDoE schools are proficient in English, the prescribed LoLT;
ii. Reflect on the extent to which proficiency in the LoLT relates to IP mathematics teachers' content delivery;
iii. Interrogate IP mathematics teachers' pedagogical content knowledge regarding language-based mathematical problems;
iv. Highlight the teaching methods currently in use in IP classrooms that may overcome language barriers in the delivery of mathematics;

v. Reflect on the aspects of IP mathematics teacher education focusing on the use of English as LoLT;
vi. Extract lessons that empower mathematics teachers to effectively overcome language barriers in content delivery; and
vii. Inform state decision making on Intermediate Phase mathematics teacher education curriculum design.

Comprehensive case studies are used to prove and disprove the extensive literature on mathematics teaching and learning in multilingual settings for the purposes of highlighting gaps that still need to be plugged. Fifty-five willing IP mathematics teachers in under-resourced schools of the Eastern Cape Province (***the key informants to the case studies referred to in this work***) were tested and observed while presenting lessons, taking into consideration the environment in which they conducted their lessons. The researcher had access to the teacher participants for the collection of quantitative data (questionnaires and teacher assessments) while the teachers attended Continuous, Professional Teacher Development (CPTD) courses lectured by the researcher. In addition, the researcher established a rapport with the teachers for the purposes of collecting qualitative data (classroom observations) at the schools where the participants are deployed.

Since teaching and learning is a two-way process, 10 teacher educators from different South African University Departments offering IP mathematics teacher education were interviewed to cross-check the information collected on the ground. Being a teacher educator, the researcher also had easy access to other teacher educators for the purposes of collecting data to cross-check the information provided by the teachers. The researcher accessed fellow teacher educators during conferences on mathematics teacher education. Data sets from the different sources were collected concurrently, analysed separately and then merged to produce the final findings and thereafter the recommendations.

Most language in education research has been conducted in well-resourced economies such as those of the United States and Europe. Although the studies conducted in well-resourced settings highlight the regularities and irregularities of language in education, it would be naïve to assume that what holds true in well-resourced settings also holds true in under-resourced settings. Language in education (like other fields) is subject to influences of national culture because people use 'rules' that derive from culture to inform decisions, to guide their behaviour, and to determine their response to other individuals (Chen, Leung & Chen, 2009; Sidle, 2009). Assuming that what is found in one research context does not necessarily apply to other contexts, researchers need to acknowledge an increase in the number of connections between individuals, institutions and countries across different local, national and international contexts – globalisation (Lambell, Ramia, Nyland & Michelotti, 2008). To understand how language influences IP mathematics instruction in the Eastern Cape Province of South Africa, local, regional and international contexts are referred to in this book.

The time at which research is performed is of great relevance because different timeframes have their specific dynamics that shape educational practices. This book was written more than 20 years after South Africa's democratic dispensation. In educational settings, 20 years is a relatively long time to plan, implement and evaluate educational policies. This project therefore interrogates whether the language practices in multilingual IP mathematics classrooms (together with the current linguistic preparation of IP mathematics teachers), are in line with producing learners who are job- and future-ready in the succeeding 20 years.

Language in education research has several audiences, and these audiences are not all interested in the same aspects of education. Different audiences require different problem solutions ranging from very abstract to very concrete; (McKelvey, 2003). The following is a non- exhaustive list of the target audiences in mathematics education:

- pre-service teachers;
- in-service teachers;
- teacher educators;
- didacticians,
- curriculum developers;
- researchers (including institutional researchers); and
- policy makers.

What these audiences have in common is their quest to know about language issues in mathematics education. The difference between the audiences is that they do not have a shared criterion or set of shared criteria against which they evaluate a mathematics teacher education institution on how it is promoting language use or what is relevant to know about teacher linguistic practices in under-resourced multilingual mathematics classrooms in South Africa. In other words, the audiences look at language in mathematics education in multilingual settings for different reasons and for solutions to different challenges.

Scope demarcates the boundaries of what we are studying about language in mathematics education. Scope refers to two issues: the level issues and the type of system under study. Level largely refers to micro (small units that can be studied, such as individuals) and macro (large aggregates such as institutions or networks of institutions). This book's key study participants are individuals: IP mathematics teachers and IP mathematics teacher educators. The book is limited to IP mathematics education in South Africa, specifically focusing on grades 4 – 7. The work also draws on the current university-level preparation of Intermediate Phase mathematics teachers in South Africa. The participants (teachers and teacher educators) ranged from newly qualified, mid-career to highly experienced professionals/experts in mathematics education.

The book is divided into ten chapters:

Chapter 1 presents the quality of South African mathematics education by first exposing the historical perspectives of South African education before detailing a contemporary perspective of South African mathematics education. The quality of South African mathematics education is viewed through a global lens by analysing international assessment on education as well as through a local lens by analysing learner performance trends in mathematics, specifically in the Eastern Cape Province.

Chapter 2 portrays language as a transparent resource in mathematics pedagogy for better serving English Language Learners (the 90% of South African learners learning mathematics in English, when English is not their first language) who face a triple challenge: learning English, learning mathematics and learning the mathematics register. The chapter briefly describes language dynamics in South Africa; identifies linguistic features influencing mathematics instruction; and also offers specific pedagogic strategies for managing ambiguity.

Chapter 3 raises the issue that since both mathematics and the teaching and learning of mathematics are symbolic practices that include inventing and use of symbols, the main function of these symbolic practices is to facilitate the communication of mathematical procedure by both the teacher and the learners. Semiotic resources are presented as contemporary tools for improving communication and enhancing meaning making, which are not a replacement but an extension of natural language.

A: Intermediate Phase Mathematics Teachers' Linguistic Competencies in the Language of Learning and Teaching: The Eastern Cape Province

---

Chapter 4 focuses on word problems which belong to the most difficult and complex problem types that learners encounter during their primary school level mathematical development. In the classroom setting, word problems are often viewed as mere calculations; however, recent research shows that a number of linguistic verbal components not directly related to arithmetic contribute greatly to their difficulty. This chapter presents word problem difficulties and provides an instructional model for solving word problems.

Chapter 5 presents mathematics teaching and learning in a second language. In contemporary South Africa, pedagogical translanguaging has recently overtaken code switching, code mixing and borrowing as pedagogical strategies used in multilingual settings. The chapter also explores different theories of bilingualism and second-language acquisition including how culture and socio-political issues impact on mathematics teaching and learning in a second language.

B: Intermediate Phase Mathematics Teachers' Pedagogical Content Knowledge Regarding Word Problems: The Eastern Cape Province

---

Chapter 6 highlights language competency levels that are necessary to support meaningful mathematics instruction (specifically in multilingual settings), by unpacking the relevant language theories. The chapter outlines various language theories including language development, acquisition and usage, and attempts to determine the effect of these language theories in developing language competency levels that support mathematics teaching and learning.

Chapter 7 begins by defining a language of learning and teaching (LoLT) before reviewing perspectives on LoLT based on studies conducted globally, in sub-Saharan Africa, in South Africa and in the Eastern Cape Province of South Africa. The chapter reviews studies that investigate the language of teaching and learning across the curriculum, before zooming in on studies reviewing the LoLT in relation to mathematics education.

C: Linguistic Practices in Intermediate Phase Mathematics Classrooms: The Eastern Cape Province

Chapter 8 examines Multilingual Based Bilingual Education (MTBBE) for Mathematics in the Eastern Cape Province and specifically focuses on a current project taking place in more than 50 schools of the Eastern Cape Department of Education where isiXhosa is used to teach mathematics to improve the recurrent poor performance. This chapter highlights the advantages of the initiative and interrogates issues of impact, sustainability, effectiveness, efficiency, relevance and replication.

Chapter 9 flips the coin by focusing on mathematics teacher education for multilingual settings. This chapter provides an overview of practices in teacher education institutions in some European and African countries that have policies and programmes preparing teachers meant for multilingual classrooms. The chapter also reviews teacher education's efforts in supporting pre- and in-service teachers.

D: University Education Departments' guidance on code switching and pedagogical translanguaging: South Africa

Chapter 10 is a consolidation of the scene-setting section: pertinent issues regarding teacher language competency for mathematics instruction versus the empirical evidence from the comprehensive cases section. Consolidation of the two informs the lessons learnt and suggests and proposes areas for further research in the broad field of language in mathematics education, specifically focusing on the IP mathematics teacher. The thrust of the chapter is on developing mathematics teaching as a way of improving the quality of mathematics education in South Africa.

This book is not a systematic exposition of the aforementioned challenges. It can be better understood as a modest continuation of previous research undertaken by numerous scholars who, directly or indirectly, have shown the potential of language proficiency in the broad field of language in mathematics education, specifically in multilingual settings. It is also a platform for future research in the field.

The book intends to broaden and widen the understanding of the relationships between teacher language competency and IP mathematics instruction in under-resourced communities of the Eastern Cape Province of South Africa. If in contemporary South Africa there still exist teachers who are not fluent in English but use English as a language of instruction, there is a deficit in the delivery of mathematics content in the IP years – and this book intends to address that deficit. Furthermore, if the linguistic challenges faced by these IP mathematics teachers are not addressed this too will contribute to far-reaching consequences, not only for the education system but also for the learners produced by these teachers as well as for the general economic growth of the country.

There is a dire need for further dialogue (***and apolitical action***) on appropriate interventions to facilitate effective delivery of mathematics content in the prescribed LoLT in the IP years. However, ***Language is political!*** (*One may say*). Although an apolitical approach is ideal, the reality on the ground reveals a mutual relationship between language and politics. This relationship entails how language issues affect political issues, and how political decisions affect linguistic developments – such as the decisions about the LoLT in secondary schools in the former Bantu Education system of the 1970s. If an apolitical approach is what we need, yet a political approach is what we find ourselves in, maybe it is time to pursue a ***semi political*** approach to language specifically designed to shrink the gaps in the delivery of IP mathematics content in a language foreign to both the teacher and the learner.

CHAPTER 1

THE QUALITY OF SOUTH AFRICAN MATHEMATICS EDUCATION

INTRODUCTION

What is the quality of South African mathematics education?

The teaching of mathematics in South Africa is amongst the worst in the world (Spaull, 2013) and research reveals increasing enrolments into private extra mathematics tuition classes partly in response to poor mathematics teaching and learning in public schools. However, these additional efforts are not enough to plug the gaps in mathematics instruction. There is a need for drastic improvements on mathematics teaching and learning – improvements which will benefit learners in public schools as well as the country's economic growth. A deeper understanding of the quality of South African mathematics education, in particular, requires some foresight into the history of the country's education system in general.

SOUTH AFRICAN EDUCATION: A HISTORICAL PERSPECTIVE

The current education crisis must be understood in the context of South Africa's history. In its third decade of democracy, South Africa's transformation of basic education is still underway and the vestiges of apartheid still hamper learners' ability to learn. Although it is in the colonial era that universities were established, the apartheid era effectively divided education, had a language policy built for separate development, unequal resources and a cognitively impoverished curriculum that resulted in the majority of the population being illiterate and innumerate (Nelson Mandela Foundation, 2004; Heugh, 2000; Navsaria, Pascoe, & Kathard, 2010).

After apartheid, one of the key changes that occurred as a direct consequence of the Schools Act in South Africa was racial desegregation, which resulted in upward social mobility of some households and consequently the migration of learners. The flow of learners resembled the apartheid racial hierarchy: Black learners migrated to schools that had previously been open only to Indian, white or coloured children, while coloured learners migrated to Indian and white schools, and Indian learners moved to white schools (Plüddemann, Xola & Mahlahela-Thusi, 2000). Classrooms transformed and became linguistically diverse – a cause for celebration – yet with no deployment of appropriately qualified African-language-speaking teachers to the relevant schools, communication difficulties between teachers and learners increased (Plüddemann et al., 2000). In the community in which this research was conducted, African learners had migrated to previously coloured schools. As a consequence, learners as well as teachers frequently came from different language backgrounds, resulting in diverse languages in the classroom (Walton, 2002). However, the language of learning and teaching (LoLT) remained English, resulting in many learners learning in a foreign language, i.e. a language that is not their mother tongue and which is often unknown to them as they had little exposure to English outside of school. This situation created numerous teaching and learning challenges contributing to low learner achievement.

In addition to the communication challenges, many of these previously black and coloured schools continued to be overcrowded and under-resourced (human and capital). Infrastructure remained inadequate, while previously white-only schools continued to be relatively well resourced (Soudien, 2008). In under-resourced communities, such as the Eastern Cape, teachers continued to encounter a large number of learners with many socio-economic problems that contributed to their learning difficulties. In both rural and poor urban schools there is a shortage of classrooms, teachers, and basic teaching and learning resources, such as stationery and textbooks, combined with poor basic needs like water and electricity (Nelson Mandela Foundation, 2004).

SOUTH AFRICAN MATHEMATICS EDUCATION: A CONTEMPORARY PERSPECTIVE

Decades after the apartheid era, the quality of South African mathematics education in particular is still significantly low (Global Competitiveness Report for the World Economic Forum 2013/2014; Trends in International Mathematics and Science Study (TIMSS), 2013) and this low quality is well documented. Reasons cited for the low quality of mathematics education are both educational and non-educational. Non-educational reasons include school management and socio-economic factors (Taylor & Taylor, 2013; Spaull, 2013). The key academic reasons include: lack of content knowledge of in-service teachers, lack of quality resource material support, mathematical skills deficit of learners as they progress through the school phases, misguided foci of assessment, curriculum not being fit for purpose, mathematics being a difficult subject, negative societal attitudes towards mathematics, the quality of teaching being substandard and poor competency in the Language of Learning and Teaching (LoLT). According to Van Staden and Bosker (2014: 2), the poor performance of South African learners as shown in international comparative studies stems from, among other things, poor communication between learners and teachers in the LoLT.

Of the various reasons cited, this book highlights mathematics teachers' competency in English, specifically in the Intermediate Phase years, as one of the most significant predictors of performance in mathematics in relation to the future development of the quality of the South African education system. The book analyses the relationship between language competency and the delivery of mathematics content; this analysis is intended to inform curriculum design so that primary school mathematics teachers are linguistically equipped with skills to effectively deliver content in the prescribed language of instruction.

The five major components of the South African education system are early childhood development, ordinary school education, special needs education, further education and training colleges, and universities. Of these, the largest is ordinary school education, which extends from Grade R (reception year) to Grade 12. Between Grade R and Grade 12, there are four phases: Foundation Phase (FP) consisting of Grades R to 3, Intermediate Phase (IP) consisting of Grades 4 to 6, Senior Phase (SP) consisting of Grades 7 to 9, and the Further Education and Training (FET) Band consisting of Grades 10 to 12. Figure 1.1 illustrates the distribution of learners in each of these phases in 2013.

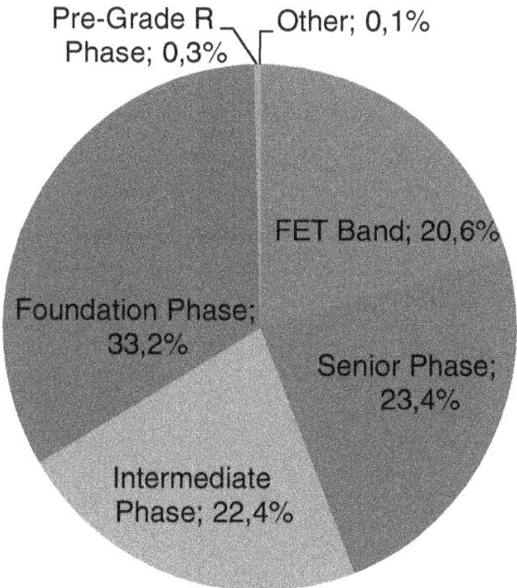

Figure 1.1: Percentage distribution of learners in ordinary schools by phase in 2013 (DBE, 2015)

Since 2000 there have been several initiatives internationally to improve the quality of education, but still, education for all has not yet been achieved (Adesina, 2017). Similarly, various stakeholders in the South African education system have made efforts to improve the quality of learning and teaching in the FET and FP years, with little attention being devoted to the IP years, although it is during these years that a number of significant changes take place, the foremost being the change in the LoLT. According to Tshabalala (2012: 22), "The majority of teachers and learners in South African schools are not first-language speakers of English, and ... many learners and teachers are not fluent in English". However, from the IP level upwards, the LoLT in 80% of the ordinary schools in all subjects, including mathematics, is English.[1] One of the negative effects of poor mastery of the English language by both learners and teachers is that teaching and learning as well as assessment are compromised.

[1] English- and Afrikaans-speaking learners learn in their home language from Grade 1 to Grade 12 and do not switch in Grade 4.

INTERNATIONAL ASSESSMENTS OF EDUCATION IN SOUTH AFRICA

The education evaluation function in South Africa is regulated in terms of section 4 of the Education Act of 1996, which provides for the national minister to determine national policy, for inter alia, monitoring and evaluation of the wellbeing of the education system (JET, 2010: 6). This provision is given effect in a number of other education policies and programmes. Examples of these are the WSE, IQMS, Developmental System (DAS) and Assessment Policy for the General Education and Training Band, which provides a systemic evaluation to be conducted on a national representative sample of learners.

The South African education system is characterised by low levels of academic performance on international assessments of education. Figure 1.2 is a representation of the quality of South African mathematics education in 2011, in comparison to other middle-income countries.

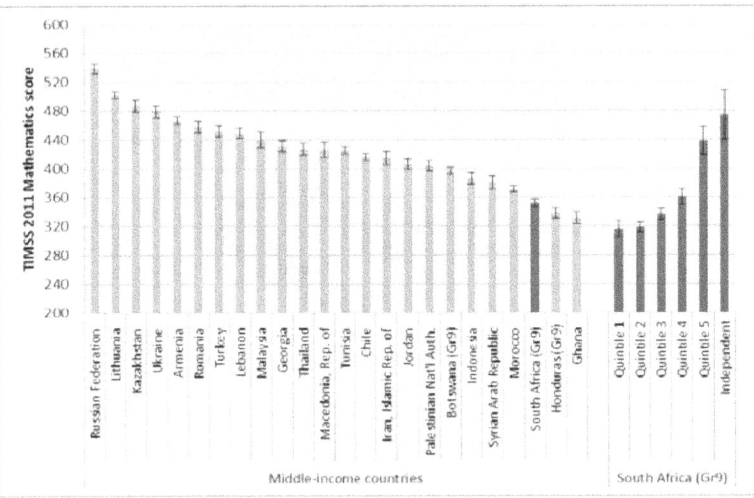

Figure 1.2: Average Grade 8 (South African Grade 9) mathematics achievement in 2011 in comparison to other middle-Income countries (Spaull 2013)

TIMSS is an international standardised test for mathematics and science. It is intended for Grade 8 learners. However, in South Africa it was decided that the tests were too difficult for South African Grade 8 learners; in 2011 only Grade 9 learners wrote the test (Spaull 2013). As illustrated in Figure 1.2, the average mathematics competencies at Grade 9 level in South Africa on the standardised test were almost at the same level as those of Morocco, Honduras and Ghana (the least performing country). Average mathematics competencies of South African Grade 9 learners was pegged at approximately 50% of the quality of Russian learners (the best performing country). When dividing South African schools into quintiles 1 - 5 (public schools) and private schools, the private schools' Grade 9 learners did much better than the average, but were still just under 90% of the average Russian Grade 8 mathematics achievement.

Table 1.1 indicates the South African education evaluation landscape. Four of the seven programmes are international comparative assessments, while the other three are national programmes implemented at various levels of the system.

Table 1.1: International and national evaluation initiatives in South Africa

	INSTITUTIONAL FUNCTIONALITY	TEACHER PROFILES	TEACHER KNOWLEDGE	TEACHING PRACTICE	LEARNER PERFORMANCE	SES
TIMSS[1]	Yes	Yes			Yes	Yes
PIRLS[2]	Yes	Yes			Yes	Yes
SACMEQ[3]	Yes	Yes			Yes	Yes
MLA[4]	Yes	Yes			Yes	Yes
Systemic Evaluation	Yes	Yes			Yes	Yes
WSE / SSE[5]	Yes			Yes*		
DAS[6]		Yes				
Total	6	6	0	1	5	5

Adapted from JET (2010)

Key:

* Intended but never implemented

** IQMS was omitted as it is comprised of WSE and DAS

1 Trends in International Mathematics and Science Study

2 Progress in International Reading Literacy Study

3 Southern and Eastern Africa Consortium for Monitoring Education Quality (regional assessment)

4 Monitoring Learning Achievement Study

5 Whole School Evaluation/School Self Evaluation

6 Developmental Appraisal System

The PIRLS surveys (2006, 2011) as well as the TIMSS (1995, 1999, 2003 and 2011) have perpetually reiterated South Africa as least performing compared to other participating counties. The NAEP revealed that grade ability levels in mathematics and science, reading and writing positioned South Africa at the lowest score among 46 other countries that participated in the tests. TIMSS results drew attention to the lack of academic potential at primary school level contributing to the fact that only 69% of school leavers entering South African Higher Education Institutions (HEIs) are prepared to cope with the academic demands of higher education studies (SAUVCA, 2003).

LEARNER MATHEMATICS PERFORMANCE: A CASE OF THE EASTERN CAPE PROVINCE

Why the Eastern Cape Province? Why single out one out of the eleven provinces in South Africa? one may ask. According to DBE (2015), 22.3% of the nation's ordinary schools are located in the Eastern Cape Province. The Eastern Cape and Limpopo Provinces' Departments of Education were placed under administration in 2011 for poor service delivery, with Mount Frere (Eastern Cape Province) being the lowest performing district in the country. Based on the 2013 Annual National Assessment (ANA) results, the bulk of IP learners in the Eastern Cape Department of Education (ECDoE) attained below adequate achievement levels in mathematics as shown in Figure 1.3:

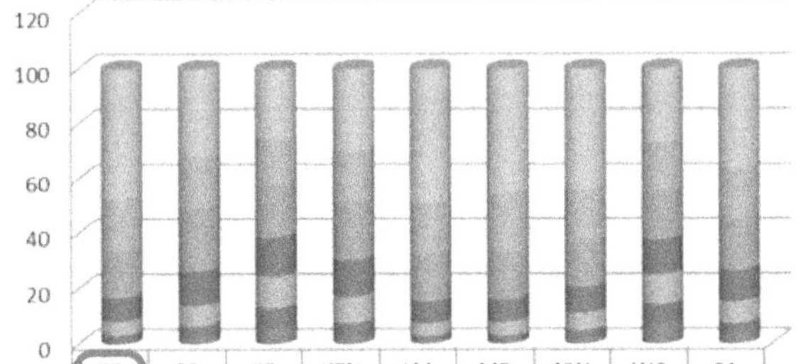

	EC	FS	GP	KZN	LM	MP	NW	WC	SA
Not achieved	47.5	31.7	25.6	30.4	49	45.6	44.1	27.1	36.2
Elementary achievement	19	19.3	16.7	18.6	18.9	21	19.1	17.9	18.7
Moderate achievement	16.7	22.5	19	20.6	16.7	17.3	16.2	17.3	18.6
Adequate achievement	8.5	12.4	14.1	13.3	7.9	8.1	9.4	12.7	11.3
Substantial achievement	5	7.7	11.5	9.5	4.7	4.9	6.3	10.7	8
Meritorious achievement	2	3.7	7.2	4.6	1.8	1.9	2.8	7.2	4.2
Outstanding achievement	1.2	2.6	5.8	3	1	1.1	2.1	7.2	3.1

Figure 1.3: Percentage of learners in various achievement levels for Grade 6 mathematics by province in 2013 (DBE, 2015)

As illustrated in Figure 1.3, if 47.5% and 20% of the learners are rated as *not achieved* and *elementary achievement* respectively, this implies that approximately 70% of the learners from the Eastern Cape (EC) province are rated below *moderate achievement* levels, and this is a cause for concern. Although this data was recorded in 2015, as of 2020, the situation has not changed much and if it has, it has deteriorated.

According to Somyo (2015) "... while there has been an improvement in the quality of education in the Eastern Cape, more needs to be done to address other education-related challenges". This book contends that IP teachers' competency in the LoLT is one of the education-related challenges that impacts on learner performance. The book also acknowledges the voluminous learner performance analyses that have been undertaken before by different educational stakeholders nationally and internationally; however, the focus will be on teacher competency in the LoLT, particularly in relation to IP mathematics instruction.

Most of the teachers referred to in this book are based in the Eastern Cape, Limpopo and KwaZulu-Natal provinces and are out-of-field, non-specialist teachers recruited into IP teaching positions to alleviate the nationwide shortage of mathematics teachers (Tshuma, 2017). The teachers enrol for in-service Advanced Certificate in Education (ACE)[2] and Advanced Certificate in Teaching (ACT) IP mathematics modules to upskill themselves. The majority of the teachers' proficiency in English, as evidenced in their written assignments (Tshuma, 2017), is below average and the book contends that this is a reflection of the quality of learning and teaching that takes place in the teachers' classrooms. Howie (2001) and Taylor (2008) believe

2 At the end of 2014 the Department of Higher Education and Training phased out the ACE programme and replaced it with the ACT programme.

that "teachers cannot teach what they do not know". Research analysing initial teacher education across five South African universities reveals that "the situation with respect to the language of learning and teaching, predominantly English, is of particular concern" (Taylor, 2015: 23).

The literature also addresses the challenges faced by teachers who are not fluent in English but use English as a language of instruction. The difficulty of communicating fluently in the prescribed LoLT leads to increased frustration for the teacher, a slow rate of learning, disciplinary problems and teacher-centred instruction (Setati, 1999; Howie, 2003). The prevailing mother-tongue versus second-language instruction debate highlights English as the gateway to global opportunities. Unless policy makers make drastic fundamental decisions, IP teachers will continue to struggle with linguistic barriers in the teaching of mathematics and other subjects. If English is increasingly used as the LoLT, as communities seem to desire, then the necessary support mechanisms – including intensive language training of teachers who will be using English as the LoLT – need to be put in place to facilitate a smooth transition from FP to IP years.

PEDAGOGICAL IMPLICATIONS: QUALITY OF SOUTH AFRICAN MATHEMATICS EDUCATION

The South African government has initiated numerous strategies addressing the crisis in the country's mathematics education:

- Mathematics is now a compulsory subject for all South African learners from Grade R up to Grade 9 level and at Further Education and Training (FET) level (Grade 10 – 12) learners are required to choose between pure mathematics or mathematical literacy as compulsory subjects.
- Curriculum Assessment Policy Statement (CAPS) developed and standardised nationally propel mathematics learning towards a subject that makes more sense to the learners, with clarified timelines and expectations for teachers.
- Investments into standardised assessments [Annual National Assessments (ANA)] administered nationally in Grades 1 – 6 and at Grade 9 between 2011 and 2014 to overcome the challenge of only having assessment data at the final exit point of the ordinary school education system indicated challenges faced in mathematics and language education.
- The Dinaledi schools project [a national mathematics (and Science) strategy currently reaching approximately 500 schools] provides additional resources to targeted mathematics and science specialist schools producing quality mathematics passes among black learners.
- The Gauteng Primary Literacy and Mathematics Strategy and the Western Cape Education Department Provincial Strategy prioritise mathematics and lead provincial interventions focusing on mathematics.

- The Fundza Lushaka national bursary scheme targeting students studying education degrees at university level promotes the teaching profession among matriculants and mathematics is identified as a priority subject.
- Overhauling of the teacher qualification framework by mandating universities to provide quality pre- and in-service teacher education recognises that the majority of South African teachers lack specialised mathematical content knowledge for teaching.
- Development and distribution of learner workbooks for mathematics and language learning at primary school level alleviate shortages of teaching and learning resources, poor pacing and lack of curriculum coverage.

Adapted from Webb and Roberts, (2017).

With all these national initiatives and investments directly or indirectly targeting mathematics education in South Africa, the crisis persists. While these efforts are rolling out nationally and in different provinces, there is a need to find lasting solutions to the current state of mathematics education, and focusing on the language aspect in mathematics education may result in the needed elevation of the country's mathematics education to national standards. The role of language and access to language skills are critical in ensuring the ability of individuals to realise their full potential to participate in and contribute to the social, cultural, intellectual, economic and political life of South Africa (Ministry of Education, 2002). As such, language can be considered to be a barrier to access and success at all levels of education.

According to Taylor and Coetzee (2013), the extent to which language factors contribute to the low performance is not clear, given the fact that language disadvantages are strongly intertwined with other pre-existing factors such as historical disadvantage, socio-economic status, geography, the quality of school management, as well as the quality of teachers. As that may be the case, there are many South African educationists who are adamant that language, and in particular, the language policy, are key determinants of education outcomes. Educationists in support of mother-tongue instruction such as Brock-Utne (2007) argue that a later transition to English is necessary, given that learners grapple with understanding the language of learning and teaching. On the other hand, English is widely viewed as a vehicle of upward mobility and this leads to a preference for using English as LoLT from as early as possible.

CONCLUSION

The chapter intends to broaden and widen the understanding of the quality of South African mathematics education against the backdrop of current trends in mathematics education worldwide. This understanding is encapsulated in Roberts (2017) who declares that the quality of mathematics education in South Africa is a mirror image of the quality of the country's education system. Performance in school mathematics is very poor and unequal. The highly unequal education system from the apartheid era spilled over to the contemporary post-apartheid education system. To be more specific, the following were passed on from one era to another – who can access mathematics and who is capable of teaching the subject. Although the contemporary post-apartheid education system attempts to equalise the playing field by increasing access to all South Africans, improving educational equality and quality remains a challenge. In international assessments, South Africa still dominates the least performing category, performing worse than Egypt, Botswana and Morocco – countries whose budget allocations for education are fractions of South Africa's allocation. Actually, South Africa's allocation is the largest while the country's performance is the poorest. Some of the explanations for this incongruity are:

- Poor performance in mathematics is a direct result of poor competence in the language of instruction;
- Teachers lack the necessary pedagogical content knowledge for teaching mathematics; and
- Schools lack human and capital resources.

Although all these explanations are valid and plausible, they reek of the consequences of the apartheid era. Without dismissing the long-lasting effects of the era (including unequal resources and white privilege), for how long is the South African education system (and mathematics education) going to remain in this stupor?

Despite some efforts and improvements, South Africa is still significantly underperforming in mathematics education. One of the key factors limiting the quality of mathematics education is the poor quality of mathematics teachers, specifically with regard to numeracy and mathematics teaching at lower grade levels, FP and IP levels. The country's development as a knowledge economy relies partly on improving the teaching and learning of mathematics in the majority of public schools (Spaull, 2013). What compounds the situation is that the relatively high number of young people not in employment or education and training (NEET) is closely linked to the quality of education, specifically numeracy and mathematics competency. One of the ways of turning around the situation is to raise both learners' and teachers' mathematical competency to international levels.

CHAPTER 2

LANGUAGE AS A RESOURCE IN MATHEMATICS TEACHING AND LEARNING

INTRODUCTION

How can language be used as a resource in mathematics education?

This chapter presents specific background knowledge on how language can be used as a resource in mathematics instruction to enhance an understanding of the connection between language and mathematics for the purpose of meeting the needs of English Language Learners (ELLs) more effectively. The chapter briefly describes mathematics as a language, the mathematics register and the linguistic features associated with mathematics teaching and learning. The chapter also outlines different strategies that can be used to teach the mathematics register in a multilingual setting and concludes by presenting other common factors that are associated with mathematics teaching and learning in a second language, including the use of code switching.

MATHEMATICS AS A LANGUAGE

Durkin and Shire (1991: 3) state that mathematics education begins and proceeds in a language, it advances and stumbles because of language, and its outcomes are often assessed in language. Although this observation can be applied to most of the ordinary school curricula, the interweaving of language and mathematics is particularly intricate and this chapter accounts for some aspects of this complex interrelationship as a way of addressing one of the many challenges posed by mathematics to learners and their teachers.

Language and communication are essential elements of teaching and learning mathematics, and this is evident from research carried out in bi/multilingual settings (Gorgorió & Planas, 2001). Language is employed as a communication tool and facilitates the transmission of (mathematical) knowledge, values and beliefs, as well as cultural practices. Language is also the channel of communication in a mathematics classroom as language provides the tool for teacher-learner interaction (Smith & Ennis, 1961). Competence in the language of communication/interaction is a prerequisite for engagement in the learning process. In South Africa the English language plays a significant role in the transition from mother-tongue to English-medium mathematics instruction. For learners of mathematics, this has two dimensions in that they are required to have competence in the language of instruction and in the language of mathematics (specifically the mathematics register). The development of mathematical learning and understanding is therefore interrelated with language capability.

Language is defined as the formal organisation of symbols, sounds and gestures used to communicate ideas, thoughts and feelings to create meaning (Ellerton & Wallace, 2004: 1). Thus mathematics appears to be a type of formal language as it too consists of symbols, sounds and gestures that are used to communicate and formulate mathematical concepts. Sternberg (2004) identifies six distinctive properties of language:

1. **Communicative**: language permits us to communicate with one or more people who share our language;
2. **Arbitrarily symbolic**: language creates an arbitrary relationship between a symbol and its referent: an idea, a thing, a process, a relationship, or a description;
3. **Regularly structured**: language has a structure, only particularly patterned arrangements of symbols have meaning, and different arrangements yield different meanings;
4. **Structured at multiple levels**: the structure of language can be analysed at more than one level (such as in sound, in meaning units, in words or in phrases);
5. **Generative, productive**: within the limits of a linguistic structure, language users can produce novel utterances, and the possibilities for creating new utterances are virtually limitless;
6. **Dynamic**: languages constantly evolve.

The above properties are applicable to mathematics as well and to the concept of mathematics as a language, because mathematics allows people to communicate. It uses symbols; it is structured, generative and dynamic.

BUT IS MATHEMATICS TRULY A LANGUAGE?

Perhaps because mathematics is associated with written work, involves symbols as opposed to words and is communicated on paper rather than orally, it is difficult to perceive it as a language (Pimm, 1987). Pimm (1987: 75) argues that it is not, since there is no one group of people for whom mathematics is their first language. Moschkovich (2012) defines the language of mathematics as the communicative competence necessary and sufficient for competent participation in mathematical discourse practices.

One way of describing the relationship between a natural language like English and mathematics is in terms of the linguistic notion of register (Durkin & Shire, 1991: 17). Therefore, mathematical language is considered a distinct 'register' within a natural language like English, which is described as "a set of meanings that is appropriate to a particular function of language, together with the words and structures which express these meanings" (Halliday, 1975: 65). Therefore, one function of mathematics as a language can be described as the expression of mathematical ideas and meanings leading to the development of a mathematical register.

THE MATHEMATICS REGISTER

Mathematics has a particular way of using language and its own particular way of expressing ideas, which is termed the mathematics register (Lee, 2006; Pimm, 1987).

The development of the mathematics register in English has been taking place since the sixteenth century, when Robert Recorde's mathematical textbooks were among the first to be printed in English rather than in the prevailing languages of that time, Latin or Greek. Similarly, democratic countries on the African continent that have been changing their language of instruction from English, French or Portuguese to indigenous languages are still in the process of developing mathematical meanings in those languages so that the new registers developed are not merely compilations of newly introduced words.

One aspect of the mathematics register consists of the special vocabulary used in mathematics (Gibbs & Orton, 1994) and it is the language specific to a particular type of situation (Lemke, 1989). But it is more than just vocabulary and technical terms. It also contains words, phrases and methods of arguing within a given situation, conveyed through the use of natural language (Pimm, 1987). The grammar and vocabulary of the specialist language are not a matter of style, but rather methods for expressing very diverse things (Ellerton & Wallace, 2004). Each language will have its own distinct mathematics register and ways in which mathematical meaning is expressed in that language. For example, the English language mathematics register includes:

> "the use of common words with specialised meanings; syntax characteristics include increased use of logical connectives, while discourse characteristics include increased density of meaning, increased use of passive voice, and the need for multidirectional reading."
>
> Barton & Neville-Barton (2003: 4)

Within the mathematics register different forms of mathematical language can be found as illustrated in Figure 2.1.

Figure 2.1: Types of mathematical language

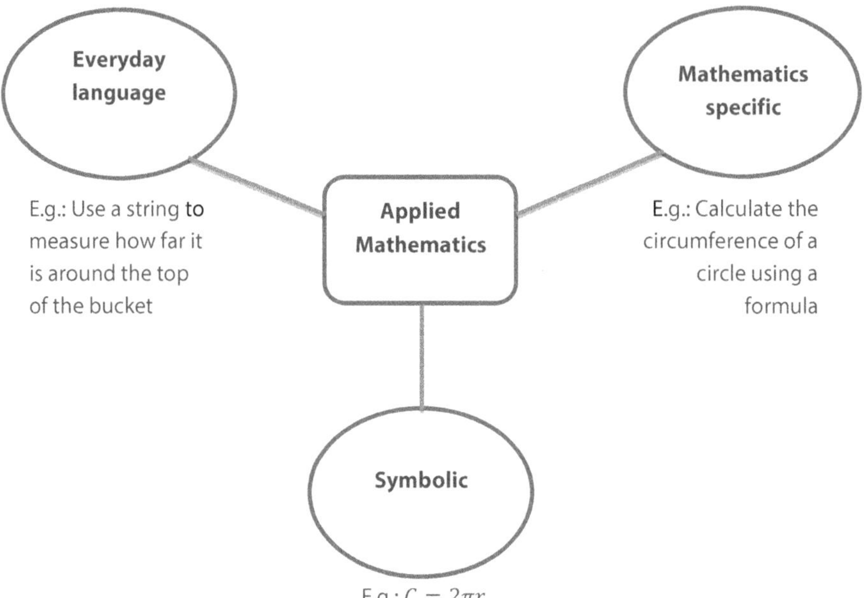

Adapted from Ní Ríordáin (2011); Meaney (2005); Bubb (1994)

As is evident, the complex register of mathematics is similar to a language and requires skills of learning similar to those used in learning a language. This adds another dimension to mathematics learning and reinforces the view that the content of mathematics cannot be taught without language. The process of learning mathematics involves mastery of the mathematics register (Setati, 2005). This allows learners to communicate their mathematical findings in a suitable manner, but "without this fluency, learners are restricted in the ways that they can develop or redefine their mathematical understandings" (Meaney, 2005: 129). By developing a learner's mathematical register, the process provides them with analytical, descriptive and problem-solving skills within a language and a structure so that they can explain a wide range of experiences. Once the register is mastered, learners will have the ability to listen, question and discuss, together with an ability to read and record.

Similarly, classroom discourse has an intricate structure, consisting of units of language such as those used in conversations, lectures, stories, essays and textbooks (Sternberg, 2003). Sentences are structured according to systematic syntactical guidelines – episodes of a discourse are also systematically structured. The context of learning is multidimensional, as understanding discourse does not rest solely on the interpretation of words written in textbooks and spoken by the teacher, but also on the knowledge of the physical, social or cultural context within which the discourse takes place (Sternberg, 2003). For example, the functions assigned to learners and teachers in the given environment, modes of communication, intentions, linguistic choices, and contexts of communication/learning are all influential in the interpretation of meaning (Georgakopoulou & Goutsos, 1997). By taking into consideration the speaker who is producing the spoken or written mathematical text, the context within which it is being produced, and the medium through which it is expressed, it becomes possible to evaluate the learners' mathematical understanding and the relationship between mathematics and language. However, mathematics learning/understanding and its relationship with language is an extremely complex and diverse area of research, which will be examined in the subsequent sections.

LINGUISTIC FEATURES ASSOCIATED WITH MATHEMATICS TEACHING AND LEARNING

Mathematics is not language-free. In other terms, language is only one component of mathematics. Furthermore, as has been exposed in preceding sections, mathematics *is* a language with its own particular vocabulary, syntax and discourse, which can cause problems for learners learning it in a second language (Barton & Neville-Barton, 2003). While many learners who learn mathematics in their mother tongue have difficulty in acquiring the mathematics register, this problem is heightened for those who must learn it in a second language, such as English. Learners have to cope not only with the new mathematics register, but also the new language in which the mathematics is being taught (Setati & Adler, 2000). Some of the language features that may impede mathematical learning are discussed in the following paragraphs.

Mathematical vocabulary

The National Numeracy Project (2000) defines mathematical vocabulary as a collection of words and phrases that make up the mathematical register. Teachers and learners need to understand these words and phrases in order to make good progress in mathematics. There are two main ways in which an ELL's (English Language Learner's) failure to understand mathematical vocabulary manifests itself, namely in an inability to perform a task as instructed by the teacher, or an inability to respond to questions during a lesson or in a test. Lack of response may occur for the following reasons:

- Lack of understanding of the spoken or written instructions (such as draw a line between ..., or find two different ways to ...);
- Unfamiliarity with the mathematical vocabulary (words such as ***difference***, ***subtract***, ***divide*** or ***product***);
- Confusion about mathematical terms (such as ***odd*** or ***table***, which have different meanings in everyday English; or
- Confusion about other words (like ***area*** or ***divide***, which are used in everyday English and have similar though more precise meanings in mathematics).

National Numeracy Project (2000)

It is therefore essential that teachers assist learners to acquire the appropriate vocabulary in order to increase participation in activities, lessons and assessment that are part of classroom life. An even more important reason for teaching mathematical vocabulary is that mathematical language is crucial to the development of learners' mathematical thinking skills. If a learner does not have the vocabulary to talk about division, perimeter or numerical difference, the learner's progress in understanding these areas of mathematics is impeded.

Developing mathematical vocabulary in the classroom

Direct teaching of vocabulary builds essential knowledge and when effective vocabulary instruction is built into a mathematics curriculum, learner achievement is likely to improve in mathematics assessments (Gunning, 2003; Vacca, Cove, Burkey, Lenhart & McKeon, 2008). Opportunities for the development of mathematical vocabulary must be regular and planned (National Numeracy Project, 2000). Thus, it must not be assumed that mathematical vocabulary development will occur unplanned or naturally, but opportunities for the development of the vocabulary should be systematically built into the teaching and learning content. Developing a better understanding of mathematical vocabulary can be done systematically by first introducing the new words, then creating opportunities for listening to the teacher using the words, followed by prompting the learners to respond to questions using the new words, followed by fluently reading written texts that include the new word, and finally writing the mathematical vocabulary in a range of contexts. Such focused and strategic development of mathematics vocabulary may benefit the ELLs (who are taught by the majority of the teachers participating in this book) better than when the development is not systematic.

- **Introducing new mathematical vocabulary**

  Although teachers often use informal language in mathematics lessons before or alongside specialised mathematical vocabulary to help learners to grasp the meaning of different words and phrases, the teacher also needs to adopt a more structured approach to the introduction of new vocabulary. Such a structured approach encourages learners to begin using the correct mathematical terminology as soon as possible (Lee, 2006; National Numeracy Project, 2000; Pimm, 1987).

- **Listening**

  Learners of all ages develop an understanding of new mathematical vocabulary by listening to the teacher or peers using the words in a suitable context, before they can copy the practice and use the words themselves. Initially learners listen to the teacher using, repeating and emphasising important words during practical activities with real objects, when playing games or when discussing pictures or diagrams. The teacher can also make use of listening comprehension exercises (comprising of pre-listening, while-listening and post-listening activities) to expand the learners' vocabulary. Listening activities guide the ELLs to listen for specific works and discover how they are used in the comprehension passage. In time, the learners pick up the words and their meanings and try them out too, particularly if the teacher, by means of probing and questioning, encourages the learners to use the new mathematical vocabulary (Lee, 2006; National Numeracy Project, 2000; Pimm, 1987).

- **Speaking**

  Initially, speaking opportunities can be combined with practical work so that learners have visual images and tactile experiences of what mathematical language means in a variety of meaningful contexts. At a later stage, using new mathematical vocabulary in oral work alone helps the learners to become less dependent on real objects and begin to visualise and work mentally. The teacher can develop the learners' understanding of terminology as well as assist the learners in understanding ambiguities and misconceptions through a range of open and closed questions. Oral work alone provides opportunities to:

  - Acquire confidence and fluency in speaking by being encouraged to use single words and phrases, then compose complete meaningful sentences that include the new mathematical words;
  - Describe, define and compare mathematical properties, positions, methods, patterns, relationships or rules;
  - Discuss ways of solving a mathematical problem, collecting data or organising the work;
  - Make predictions or hypothesise about possible results;
  - Present, explain and justify their methods, results, solutions or reasoning in pairs or small groups to the whole class; and
  - Generalise or describe examples that match a general statement.

  (Lee, 2006; National Numeracy Project, 2000; Pimm, 1987)

- **Reading**

  The teacher can provide a variety of texts with new mathematical vocabulary to be read aloud or silently. The teacher can ask learners to bring relevant texts such as newspaper and magazine extracts (with some bearing on the mathematical concepts to be taught) to be read in class. The texts can be read individually, as a whole class or through chorus reading. The teacher can provide reading opportunities in the form of:

  - Numbers, signs and symbols, expressions and equations written on the chalk board;
  - Instructions and explanations in work books, textbooks or eBooks;
  - Prose in library books, including reference books, story books or books of rhymes;

- Labels and captions on classroom displays, on diagrams, graphs, pictographs, charts or tables; or
- Definitions in illustrated dictionaries, ready-made dictionaries or dictionaries made by the learners in order to discover synonyms, origins of words or words beginning with the same prefix such as 'triangle' and 'trisect'.

(Lee, 2006; National Numeracy Project, 2000; Pimm, 1987)

Just like the variety of listening activities listed above, the teacher can incorporate pre-reading, while-reading and post-reading activities that improve the learners' awareness of mathematical vocabulary within the text.

- **Writing**

  The teacher can provide writing and recording opportunities in a variety of ways progressing from words, phrases and short sentences to poems, paragraphs and longer pieces of writing such as:
  - Writing prose in order to describe, compare, predict, interpret, explain or justify;
  - Writing formulae, first using words, then symbols;
  - Sketching and labelling diagrams in order to clarify their meaning; and
  - Drawing and labelling graphs, charts or tables or interpreting and making predictions from the data in graphical presentations.

(Lee, 2006; National Numeracy Project, 2000; Pimm, 1987)

Questioning techniques for developing mathematical vocabulary

Since the meaning of words cannot be learnt in isolation, the teacher's use of questions is crucial in helping learners to understand mathematical ideas and use mathematical terms correctly. According to the National Numeracy Project (2000), teachers often find it easy to ask simple recall questions – those that require learners to recall facts rather than asking questions that require learners to engage higher levels of thinking. Therefore, teachers are encouraged to use the full spectrum of question types to allow learners to respond with more complex answers in which they explain their thinking. Six types of questions that can be used by teachers to improve vocabulary learning are indicated in Table 2.1.

Table 2.1: Types of questions for developing mathematical vocabulary

TYPES OF QUESTIONS	EXAMPLES
Recalling facts	▪ What is 3 add 7? ▪ How many days are there in a week? ▪ How many centimetres are there in a metre? ▪ Is 31 a prime number?
Applying facts	▪ List two numbers that have a difference of 12. ▪ What unit would you choose to measure the width of a table? ▪ What are the factors of 42?

TYPES OF QUESTIONS	EXAMPLES
Hypothesising/ predicting	▪ Estimate the number of counters in the jar. ▪ If the survey is conducted again on Friday, how likely is it that the graph would be the same? ▪ Roughly, what is 51 times 47? ▪ How many rectangles will be in the 5th diagram?
Designing & comparing procedures	▪ How might we count this pile of sticks? ▪ How can you subtract 37 from 82? ▪ How can you test if a number is divisible be 6? ▪ How can we find the 20th triangular number? ▪ Are there other ways of solving the problem?
Interpreting results	▪ So what does that tell us about numbers that end in five or zero? ▪ What does the graph tell us about the most common shoe size? ▪ So what can we say about the sum of the angles in a triangle?
Applying reasoning	▪ The seven coins in my purse total 23 cents, what could they be? ▪ In how many different ways can four children sit at a round table? ▪ Why is the sum of two odd numbers always even?

Adapted from National Numeracy Project (2000: 4)

Different types of questions can be used by the teacher to promote good dialogue and interaction in a mathematics classroom. Questions can either be closed, with only one correct answer, e.g. 'Is 16 an even number?', or open, with a number of different correct answers, e.g. '*List the even numbers between 10 and 20*'. Open questions provide more learners with a chance to respond and they often provide a greater challenge for the most able learners, who can be asked to think of alternative answers, and in suitable cases, to count all the different possibilities. Questions can be used at any point during the lesson to extend the learner's thinking. When learners are in the entry phase of an activity, the teacher may ask the learners a question such as: '*What information do you have?*' To make positive interventions for the purposes of checking progress while learners are working, the teacher may ask: '*Can you explain what you have done so far?*' To assist learners who are stuck, the teacher may ask: '*Can you describe the problem in your own words?*', and during the feedback session, the teacher may ask, '*How did you get the answer?*' As learners provide answers to the teacher's questions, the teacher needs to encourage the correct use of newly learnt words as well as other relevant mathematical vocabulary, and reward or positively reinforce correct usage of the mathematical terms in the dialogue. Such practices would help to develop the learners' mathematical vocabulary.

A study conducted by Moloi, Morobe and Urwick (2008) in Lesotho primary schools identified three major challenges in teachers' questioning. The first is that most questions asked simply require learners to recall what the teacher has said; these are closed questions with one correct answer. Learners are therefore restricted in their thinking skills. Probing, analytical, general interpretation and open-ended questions are rarely asked. While there is some element of multiple representations of mathematical concepts in the primary classrooms, generally questions used in exercises only require recall, with no room for critical thinking. Learners are not afforded the opportunity to imagine or to interpret. Similar observations were made by Ntoi and Lefoka (2002: 280), who noted that teacher education lecturers' questions "do not lead to much discussion because of their restrictive nature". Thus the answers called for are safe and simple for the student-teacher.

Secondly, questions are not always distributed around the whole class, especially in large classes. The majority of the learners get an opportunity to respond to a question only through chorus answers; otherwise they may go through the whole lesson without getting the opportunity to demonstrate their understanding or lack thereof. The learners who get a chance to respond to questions feel involved in the lesson.

The third challenge is lack of feedback. Generally, teachers correct learners' inaccurate answers and redirect or rephrase questions, but they do not reinforce the correct answer. Either they underestimate the importance of reinforcement for learners' motivation, or they are not very interested in the matter. This practice of providing inadequate feedback defeats the whole purpose of using questioning techniques to improve language competency in the classroom.

Ambiguous terms

A key issue that creates significant hurdles for ELLs (as well as monolingual learners) is the number of words *borrowed* from everyday English (Pimm, 1987). These words tend to be ambiguous as they have one very specific meaning in the mathematics register and a range of possible other meanings in their everyday uses (Yushau & Bokhari, 2005). Table 2.2 is a compilation of the more common ambiguous words found in mathematics education. The non-mathematical meanings of these terms can interfere with mathematical understanding, as well as being a source of confusion.

Table 2.2: Some ambiguous words used commonly in school mathematics

above, altogether, angle, as great as, average, base, below, between, big, bottom, change, circular, collection, common, complete, coordinates, degree, difference, different, differentiation, divide, down, element, even, expand, face, figure, form, grid, high, improper, integration, leaves, left, little, low, make, match, mean, model, moment, natural, odd, one, operation, overall, parallel, path, place, point, power, product, proper, property, radical, rational, real, record, reflection, relation, remainder, right, root, row, same, sign, significance, similar, small, square, table, tangent, times, top, union, unit, up, value, volume, vulgar

Durkin and Shire (1991: 74)

Furthermore, Rudner (1978) found that the following are sources of difficulty and hinder learners' interpretation and understanding of mathematical word problems:

- conditions (if, when);
- comparatives (greater than, the most);
- negatives (not, without);
- inferentials (should, could, because, since);
- low-information pronouns (it, something); and
- lengthy passages.

Lexical ambiguity in mathematics can be classified under homonymy, polysemy, homophony, and shift in application or imprecision (Durkin & Shire, 1991: 71 – 75). Table 2.3 summarises vocabulary misconceptions commonly found in mathematics classrooms.

Table 2.3: Summary of vocabulary misconceptions

MISCONCEPTION	EXAMPLES
Homonyms: words shared by mathematics and standard English but have different meanings	▪ ***Right*** angle ▪ ***Right*** answer ▪ ***Reflection***: flipping over a line ▪ ***Reflection***: thinking about something ▪ ***Table***: organising information ▪ ***Table***: piece of furniture
Polysemy: mathematical words shared with standard English and have comparable meanings, but with a more precise mathematical meaning	▪ ***Difference***: answer to a subtraction problem ▪ ***Difference***: general comparison ▪ ***Even***: divisible by 2 ▪ ***Even***: smooth
Mathematical words **shared with other disciplines** and have different technical meanings in the two disciplines	▪ ***Divide*** in mathematics means to separate into parts, but the continental ***divide*** is a geographical term referring to a ridge that separates eastward/westward flowing waters. ▪ ***Variable*** in mathematics is a letter that represents possible numerical values, but ***variable*** clouds in science describes a weather condition.
Shift of application: words with more than one mathematical meaning	▪ ***Round***: a circle ○ ▪ ***Round***: a number rounds off… **243** rounds off to **240** ▪ ***Square***: a shape □ ▪ ***Square***: a number times itself $2^2 = 4$ ▪ ***Second***: a measure of time ▪ ***Second***: a location in a set of ordered items 1st 2nd 3rd 4th …
Homophones: mathematical terms sharing the same pronunciation with standard English terms	▪ ***Sum***/some ▪ ***Arc***/ark ▪ ***Pi***/pie
Irregularities found in English spelling and usage	▪ ***Four*** has a 'u' ▪ ***Forty*** does not
Specialist terms: mathematical terms only found in mathematical contexts	▪ ***quotient*** ▪ ***decimal*** ▪ ***denominator*** ▪ ***isosceles***
The same mathematical concepts **expressed in more than one way**	▪ ***One quarter/one fourth/¼***
Related mathematical words whose distinct meanings may be confused	▪ ***factors*** and ***multiples*** ▪ ***hundreds*** and ***hundredths*** ▪ ***numerator*** and *denominator* ▪ ***mode, mean*** and ***median***
Imprecision: adapting an informal term and using it as if it is a mathematical term	▪ ***Diamond*** for rhombus ▪ ***Corner*** for vertex ▪ ***Cross*** for intersection ▪ ***Turn*** for rotate

Adapted from National Numeracy Project (2000)

Homonymy denotes the property of some words that share the same form but have distinct meanings. A standard example is *bank* – meaning a financial institution, and *bank* meaning the area of land beside a river. An example relevant to mathematic is *leaves* which in everyday use refers to the outgrowths of a tree, but has a different meaning when referring to the process of subtraction (3 from 7 *leaves* 4).

Polysemy refers to the property of some words that they can have two or more different but related meanings. A standard example is *mouth*, which can refer to a facial aperture or the place where a river meets an ocean. These are different but related meanings, which may be apparent to the native speaker of English in that they contain some shared sense of an opening. A mathematical example is *product*, which in everyday use refers to an item which has been made, whereas in mathematics it refers to a quantity obtained by multiplication. Again, there is a shared sense of that which is produced in both the everyday and specialised meanings; however, this shared sense may not be so obvious to a second-language learner of English.

Homophony refers to the phenomenon wherein two distinct words have the same pronunciation. Standard examples include *bare/bear*; *flour/flower*. These are linguistic coincidences which may lead to misconceptions. Mathematically relevant examples include *two/too/to; sum/some; pi/pie*.

Shifts of application refer to occasions where the same sense can be considered from different perspectives. Pimm (1987) suggests *wall* as a standard example which has different aspects according to whether it is discussed with reference to its composition (brick, stone) or its function (from the perspective of brick layers, architects, residents). A mathematical example is the word *number*, which is subject to shift of application, for example, when used to describe *nominal* (the number 5), *ordinal* (the second number she said) or *visual* (the number 7 is crooked) properties.

Imprecision and use of **informal** terms occur in any area of vocabulary use; however, it is worth mentioning that these are frequent in mathematical contexts and contribute to ambiguity. Durkin and Shire (1991) explain how the word *share* refers to division satisfactorily for some problems, but is misleading for others. Pimm (1987), Durkin and Shire (1991) and Hanley (1978) provide the following examples: *average* when *mean* is intended; *data* to indicate a singular referent, or *diamond* to refer to a square in a particular orientation. Some words may be used more informally in ways which conflict with their precise meanings in mathematics and logic, for example, the phrase '*Let's move the tables to make more space*' is a description which would fail the conservation of area assessment (Hobart, 1980).

As stated in the earlier section of this chapter, since languages are dynamic, most language use involves shifts of application, otherwise nothing new would be said, and thus even mathematics teachers who pride themselves on the precision of their subject are as capable as any other language users of imprecise and misleading descriptions (Durkin & Shire, 1991). Certain word combinations created in mathematics are intended to represent meanings which are different from the sum of their everyday senses. Such examples include: simple interest, pie chart, square root, closed figure (Shuard & Rothery, 1984: 28). Conversely, some colloquial descriptions of mathematical operations have different everyday meanings, such as *take away* (*takeaway*).

Specialist terms

The use of specialist terms can lead to misunderstanding and misinterpretation of mathematical tasks. Learners tend to encounter these terms only within the mathematics classroom (for example, *quadrilateral*, *parallelogram* and *hypotenuse*) and they are unlikely to be reinforced outside of it (Pimm, 1987). If second-language learners do not acquire the correct meaning of these terms, this can lead to difficulties within the mathematics context. Second-language learners have a tendency to translate new mathematical terms/vocabulary into their mother tongue. This translation may not exist and/or it may be done incorrectly, thus resulting in further confusion and misinterpretation (Graham, 1988).

Context

Context is also a key aspect in lexical ambiguity. Words can change their meaning depending on their context within the mathematics lesson (Gibbs & Orton, 1994: 98). In terms of language analysis, this is referred to as semantics: establishing meaning in language, or the relationship between and representation of signs and symbols. Because of the multiple meanings that various words can have, the context is vital in determining the correct interpretation. A review of the literature showed that learners experience more difficulties with the semantic structure of word problems than with other contributing factors such as the vocabulary and symbolism of mathematics and standard arithmetic (Ellerton & Clarkson, 1996).

Symbolism

Symbolism, the use of symbolic expressions or symbols, is one of the most distinctive features of mathematics. It is crucial for the construction and development of mathematics. Unfortunately, symbolism can accordingly cause considerable difficulties to those whose mother language has different structures. One of the requirements for mathematical learning is that learners should be able to interpret the mathematical text and convert it to an appropriate symbolic representation, and perform mathematical operations with these symbols (Brodie, 1989). Thus, if learners cannot understand the text because of the language of instruction, they will be unable to convert it to the appropriate mathematical constructions needed to solve the problem. Symbols provide structure, allow manipulation, and provide for reflection on the task completed. Some lexical ambiguities are compounded by the fact that they relate to mathematical symbols which themselves are described by different words in different contexts; for instance, the symbol = can mean *equals*, *means*, *makes*, *leaves*, *the same as*, *gives*, *results in*, any one of which itself has multiple meanings. Although there may be other conceptual factors involved, this linguistic diversity seems likely to be implicated in the findings that learners experience difficulties in interpreting the equals sign as they proceed further with their school years (Baroody & Ginsburg, 1983; Cobb, 1987).

Registers exist in other disciplines such as science or technology, but Standard English language can also be classified as a register. In the teaching and learning process, mathematics and ordinary language registers can interfere with each other. Thus learners need to recognise each of these registers so as to identify which is being used at any given time (Sierpinska, 1994), and this is a challenge many IP mathematics teachers face when managing the transition to teaching mathematics through the medium of English.

This chapter contends that mathematics classrooms are among the *few* places where learners engage with mathematical vocabulary; therefore, teachers *must* create opportunities for learning mathematical vocabulary and its applications; this applies specifically to IP mathematics teachers, for the purposes of facilitating a smooth transition from mother-tongue instruction to instruction using English as the LoLT.

Vocabulary teaching and learning strategies and the role of vocabulary

From Schmitt's (1997: 207 – 208) taxonomy of vocabulary learning strategies, some key vocabulary teaching strategies that can be inferred and are relevant for the second-language IP classrooms include:

- provision of a mother-tongue equivalent;
- guessing word meaning from contextual use;
- teaching word lists;
- using flash cards;
- using a new word in a sentence;
- giving the meaning or synonym of a word;
- practising word meaning with peers for consolidation;
- representing a word in pictorial form;
- making connections between words and their synonyms or antonyms;
- using semantic maps;
- drilling an aspect of a word, for example, pronunciation, spelling;
- using physical action e.g. demonstration or gestures;
- repeating a word verbally and/or in written form; and
- labelling objects in the classroom.

Long and Richards (2001) found that strategies of drawing attention to words and defining them lead to vocabulary gains. Attention-drawing strategies include activities such as presentation of words to learners prior to the reading of a particular selection and pre-teaching them as well as highlighting them. The manifestation of such strategies in teacher practices would be indicators of the potential for teacher practices to impact on vocabulary learning. Learners can only learn those words that they are exposed to.

Hart and Risley's (1995) study on the discrepancies in vocabulary knowledge of learners according to their different socio-economic statuses attests to the significance of word exposure for word acquisition and knowledge. There is, however, no consensus in the research on the average number of words average learners of an average school age learn per day or per week. Daily word acquisition estimates are extremely varied, ranging from three words a day (Joos, 1964), through seven (Beck, McKeown, & Kucan 2002) to twenty (Miller, 1978). Beck et al. (2002) even found that among what they called 'at-risk' learners the figure was as low as one or two words a day.

For first-language speakers of English, Biemiller (2005) found that the lowest quartile learners learnt between 500 and 600 words a year. According to Nagy and Herman (1985: 16), the Vocabulary Learning Hypothesis states that most vocabulary is learned gradually through repeated exposure to new and known words, in various contexts. They note that a single word encounter gives the learner only a 5 to 10% probability of knowing and retaining it. They peg the ideal number of word encounters for the entrenchment of a word in memory in varied contexts at 10 to 12 exposures. More and varied contextualised exposures would be needed to ensure transference of words from passive to active vocabulary. Hinkel (2007: 6) notes that "a large passive vocabulary does not necessarily result in a better active vocabulary. Active vocabulary is one a language user can easily and readily draw on for productive language use, whereas passive vocabulary is somehow distant and slightly hidden". The frequency of word usages required for learners to 'own' a word underscores the importance of strategies that will ensure vocabulary instruction. Sacrificing the contextual usage of words for the number of words learnt is counter-productive. Inasmuch as having a million bricks does not mean having a house, the meaningfulness of the context in which vocabulary is encountered is as important as, if not more important than, mere exposure of the vocabulary, especially for developing active vocabulary.

Lessons focusing on a specific topic or theme allow for the recycling of vocabulary within a meaningful context, which allows for sufficient practice with the related vocabulary. There is also a need for multiple exposures to complement such practices for the purposes of developing learners' vocabulary. Mere word repetition or word drills will not achieve much (Stahl, 2005). Possibly the most consistent finding in vocabulary instruction is that varied, multiple and meaningful word exposure is essential for vocabulary development. The more concrete the learners' experiences with new words, the better they will be positioned to comprehend and retain them.

Biemiller (2005) documented the meagre percentage of vocabulary instructional time and the absence of systematic, explicit vocabulary instruction in schools within second-language contexts. Traditionally, word instruction was delegated to the glossary and the dictionary, or took the form of a quick oral definition. This accorded learners an on-the-fly word exposure, which hardly translated to word learning on a long-term basis, seeing that learners needed multiple word exposures in multiple contexts for them to understand, remember and apply new words appropriately (Nagy, 2005).

In the study by Saragi, Nation and Meister (1978) learners learned 93% of the words after six or more exposures to the words, while words presented fewer than six times were learned by only half the learners. The study by Jenkins, Stein and Wysocki (1984) found only 25% of the learners learning a word after 10 encounters. Although such discrepancies characterise research on the amount of exposure that triggers word learning, what is held in common is the idea that the higher the frequency of word exposure, the greater the likelihood of it being internalised. The discrepancies in the estimates emanate from a host of other confounding variables, including the nature of the word exposure and the word testing, and the proficiency levels of the learners, among others.

PEDAGOGICAL IMPLICATIONS: LANGUAGE AS A RESOURCE IN MATHEMATICS TEACHING AND LEARNING

The previous sections explored the link between language and mathematics achievement, including vocabulary misconceptions prevalent in mathematics classrooms. The literature has revealed that for learners of mathematics, especially in a second language, there are some linguistic constraints that might hinder the mathematics learning and understanding. This section looks at some strategies that may help in making mathematics meaningful for bilingual learners in spite of their linguistic constraints. A study conducted in Botswana by Kasule and Mapolelo (2005) summarises common teachers' strategies of teaching primary school mathematics in a second language, as illustrated in Table 2.4.

Table 2.4: A typology of strategies used in teaching mathematics in a second language

LINGUISTIC STRATEGIES	GAMES & OTHER STRATEGIES	ORGANISATIONAL STRATEGIES
▪ Code switching from English to first language	▪ Using concrete objects	▪ Group work based on ability so that the weaker learners can be helped
▪ Inviting questions from learners	▪ Using flash cards and playing cards	▪ Giving large amounts of homework daily
▪ Teaching the language of mathematics	▪ Using jigsaw puzzles	▪ Providing variety in teaching methods and classroom activities
▪ Translating the textbook from mother tongue to English	▪ Snakes and ladders	▪ Forming Maths clubs
▪ Using simpler English words wherever possible	▪ Using multiple tables	▪ Asking one learners to solve a problem while others listen
▪ Doing 'oral' maths	▪ Role play	▪ Using the solution to check the meaning of the problem
▪ Using discussion	▪ Asking learners to convert number to story/word problems	▪ Involving the learners when working out a problem on the board
▪ Encouraging learners to speak English	▪ Regular participation in mathematics ▫ quizzes, ▫ fairs and ▫ other competitions	▪ Providing different ways of getting to the answer.
▪ Relating a new to an old topic/ concept		

Adapted from Kasule and Mapolelo (2005).

To minimise lexical ambiguity in mathematics classrooms, Durkin and Shire (1991: 74 – 78) propose several strategies that can be adopted by teachers. These strategies include monitoring lexical ambiguity, enriching contextual cues, exploiting ambiguity to the learners' advantage as well as confronting ambiguity. In addition to minimising lexical ambiguity, the literature reveals other pedagogical strategies that may help in making mathematics meaningful for bilingual learners in spite of their linguistic constraints. These are: appraising mathematics ability, employing bilingual approaches, contextualising mathematics, localising mathematics, employing linguistic approaches, removing reading difficulties, monitoring language in the classroom, and using analogy and metaphor.

- **Monitor lexical ambiguity**

In preparing materials for use in mathematics lessons, the teacher must consider whether any words might have a different meaning for the learner from that intended or assumed in the specialist context. If the teacher is aware in advance of potential misconceptions, sensitivity to the learners' needs is increased.

- **Enrich contextual cues**

Successful decoding of a specific use of an ambiguous word is dependent upon the context in which it is used. The contexts provided by mathematics lessons may not always be sufficiently familiar or secure for the learner to readily access the appropriate mode of discourse and terminology. For the most able learners in mathematics, this may not be a problem, but for weaker learners misconceptions over the specialist use of terminology can only compound the confusion. Structured and sensitive preparation by the teacher can help guide the learners to the context-specific meaning of the target words. Shire and Durkin (1989) found that providing learners with visual support for judgements about rank ordering facilitated the learners' performance. When mathematical definitions need to be differentiated from everyday or scientific meanings, the teacher should allow the learners to discuss the differences and draw pictures or construct meaningful sentences to contrast the two meanings such as:

- The *difference* between my two bags is that one is red and the other is blue.
- The *difference* between 12 and 7 is 5 because *12 – 7 = 5.*

In addition, keeping a plain word wall, or journal entries of new words, or using published dictionaries to search for word meanings might not be as enriching and creative as posting new words alongside their definitions and illustrations to make the new words more meaningful. Compared to published dictionaries, learner-made dictionaries, also called *vocabulary journals*, allow learners to draw on their creativity, which could be advantageous in aiding memory. Using illustrative word walls as shown in Figure 2.2 provides visual cues that can help learners with word recognition, automaticity, decoding and spelling (Browne, 2002; Peregoy & Boyle, 2005).

Figure 2.2: Examples of illustrative journal entries

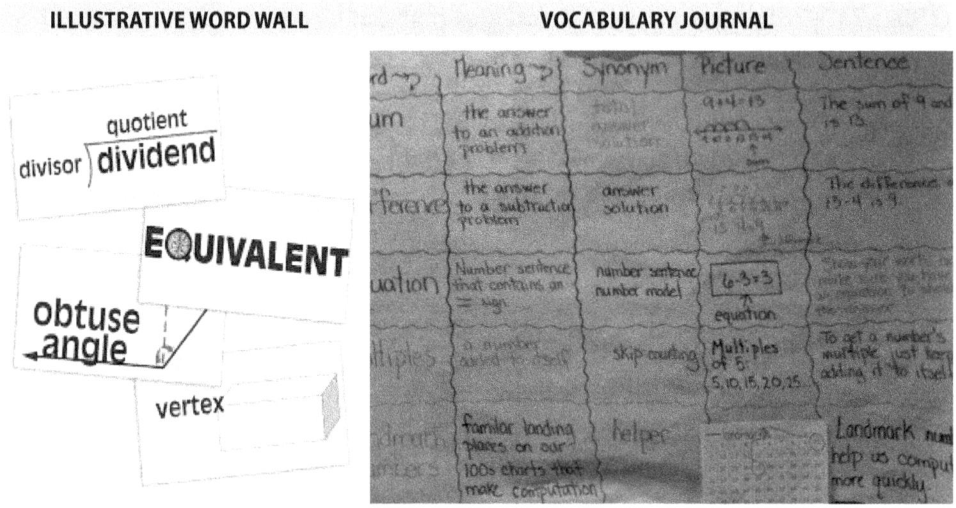

ILLUSTRATIVE WORD WALL **VOCABULARY JOURNAL**

Adapted from Virginia Department of Education, (2006).

Barwell et al. (2002) note that it is not enough for learners to hear a few examples of a word being used or to be given a formal definition. Learners need to explore the concepts involved, push at the limits of definitions and most of all, make the meaning their own as they learn to talk mathematically.

- **Exploit ambiguity to advantage**

In many cases, particularly with polysemous words, purposeful use of the contexts in which the everyday and specialist meanings of the terms coincide can provide learners with a secure base from which to familiarise themselves with the new extensions of their vocabulary before venturing into more demanding uses. For the most able learners as well as in the upper levels of the IP, teachers may use strategies that include analyses of word origins as a way of exploiting ambiguity to advantage. Analysing word origins allows teachers to share with learners the '**words behind the words**', so that learners connect terms that sound foreign to the words they already know:

- *Parallel* comes from '*alongside*' ... (*para-*) as in a paramedic who works alongside medical professionals.
- *Percent* comes from '*for each hundred*' (*per* dozen – for each dozen) from which the term *divided by 100* is derived.

Word origins can also be used to clarify distinct meanings of mathematics word pairs that are commonly confused such as (numerator/denominator). The root *nom* means to name, hence the *denominator* names the fraction, while the *numerator* tells the *number* of parts of interest. Thus ³⁄₅ means that something is divided into *five* equal parts, and we are referring to *three* of those parts.

- **Confront ambiguity**

In terms of linguistic economy and organisation, there are good reasons why homonymy and polysemy exist and learners' appreciation of the varieties of meanings can be promoted by developing teaching materials which bring potential conflicts into focus. This calls for careful monitoring of learners' responses by the teacher, and learners can only benefit from such encounters, particularly from middle childhood, approximately at the IP level of the South African education system, the period during which learners generally become more sensitive to linguistic ambiguities. Shultz and Horibe (1974) describe this period as characterised by an increasing interest in word play, jokes, puns and riddles. To this effect, activities such as quizzes, word searches, crossword puzzles and creative writing (discussed later in this section) in mathematics become relevant.

While providing learners with opportunities to gain access to the resources implicit in natural language can be viewed as a common aim of all teachers, Durkin and Shire (1991) view a particular aim of mathematics teachers as providing their learners with a means to make use of the mathematics register for their own purposes. As such, this chapter values IP mathematics teachers' knowledge of the language forms and structures that comprise aspects of that register, so that part of learning mathematics reinforces gaining control over the mathematics register, which in turn means that learners are able to talk like mathematicians and express their understanding of mathematics coherently.

- **Appraise mathematics ability**

Individual learner strengths and needs vary. Bilingual learners may have high ability in maths and yet not be able to communicate that ability due to lack of English proficiency, or a lack of communication skills in either language. Therefore, it is crucial to appraise mathematics ability as early as possible, and on the basis of cognitive ability and not assume that it exists on the basis of learner proficiency in English.

- **Bilingual instructional approach**

The level of bilingual language proficiency should be measured to ascertain the level of language proficiency in both languages. If the learner is stronger in their mother tongue than English, then instruction should continue in the mother tongue before transition to English (Cuevas & Beech, 1983; Setati, 2003).

> If it is determined that the native language will be used for instruction, then a teacher who is fluent at the academic level of that language should provide instruction. Just as the language proficiency of bilingual learners can range dramatically, so can the language proficiency of bilingual teachers. If the teacher is highly proficient in the academic language of mathematics, then the learner will be more likely to learn that higher level of their native language. A teacher who is only fluent in the social aspects of the native language may struggle to communicate precisely with mathematical terms and expressions in that language,

> and may not be able to take their native language learners to a higher level of learning.
>
> Raborn (1995)

However, if the language of instruction is the learners' second language, a translation of key words into the learners' first language can help (Yashau & Bokhari, 2005; Secada & Cruz, 2000; Setati & Adler, 2001). The teacher may also request a strongly bilingual/multilingual learner to translate for the weaker bilinguals and multilinguals, so that the rest of the class may have an opportunity to participate in the classroom discussions. Teachers might request having some of their teaching materials translated into the learners' native language, or to have the material rewritten in simplified language to make it more accessible to the learners.

- **Contextualisation of mathematics**

One of the major constraints faced by learners is that mathematics has remained inaccessible to teachers and hence to the learners.

> If the young learner receives only a quick and abstract initial encounter with variables followed by practice in the essentially syntactic skills of manipulating algebraic expressions, the semantics component of the new language may never be realised.
>
> Burton (1988: 6)

In this case the learner will have difficulties in making sense of the problem; therefore, teachers should always try to use a context or theme that learners are familiar with to serve as a matrix out of which they generate mathematical activities and problems. Since mathematical language, like language in general, develops in context to support communication, the best way to increase learners' use of precise language is to use contexts that require such precision (Secada & Cruz, 2000).

- **Localising mathematics**

Effective bilingual teachers use their learners' home cultures to support classroom management and learning. All indigenous societies have their own forms of mathematics, which are different from the internationalised mathematics (Ellerton & Clarkson, 1996: 1014). Localising educational theories and practices allows the curriculum to be directed to those for whom it is being developed. Such a curriculum can be built upon the natural cognitive modes of the learners in question, which are determined by their language (Brodie, 1989: 51).

> The cultural perspective requires us to culturalise the curriculum at each of the levels, and demonstrate that no aspect of mathematics teaching can be culturally neutral.
>
> Ellerton & Clarkson (1996: 1017)

- **Linguistic approach**

Two linguistic strategies are suggested. The first strategy should be to aim at improving the linguistic skills and fluency in the language of instruction, while at the same time developing the skills and support for the mother tongue. Thus, developing proficiency in the language of instruction should not be at the expense of the development of the mother tongue, but should be encouraged in such a way that proficiency in both the languages is developed. The second strategy is to involve language teachers who would spend a substantial portion of their teaching time concentrating on linguistic skills that are relevant to mathematics (Brodie, 1989: 49).

- **Removing reading difficulties**

> Whether the text is written in the learner's first or second language, an important factor is readability.
>
> Brodie (1989: 50)

To make mathematics accessible to the majority of the learners, it is imperative to develop the reading skills of the learners and to make mathematics textbooks simple and directly to the point. Thus, teachers should avoid using dense stylised mathematical writing, which tends to redirect the reader's attention to the correctness of what is written instead of concentrating on the richness of the meaning.

- **Language in the classroom**

The classroom conversational language between teachers and learners, and amongst the learners themselves, has a great impact on mathematics learning. The language of the classroom should be simple and straightforward to avoid communication gaps. ELLs usually receive little encouragement to talk about their ideas, in part because of the belief that they will find it too difficult to express themselves (Secada & Cruz, 2000). Therefore, teachers should initiate discussion in the class and should encourage weak language learners to participate in classroom discussions. Such practices will enable the teachers to identify the learners struggling with new vocabulary in mathematics and thereby provide the necessary support and guidance.

> In addition to simplifying oral language, teachers should expand on learner responses and build on those when posing their next question. Some learners pass through a 'silent period' when learning a second language, during this time, they focus on listening and trying to make sense of the rules for conversation in the classroom as well as in the larger world. Teachers will need to give learners time and look for nonverbal cues as to whether they understand the gist of the lesson.
>
> Secada and Cruz (2000)

It is important to learn both vocabulary and the way that vocabulary is used. Thus, learners need to use the mathematical language themselves in order to get used to the way that the expressions are used, and to begin to use the mathematical terms to express the web of concepts and ideas that are encompassed by those terms (Lee, 2006: 36).

- **Analogy and metaphor**

> Other texts, especially literary ones, convey ideas through imagery and metaphor, and often restate them in different ways in order to present a more complete picture.
>
> Brodie (1989: 49)

The use of analogy and metaphor is also suggested and shown to be a useful instructional tool in mathematics. The use of analogy and metaphor in mathematics can address the language constraints of ELLs by simplifying contextualised mathematical concepts and by strengthening vocabulary learning. Analogy and metaphor can also be incorporated into creative writing activities (Newby & Stepich, 1987; Dickmeyer, 1989; Wessels, 1990).

Creative writing strategies appropriate for mathematics teaching and learning include short stories, songs and poems. A poem based on a mathematical concept can easily be composed by learners of varying abilities. Once learners formulate the first two sentences, they build up more impressive lines and they begin to play with the rhythms created by the lengthening of the sentences. Poetry sessions can be done impromptu, within specified times and without warning, or they can be done over a longer period of time to allow learners to find out more information on a mathematical concept. The poem below by Baartman (2015) is an example of creative writing composed by completing two given statements:

- *If I were a shape, I would be a ______ Because ______*
- *If I were a shape, I would* **not** *want to be a ______ Because ______*

If I were a shape, I would like to be a *circle*
I would have neither end nor beginning
I would be *infinite*
You would see me everywhere
You would see me every day
On your watch, the shape of the Sun
Even when you eat
My *circumference* loves *pi diameter*
But my *radius* loves *pi half* as much as my *circumference*

If I were a shape I would *not* want to be a *square*,
Everything would always be the same
Everyone would know me wherever I go
Even when I am stretched,
I would still just be a *special square*
You would find me in any *area*... how boring!

Althea Baartman (2015)

SP Mathematics Teacher at Aloe Junior High School, Cape Town[1]

1 Ms Baartman composed the poem during an activity conducted in one of the author's lectures with in-service mathematics teachers.

Writing about mathematics allows learners to understand mathematical vocabulary, provides them with the means to make use of the mathematics register for their own purposes, and gives teachers the opportunity to assess learners' understanding of the terms they use. Thus creative writing enforces control over the mathematics register, so that learners are able to talk like mathematicians and express their understanding of mathematics coherently (Durkin & Shire, 1991).

> While teachers may consider mathematics to be universal, there are factors related to language, culture and cognition that must be taken into consideration in mathematics education. With careful assessment planning and implementation, learners with diverse learning characteristics can be successful in mathematics.
>
> Raborn (1995)

To achieve this, the following recommendations of the National Council of Teachers of Mathematics (NCTM) are imperative to address mathematics teaching and learning using a second language holistically, including other stakeholders involved in the teaching and learning process, other than just the teacher and the learner.

- Schools should provide second-language learners with support in their home language and in English language while learning mathematics.
- Teachers, curriculum advisors and other professionals who have the necessary expertise should carefully assess the language and mathematics proficiencies of each learner in order to make curricular decisions and recommendations.
- Mathematics teaching, curriculum and assessment strategies should be based on best practices and built on the prior knowledge and experiences of learners and their cultural heritage.
- To verify that barriers have been removed, teachers should monitor enrolment and achievement statistics to determine whether second-language learners have gained access to and are succeeding in mathematics courses. Reviews should be conducted at school, district, provincial and national levels.

It is worth pointing out that mastering the mathematical register alone or accumulating a collection of mathematical vocabulary does not constitute better performance in mathematics. There are other factors that are associated with mathematical learning and teaching. The section below details a number of key factors that need to be considered when using language as a resource in mathematics teaching and learning.

CONCLUSION

This chapter sought to explain how language can be used as a resource in mathematics instruction and concurs with research that deems **language as a transparent resource in teaching IP mathematics**. This chapter highlights some aspects of the English Mathematics Register that may be sources of difficulty, specifically for

IP learners in the transition from mother-tongue to English-medium education in the South African education system – and some pedagogical strategies that may be useful in managing that transition.

In multilingual communities, when using an LoLT that is not the home language of the learners and the teachers, it is important to equip teachers with the necessary linguistic skills to facilitate mathematics instruction. This chapter does not seek to enshrine English at the expense of other official languages in South Africa, but endevours to cater for ELLs within the education system today who are supposed to be taught and assessed in English, as stipulated in the current Language in Education Policy (LiEP), so that those learners are not written off as a doomed generation. Even if policies change and promote mother-tongue instruction throughout the entire education system, proficiency in English will still be a prerequisite for ELLs to access the global village. To that effect, the following rhetorical questions are posed:

- For how long are we going to have the same conversation addressing language as a barrier in mathematics?
- If identifying the need to use language as a resource in mathematics instruction is only the first step, what should be the second, third and fourth steps?
- Under the current LiEP, how easily can language as a resource in mathematics classrooms facilitate a smooth transition from mother-tongue instruction in (FP) to using English as LoLT in the (IP)?

Wolff (2006) in Alidou, Boly, Brock-Utne, Diallo Heugh & Wolff (2006:9) sums it up best by stating that 'Language is not everything in education, without language, everything in education is nothing'; and I take the liberty of extrapolating thus: **'Language is not everything in mathematics education, but without language, everything in mathematics education is nothing'**.

***

* Modifications of this chapter are also available in the following publications:

1. (Afrikaans version): http://www.scielo.org.za/scielo.php?script=sci_arttext&pid=S0041-47512017000400003.
2. (English version): Tshuma, L. in Webb, P. & Roberts, N. (2017). *The pedagogy of Mathematics in South Africa: Is there a unifying pedagogy?* Johannesburg: Mapungubwe Institute for Strategic Reflection (MISTRA) and Real African Publishers.

CHAPTER 3

MEANING MAKING IN MATHEMATICS TEACHING AND LEARNING THROUGH SEMIOTIC RESOURCES

INTRODUCTION

What is semiotics?

Semiotics is the study of signs. The signs referred to in this chapter are more than just the everyday signs such as road signs; they include drawings, paintings and photographs. Furthermore, these signs also include words, sounds and body language. To encapsulate all these 'signs' into one, the Italian semiotician Umb*erto Eco* described semiotics as concerning everything that can be taken as a sign. It is a theory of language that offers more than an understanding of a word as simply standing for something else. Semiotics is a general science of signs.

Does semiotics have anything to offer in mathematics education?

Since mathematics relies on an intensive use of different kinds of signs (letters, signs for numbers, diagrams, formulas) semiotics is indeed significant in mathematics teaching and learning.

What can semiotics offer to mathematics education?

Historically, mathematics was viewed as a system of mathematical concepts and procedures, before mathematics education research widened to include the role of language in mathematics mediation. Contemporary mathematics education research has further widened to include the use of both language and visuals to make meaning of the mathematical concepts and procedures. Thus semiotics describes how to construct, visualise and communicate mathematical concepts while appreciating culture and history as aspects that play an important role in mathematics meaning making.

Adapted from Radford, (2001).

This chapter begins by exploring semiotic systems before detailing the different types of semiotic resources that contribute to meaning making in mathematics teaching and learning, specifically in under-resourced, multilingual settings. Amongst many theories in semiotics, Goldin's theory of representations explains how external and internal representations (emanating from visuals) contribute to meaning making in mathematics. The chapter concludes by suggesting possible pedagogical implications for enhancing meaning through the use of signs, visuals, gestures and representations in mathematics teaching and learning.

SEMIOTIC SYSTEMS

In mathematics, it is not enough to be able to work with the language alone; mathematics draws on multiple semiotic (meaning-creating) systems to construct knowledge: symbols, oral language, written language, and visual representations such as graphs and diagrams; Schleppegrell (2007). In addition, it uses features such as order, position, relative size, and orientation in meaningful ways. Because concepts that mathematics construct are often difficult to articulate in ordinary language, mathematics symbols express meanings that go beyond what ordinary language can express. For example, mathematics symbols can be used to describe relationships of parts to whole, and to construct trends and patterns of continuous covariation that cannot be presented as precisely in natural language. Visual displays, in the form of graphs and diagrams, can represent the information presented in the mathematics symbolism in ways that language cannot.

There are advantages for each type of display of meaning, whether in natural language, mathematical symbols, or the visual display of diagrams and graphs. A word problem on calculating the perimeter of a rectangular shaped vegetable garden (*sketch provided or to be drawn by the learner*) uses natural language, mathematics symbolism, and graphic representations, requiring learners to recognise the meanings in the interaction of these semiotic systems.

Language, mathematical expressions, and visual diagrams, as well as the gestures and actions of participants in the classroom, together construct meaning, and "...it is only by cross-referring and integrating these thematically, by operating with them as if they were all component resources of a single semiotic system, that meanings actually get effectively made and shared in real life" (Lemke, 2003). Thus, learning mathematics is not just a question of manipulating symbols, but of understanding how different systems for making meaning interact.

THEORY OF REPRESENTATIONS: GOLDIN 2020

The visuals and gestures present above are not only images and diagrams, but also **semiotic resources** for enhancing mathematical meaning. The section below describes semiotic resources as representations as informed by Goldin's theory of representations.

SEMIOTIC REPRESENTATIONS

According to Goldin (2020) semiotic representations or mathematical representations are visible or tangible inscriptions – such as diagrams, number lines, graphs, arrangements of concrete objects or manipulatives, physical models, written words, mathematical expressions, formulas and equations – that encode, stand for, or embody mathematical ideas or relationships. An inscription does not include an interpretation. However, a representation includes reference to some meaning or signification. There are two types of semiotic representation: external and internal.

EXTERNAL REPRESENTATIONS

External representations comprise the conventional symbol systems of mathematics, such as base-ten numeration, formal algebraic notation or the number line. Learning environments are also included, such as those that use concrete teaching and learning materials or computer-based activities. Some representations show relationships visually or graphically, such as number lines, graphs based on Cartesian or polar coordinates, and geometric diagrams; the words and expressions of ordinary language are also external representations. These representations may describe material objects, physical properties, actions and relationships, or objects that are much more abstract (Goldin, 2020).

INTERNAL REPRESENTATIONS

Internal representations comprise learners' constructs of meaning to mathematical concepts such as learners' natural language, visual imagery and spatial representation, problem-solving strategies, and their significance in relation to mathematics. Internal mental representations may be constructed from external representations. They cannot be directly observed, but are inferred from observable behaviour.

Types of cognitive representations include:

- Verbal or syntactic: the use of natural language by individuals, mathematical and non-mathematical vocabulary, including the use of grammar and syntax.
- Figural (imagistic) and gestural systems, including spatial and visual cognitive configurations, or mental images, gestures and body language.
- Mental manipulation of formal notations (numerals, operations, visualisation of symbolic steps to solve an equation).
- Strategic and heuristic processes: trial and error, breakdown into stages.
- Affective systems of representation: emotions, attitudes, beliefs and values with respect to mathematics, or about themselves in relation to mathematics.

Adapted from Goldin, 2020.

INTERACTION BETWEEN EXTERNAL AND INTERNAL REPRESENTATIONS

The interaction between external and internal semiotic representations is essential for teaching and learning. The teaching process is grounded in the nature of the internal representations being developed by learners. The connection between external and internal semiotic representations manifests itself in the use of analogies, images and metaphors, as well as in contrast of similarities and differences.

REPRESENTATION IN MATHEMATICS TEACHING AND LEARNING

In mathematics teaching and learning representations are interpreted as follows:

- An external, structured physical situation, or structured set of situations in the physical environment, that can be described mathematically or seen as embodying mathematical ideas;

- A linguistic embodiment, or a system of language, where a problem is posed or mathematics is discussed, with emphasis on syntactic and semantic structural characteristics;
- A formal mathematical construct, or a system of construct, that can represent situations through symbols or through a system of symbols, usually obeying certain axioms or conforming to precise definitions – including mathematical constructs that may represent aspects of other mathematical constructs;
- An internal, individual cognitive configuration, or a complex system of such configurations, inferred from behaviour or introspection, describing some aspects of the processes of mathematical thinking and problem solving.

Godino & Font (2010)

The relationship of representation (symbols/visuals) between two systems is reversible. Mathematical representations cannot be understood in isolation. A specific formula only has meaning as part of a wider system with established meanings and conventions. Multiple representations of a mathematical concept implies working with different representations of the mathematical concept, leading to meaning making.

MATHEMATICS AS MEANING MAKING

Signs do not operate in isolation; they are always part of an on-going process of semiosis or meaning making. People solve problems, communicate with one another, or get some practical task done, and in the process mobilise semiotic resources, which are not simply isolated signs, but are always part of organised systems of signs, in which each sign has some specific meaning relationship to every other sign in the system. Mathematics is a system of related social practices, a system of ways of doing things. Mathematics has its characteristic ways of doing things: calculating, symbolising, deriving, analysing. Sometimes mathematics exists without any algebraic symbols at all, except arguments written in words, accompanied by diagrams. Mathematics is identified by the kinds of meanings it makes: meanings about addition, subtraction, multiplication, and division; about numerical difference and equality; about geometrical relationships of parallelism, orthogonality, similarity, congruence, tangency, and many other endeavours in mathematical history. It is distinguished by such meanings, whether they are made by writing natural language, by drawing diagrams, or by formulating symbolic expressions. However, mathematical sentences (even in word form) are about kinds of meanings that natural language fails to articulate.

MEANING MAKING THROUGH SEMIOTIC RESOURCES

Meaning is achieved through different semiotic resources for each literacy mode namely: spoken language, written language, visual, audio, gestural, and spatial

(Kalantzis, Cope, Chan & Dalley-Trim, 2016). Learners engage with all of these meaning making practices through a multicultural and/or multilingual lens. There is is a lot pedagogic support for teaching meaning making through spoken and written language, and some resources developed to support teaching meaning making in the visual mode, through 'viewing'. However, there are limited resources available for teaching learners (specifically at primary school level) how to comprehend and compose meaning in the other modes.

- **Written meaning**
  Conveyed through written language in the form of handwriting, the printed page, and the screen. Words, phrases, and sentences are organised through linguistic grammar conventions, register (where language is varied according to context), and genre (knowledge of how a text type is organised and staged to meet a specific purpose). In multilingual contexts learners may write words from their mother tongue using English letters (transliteration).
- **Spoken (oral) meaning**
  Conveyed through spoken language in the form of live or recorded speech and can be monologic or dialogic. Choice of words, phrases and sentences are organised through linguistic grammar conventions, register, and genre. Composing oral meaning includes choices around mood, emotion, emphasis, fluency, speed, volume, pitch, rhythm, pronunciation, intonation, and dialect. Learners may make additional choices around the use of home languages to emphasise meaning.
- **Visual meaning**
  Conveyed through choices of visual resources and includes both still and moving images. Images may include diverse cultural connotations, symbolism and portray different people, cultures and practices. Visual resources include framing, vectors, symbols, perspective, gaze, point of view, colour, texture, line, shape, casting, saliency, distance, angles, form, power, involvement/detachment, contrast, lighting, or naturalistic/non-naturalistic.
- **Visuals as semiotic resources in mathematics**
  In mathematics teaching and learning, visualisation links images and diagrams with using metaphors. The incorporation of visualisation processes is an essential element of any classroom environment and a teacher's daily practice. Ballstaedt (2003) classifies visuals particularly relevant in mathematics teaching and learning into seven main categories namely: nonrepresentational images, depictions (representational images), charts (analytic images), diagrams, pictograms, maps and complex images (infographics):
  - ***Non-representational images:*** These images do not actually represent anything in particular, but instead communicate an aesthetic experience

such as graphic elements or abstract images. They usually do not require any interpretation but they can also induce the projection of meanings.

- ***Depictions, representational images:*** A representational image falls under informative and didactive images. Such visual information is often difficult to communicate verbally.
- ***Charts, analytical images:*** Charts represent links between different elements. Comprehensive charts provide orientation support in complex structural relationships, while simple charts aid memory. There are different types of charts: flow charts, mind maps, time charts.
- ***Diagrams:*** Diagrams are used to visualise quantitative relationships. The exact values remain visible, allowing rough visual comparisons to be made quickly and easily. Diagrams are a form of visualisation, which often make use of visual manipulation: the data shown might be numerically correct, but the visualisation either weakens or strengthens certain messages. There are different types of diagrams available: pie charts, bar charts, line graphs or range diagrams.
- ***Pictograms:*** A pictogram is a simple symbol or picture that induces or suppresses a particular action. The use of pictograms is very popular for data handling.
- ***Maps:*** Maps present an image from above that usually also contains non-visible symbolised information. Maps provide spatial orientation.
- ***Complex images: infographics:*** Infographics are a combination of text and image in which verbal and non-verbal forms of representation are closely linked to each other. Their aim is to provide access to and make complex relationships understandable.

Adapted from Risku & Pircher, (2008).

Visualisation includes creating a mental picture that aids in making sense of a concept. In general, it is about communication through imagery and, in a broader sense, visualisation goes beyond images and physical artefacts to encompass the use of all senses in pursuit of meaning making.

- **Audio meaning**

Conveyed through sound, including choices of music representing different cultures, ambient sounds, noises, alerts, silence, natural/unnatural sounds, and use of volume, beat, tempo, pitch, and rhythm. Lyrics in a song may also include multiple languages.

- **Spatial meaning**

Conveyed through design of spaces, using choices of spatial resources including: scale, proximity, boundaries, direction, layout, and organisation of objects in the space. Space extends from design of a page in a book, a page in a graphic novel or comic, a webpage on the screen, framing of shots in moving image, to the design of a room.

- **Gestural meaning**

Conveyed through choices of body movement, facial expression, eye movements, dance, acting, and action sequences. It also includes use of rhythm, speed, stillness and angles. Gestures and body language may have diverse cultural connotations.

  - **Gestures as Semiotic Resources in Mathematics**

  Communication can be nonverbal such as a nod, a wave of the hand or thousands of other subtle and unsubtle elements of body language which communicate information to the next person. Gestures are part of nonverbal communication and are often used as resources that communicate information within the classroom. Learning and teaching mathematics requires the activation of *a variety of resources*, which are classified into two types of mathematics knowledge: the *body* and its activity with artefacts, and the activity with *signs*. The cognitive significance of the body in mathematics education is relatively recent, and it places the origins of all human knowledge – mathematics included – in bodily experiences and perceptions. On the other hand, knowledge formation is embedded in cultural and social contexts, and the use of signs becomes crucial in cognition. This is of particular interest in the teaching and learning of *mathematics*, a discipline that has always been considered as *abstract* yet the discipline is also based on signs.

  Mathematics education research frames teaching and learning activity within semiotic perspectives, focusing on written semiotic systems such as algebraic symbols. Only recently has attention been focused on including body expressions as *semiotic resources* in teaching and learning mathematics, and looking at their relationship to written mathematical symbolism (Radford, Schubring & Seeger, 2008).

OTHER DIMENSIONS OF SEMIOTICS IN MATHEMATICS TEACHING AND LEARNING

There are three dimensions of semiotics and mathematics education research which go beyond the visuals and representations. The first is the relationship between sign systems (such as natural language, diagrams, pictorial and alphanumeric systems) and the translation between sign systems in mathematical thinking and learning. The second one is semiotics and intersubjectivity. The third one concerns semiotics as the focus of digital teaching and learning materials (Presmeg, 2016).

THE RELATIONSHIP BETWEEN SIGN SYSTEMS AND TRANSLATION

According to Duval, (2006) the relationship between sign systems and the difficulties faced by learners when moving between semiotic registers is complex. As a result, an investigation into the meaning that learners ascribe to their first algebraic formulas expressed through the standard algebraic symbolism suggested that their emerging meanings are deeply rooted in significations that come from natural language and perception.

SEMIOTICS AND INTERSUBJECTIVITY

Vygotsky denotes semiotic mediation is replaced by intersubjective speech field. Since learners always find themselves in an intersubjective speech field, the world and language are given to them as their own. Intersubjectivity is different from intrasubjectivity (Mikhailov, 2006); however, subjectivity is a wide topic falling beyond the scope of this book.

SEMIOTICS AS THE FOCUS OF INNOVATIVE LEARNING AND TEACHING MATERIALS

Mobile phones/tablets, digital mathematics textbooks, teaching and learning materials integrating interactive diagrams, interactive visual examples and visual demonstrations, videos, animations and simulations form part of the contemporary research in mathematics education; and they incorporate visual resources in the mathematics classroom. Semiotics helps to understand the challenges arising from these materials. Examples include:

- Innovative visualisation tools for teaching and learning;
- Design of activities and tasks that are based on interactive visual examples;
- Patterns of reading, using and solving with interactive linked multiple representations;
- Roles of diagrams, animations and video as teaching and learning materials with recent technologies.

Presmeg (2016).

In the absence of ICT technologies mentioned above, cost-effective physical manipulates may be used in under-resourced settings. Examples include: self-made geoboards (made of wood and soft nails) with elastic bands – for presenting plane shapes; wooden cubes – for presenting views of 3-dimentional objects; empty egg trays – for illustrating calculation on fractions; or bottles tops – used as counters; and paper plates – illustrating analogue clocks. Jornet and Roth (2015) illustrate the different types of work learners need to accomplish to relate natural phenomena and different computer-based images and graphs that are used to stand in for natural phenomena. In support of visual- and symbolic-based teaching and learning, the National Curriculum and Assessment Policy Statement (CAPS, 2011) encourages teachers to develop learners' effective use of science and technology, and effective communication using the visual and symbolic in various modes.

THE SEMIOTIC CLASSROOM

According to Fast, (2012), the classroom environment is central to most educational experiences and the meanings made within this environment become the curriculum, whether intended or unintended. For teachers to create a learning environment that prepares learners to engage in critical thinking, the classroom itself must move away from being a 'traditional classroom'. This requires a reorganisation of the classroom in such a way as to ensure participation of all learners, creating an inviting environment that allows learners to fully engage, and therefore learn, through bringing in their own experiences, thereby examining material for its relevance to their lives. To successfully create an environment where all of this becomes possible, involves understanding the meaning that is being made by the learners when they enter the learning environment. This is a significant approach that goes beyond the prevailing practice in today's classrooms, even when the teacher intends to create an engaging environment and facilitate transformative learning sometimes using very limited resources.

PEDAGOGICAL IMPLICATIONS: MEANING MAKING IN MATHEMATICS TEACHING AND LEARNING THROUGH SEMIOTIC RESOURCES

Mathematical understanding and meaning is lost if Foundation and Intermediate Phase mathematics concepts and procedures are not taught simultaneously with language and visual representation. Figure 3.1 below is a visual representation of a simple balance scale (*made of a trouser hanger and two empty mayonnaise jars*) that may be used to introduce the concept of mass:

Figure 3.1: A model of a simple balance scale

a. Teachers should emphasise to learners how mathematical expressions and mathematised visual representations can be translated partially, but never completely, into natural language statements and questions. Conversely, natural language can be partially translated into mathematical symbols.
b. Teachers should explore the use of mathematics in contexts that either gave rise to the mathematics historically, or are among the most interesting uses of the mathematics today.
c. Teachers should expose learners to real, out-of-school settings and practices, in which real people are using mathematics for practical purposes as part of

everyday activities. Learners should observe such applications of mathematics in everyday life, simulate them in a simplified version, and appreciate a sense of their real contexts and level of difficulty.

d. Teachers should use semiotic approaches to present lessons that include practical activities and are learner centred. When teachers are aware of their learners' personalities, backgrounds and present circumstances, the teachers will most likely engage the learners in motivating activities.

These useful strategies come in handy when teaching particular topics such as fractions and ratios. The history of mathematics and diagrams assists teachers in understanding why learners have difficulties at certain conceptual points and in identifying alternative approaches to these topics. The mathematics curriculum and including mathematics teacher education need to give teachers (and learners) an insight into the historical contexts and intellectual development of mathematical meanings, and the practical connections of mathematics with natural language and visual representation. A semiotic approach reconceptualises mathematics not only as a system of signs, but as an integral component of a much larger system of semiotic resources for making mathematical meaning in real, historical contexts.

CONCLUSION

Mathematics education, like many other fields significantly draws upon semiotics. The question is:

What exactly does semiotics have to contribute to mathematics education?

The answer has both a simple and a complicated version. The simple version entails that mathematics is an intrinsic symbolic activity that is accomplished through written, oral, bodily and other signs. Semiotics therefore helps learners and teachers to understand the mathematical processes of thinking, symbolising and communicating. The complicated version entails that processes of thinking, symbolising and communicating are subsumed in more general encompassing symbolic systems. The nature of our ways of thinking and doing that is intertwined with the symbolic systems makes mathematical thinking and discourse entangled with the cultural, historical and political dimensions of life. Semiotics offers an advantage to investigate and transform the signs and sign systems which are part and parcel of our everyday lives including the mathematics classrooms in under-resourced communities. Similarly, as Chapter 2 explored the exploitation of language in mathematics teaching and learning, an understanding of what semiotics can offer in contemporary mathematics education therefore leads to an even greater question:

How can semiotic resources be exploited to improve mathematics teaching and learning, (specifically) in multilingual, under-resourced settings?

Chapters 1, 2 and 3 of this book highlight language and semiotics as some of the transparent resources in mathematics pedagogy on the backdrop of the quality of South African Mathematics Education. **Case Study A** specifically focuses on teacher language proficiency since proficiency precedes the use of a language as a resource. The purpose of the case study is to analyse the Eastern Cape Department of Education (ECDoE) Intermediate Phase (IP) mathematics teachers' language proficiency in English, the prescribed LoLT. This case study is informed by the socio-psycho-linguistics theory. A standardised teacher English language proficiency assessment piloted in five South African universities was administered on 55 Intermediate Phase (IP) mathematics teachers purposefully selected from 16 education districts in the ECDoE. Data were quantitatively and qualitatively analysed. Results show that teachers' language ability in English is very low and the IP teachers who are not proficient in the language of instruction are likely to compromise the quality of mathematics instruction. Since study participants are qualified practicing teachers, this case concludes that the onus is on teacher education institutions to linguistically prepare IP mathematics teachers adequately.

CASE STUDY: A

INTERMEDIATE PHASE MATHEMATICS TEACHERS' LINGUISTIC COMPETENCIES IN THE LANGUAGE OF LEARNING AND TEACHING: THE EASTERN CAPE PROVINCE

INTRODUCTION

Durkin and Shire (1991: 3) state that mathematics education begins and proceeds in a language, and it advances and stumbles because of language, and its outcomes are often assessed in language. Inasmuch as this notion applies to all learning areas, its crucial application in mathematics is highlighted in the example below:

Picture the scene: While presenting a continuous professional teacher development course in mathematical thinking and problem solving, a lecturer assigned a group of Intermediate Phase (IP) mathematics teachers from the Eastern Cape Province of South Africa the following word problem to solve: ***"Enzo a botte le ballon 4 fois. Nathan a botte le ballon 11 fois. Combien de fois ont-ils botte le ballon en tout?"*** On realising that the teachers did not seem to understand the problem, some clues were provided to assist the teachers and the word problem read as follows: ***Enzo*** *kicked* ***le ballon 4 fois. Nathan*** *kicked* ***le ballon 11 fois. Combien de fois ont-ils*** *kick* ***le ballon en tout?*** After fruitless deliberations among the teachers, the lecturer provided a diagrammatic representation of the problem as illustrated in Figure A.1:

Figure A.1: A diagrammatic illustration of a word problem

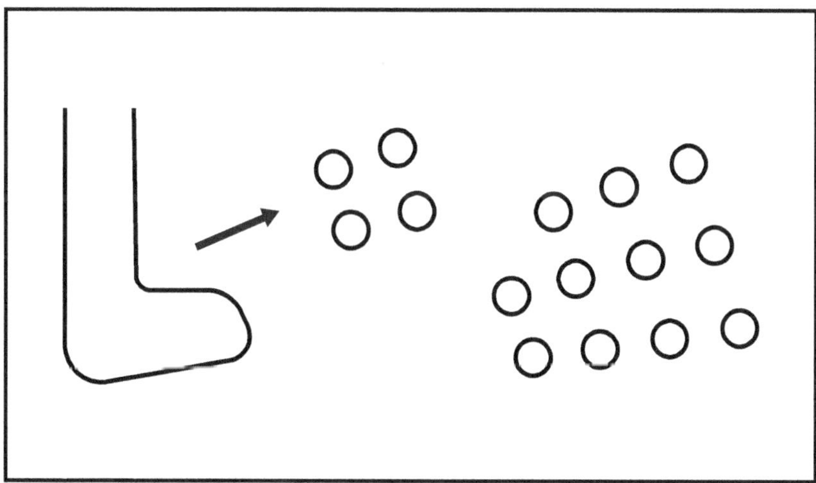

Adapted from Barwell, Leung, Morgan & Street (2002).

The added clues as presented in the diagram seemed to help the teachers and some responded to say the solution of the problem is 15. When the lecturer requested an explanation as to how they arrived at the answer 15, one teacher said it was through guesswork. The lecturer confirmed that the answer is indeed 15, and translated the French word problem into English: ***Enzo kicked the ball 4 times. Nathan kicked the ball 11 times. In total, how many times did they kick the ball?*** It was intriguing for the teachers to discover that the word problem that had seemed so problematic was actually a simple addition sum of **4 + 11 = 15**, yet the teachers had to be provided with a few clues before they could **guess** the answer.

The example above attests to Durkin and Shire's (1991) notion: in the absence of a language barrier (French is a foreign language to the participants of the study) presented by the word problem, solving the addition sum would have been an easy task for the Intermediate Phase (IP) mathematics teachers and they would not have resorted to guesswork. Since mathematics education is pivoted on one's competency in the language in which the mathematics concepts and procedures are presented, how teacher language competency impacts on mathematics education in under-resourced provinces such as the Eastern Cape Province warrants interrogation.

Mathematics has a particular way of using language and its own particular way of expressing ideas, which is termed the 'mathematics register' (Lee, 2006; Pimm, 1987). One way of describing the relationship between a natural language like English and mathematics is in terms of the linguistic notion of register (Durkin & Shire, 1991: 17). Therefore, mathematical language is considered a distinct 'register' within a natural language like English, which is described as "a set of meanings that is appropriate to a particular function of language, together with the words and structures which express these meanings" (Halliday, 1975: 65). Language competence is an individual's ability to speak, read, write, interpret and generally use a language well and efficiently. It is the quality of being adequately or well qualified to use a language for both basic communication tasks and academic purposes (DBE, 2010). Language proficiency is an individual's level of competence to use a language for both basic communication tasks and academic purposes; proficiency also refers to the quality of great facility and competence (DBE, 2010). Thus, proficiency requires higher levels of mastery than just the basics. Linguistic competence is an individual's implicit internalised knowledge of the rules of a language (DBE, 2010). In this case, the terms 'language competency' and 'linguistic competency' are used interchangeably, and the focus is on the teacher language proficiency in using English as a language to mediate meaningful mathematics instruction.

The case study provides a brief description of teacher language competency and mathematics instruction in South Africa; phase level changes in the South African school education system; the research questions; conceptual framing; research design and methodology; and findings and discussion.

Perspectives on Teacher Language Competency and Mathematics Instruction in South Africa

According to Tshabalala, (2012: 22) the majority of teachers in South African schools are not first-language speakers of English. This is not peculiar to South Africa, but common practice in several sub-Saharan countries where English is used as a lingua franca and one of the languages of instruction specifically used from upper primary school up to tertiary level. Difficulty in communicating fluently in the LoLT leads to increased teacher frustration, a slow rate of learning, disciplinary problems and teacher-centred instruction (Setati, 2008; Howie, 2013). Thus, a mathematics teacher who is not competent in the language of instruction faces challenges in imparting mathematical concepts and procedures, and these challenges compromise the quality of instruction that takes place in the classroom. When teachers do not use mathematical language fluently, learners are unable to describe mathematical ideas using appropriate language (Howie, 2001; Murray, 2012).

Phase Level Changes in the South African School Education System

South African schooling is divided into four levels: Foundation Phase (FP), from Reception to grade 3; Intermediate Phase (IP) from grade 4 to 6; Senior Phase (SP) from grade 7 to 9 and the Further Education and Training (FET) band from grade 10 to 12. Several changes take place during the change from FP to IP, such as an increase in the number of learning areas offered, and these learning areas are taught by different teachers (NEEDU, 2013). Of the several changes, this research highlights the change in the language of instruction as the most significant. 90% of South African learners are not native English speakers, and their different home languages are used in the FP while from the IP onwards English is used as the official Language of Learning and Teaching (LoLT).

OBJECTIVES

Interrogating the extent to which IP mathematics teachers in ECDoE schools are proficient in English, the prescribed LoLT seeks to:

i. Determine the extent to which proficiency in English, the prescribed LoLT, relates to English symbol obtained at matriculation.
ii. Determine the extent to which proficiency in English, the prescribed LoLT, relates to initial teacher education qualification.
iii. Determine the extent to which proficiency in English, the prescribed LoLT, relates to experience in teaching IP mathematics.

CONCEPTUAL FRAMING

The conceptual framework grounding this case is Gawned's (1990) Socio-Psycho-Linguistic Model.

Socio-Psycho-Linguistic Model: Gawned (1990)

Gawned's socio-psycho-linguistic framework is based on a model of language learning (Gawned, 1990; Ellerton & Clarkson, 1996). The current research employs the socio-psycho-linguistic framework as it best reflects the nature of interaction between natural language and the mathematics register (mathematical language), and the way that language can influence mathematics learning and understanding as illustrated in Figure A.2.

Figure A.2: A summary of the socio-psycho-linguistic model (Gawned, 1990)

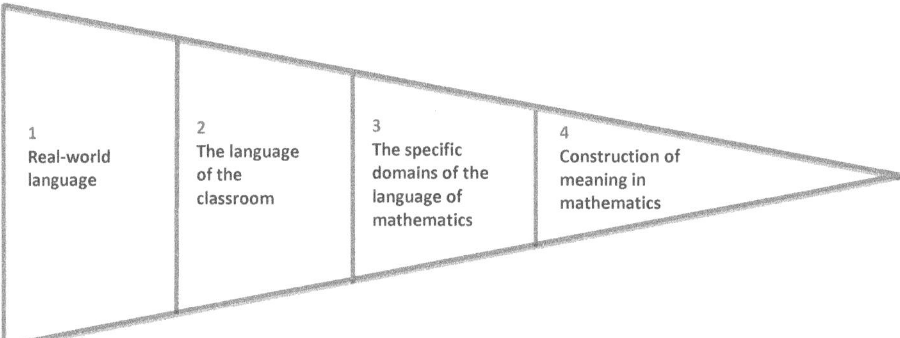

Particularly important in Gawned's model is its acknowledgement that the language of the classroom has a great influence on learners' understanding of mathematics, and that each classroom has a unique culture of its own (Gawned, 1990). Gawned's model also discusses the discourse patterns found in mathematics classrooms. This framework reflects the nature of mathematics classrooms and the way that language plays a key role in learning, particularly the language of the teacher and the textbook, while also highlighting the cultural influences on mathematics education.

Viewing IP mathematics instruction as a socio-psycho-linguistic process emphasises the construction of meaning and draws upon an individual's unique constellation of prior knowledge, experience, background and social contexts. Thus, the socio-psycho-linguistic process is holistic instead of being focused on discrete parts. If teachers understand mathematics instruction as a socio-psycho-linguistic process, they will then be in a better position to incorporate the linguistic elements of mathematics instruction into their classroom practice. As classroom teachers teach mathematics as a socio-psycho-linguistic process, a component of their learners who will end up in teaching and academic careers will also be inclined to adopt that method. As this approach becomes cumulative and widespread, it could revolutionise the way mathematics is taught and learned.

Gawned's (1990) framework demonstrates how language and the language of instruction are key areas in the learning and teaching of mathematics. This has implications for the mathematics learning and teaching process through the medium of a second language, particularly using English as a LoLT. Without discrediting the contribution made by the two frameworks in understanding language issues in mathematics education, one cannot be blind to the fact that South Africa, and the third world in general, have contextual socio-economic factors that impact on education which might have been overlooked in Western contexts.

METHODOLOGY

Based on the Eastern Cape Province's recurrent poor performance in mathematics, this investigation of mathematics teacher language proficiency in the prescribed LoLT was designed as a case study in order to zoom into, and better understand the situation in the province. Ethical clearance to collect data was obtained from University of Stellenbosch Research Ethics Committee and ECDoE Strategic Planning Policy Research and Secretariat Services. The data collection techniques included a teacher language proficiency assessment and questionnaire focusing on demographic data.

The English language proficiency assessment was provided by JET Educational Services and is a standardised teacher assessment, conforming to the South African context and piloted in five different university teacher education departments in the country. The assessment is primarily based on the National Curriculum Statement (NCS) home language level assessment standards for Grade 7 level. The rationale used is that teachers, at the absolute minimum, need to show a proficiency in a language that is two years beyond that of their learners (JET, 2015). The assessment was made up of 47 items and covered a range of topic areas including comprehension, text structures, words, grammar and writing. Test items consisted of open-ended and multiple-choice questions. The level of the assessment is quite low, as it is trying to establish the absolute minimum standards for teachers. Considering that the case study participants are qualified and practicing teachers, the pass rate was thus set at 70%.

The researcher was the sole assessment administrator; however, this was not a challenge as the case study is triangulated. Assessments were administered anonymously through the use of alpha numerical codes. Teachers were given 90 minutes in which to complete the test. On average, teachers spent 75 minutes completing the proficiency test. No teacher required the full amount of time to finish any of the tests.

The teacher demographic questionnaire was developed by the researcher to collect data on gender, age, race group, education districts, home language(s) spoken, university attended, highest qualification obtained, English symbol obtained at matriculation and experience in teaching IP mathematics.

The data collected were analysed using STATISTICA version 13 (Dell Inc. (2015) Dell STATISTICA (*data analysis software system*), version 13 software.dell.com) at University of Stellenbosch's Centre for Statistical Consultation, while JET Education Services conducted descriptive analyses of the quantitative data. Trends emerging from the two independent analyses of the quantitative data were compared and analysed qualitatively.

SAMPLING

The unit of analysis was language proficiency of 55 IP mathematics teachers from the ECDoE. For the purposes of this research, the teachers were studied not so much for their mathematics content knowledge, but specifically as to whether they were competent and proficient in using English as the LoLT in mathematics instruction. The section below details the IP mathematics teacher participants' demographic information.

Demographic data of research participants

Of the 55 teacher participants, 89% were female and 11% were male. 3% were below the age of 25, 2% were between 26 and 29, 2% were between 30 and 34, and the majority (93%) of the participants were above 35. The majority (96 %) of the participants belonged to the black race group, 2% were coloured, 2% were white and none were Indian. The distribution of participants across race groups is closely related to the participant teachers' home languages. IsiXhosa is home language to 91% of the participants, seSotho to 7% of the participants and Afrikaans to 2% of the participants. It can be inferred that teachers of the black race group constituted those whose home language is isiXhosa and seSotho, while the participants of the coloured and white race groups constituted those speaking Afrikaans as their home language.

The Eastern Cape is a relatively large province divided into 24 districts. The districts where the participant teachers were deployed is indicated in Figure A.3.

The teachers were from only 16 of the 24 education districts in the Eastern Cape Province (the full list of districts: *Bisho, Butterworth, Cofimvaba, Cradock, Dutywa, East London, Fort Beaufort, Graaff-Reinet, Grahamstown, King William's Town, Lady Frere, Libode, Lusikisiki, Maluti, Mbizana, Mt Fletcher, Mt Frere, Mthatha, Ngcobo, Port Elizabeth, Queenstown, Qumbu, Sterkspruit and Uitenhage*).

27% of the teachers studied at teachers' colleges, which have since been closed after policy changes stipulating that all Initial Teacher Education (ITE) programmes be offered by the various university departments of education; 25% of the teachers studied at University of South Africa, 10% of the teachers studied at University of Pretoria, while 5% of the teachers studied at University of Zululand. For finer details regarding the teachers' level of education, information on their highest qualification obtained in teacher education was captured. The majority of teachers (49%) only had a Diploma in Education, 15% had a Bachelor's Degree in Education and 36% had either an Advanced Diploma or an Honours' degree. According to these statistics, all the teachers were qualified including two outliers: an Adult Basic Education and Training (ABET)[1] trained teacher and a teacher currently enrolled for a Masters degree in Education.

Since the current research is based on interrogating the teachers' proficiency in English, the prescribed LoLT in most South African Schools, it was necessary to

1 ABET is now called Adult Education and Training (AET).

Figure A.3: Histogram of teacher participants' school districts

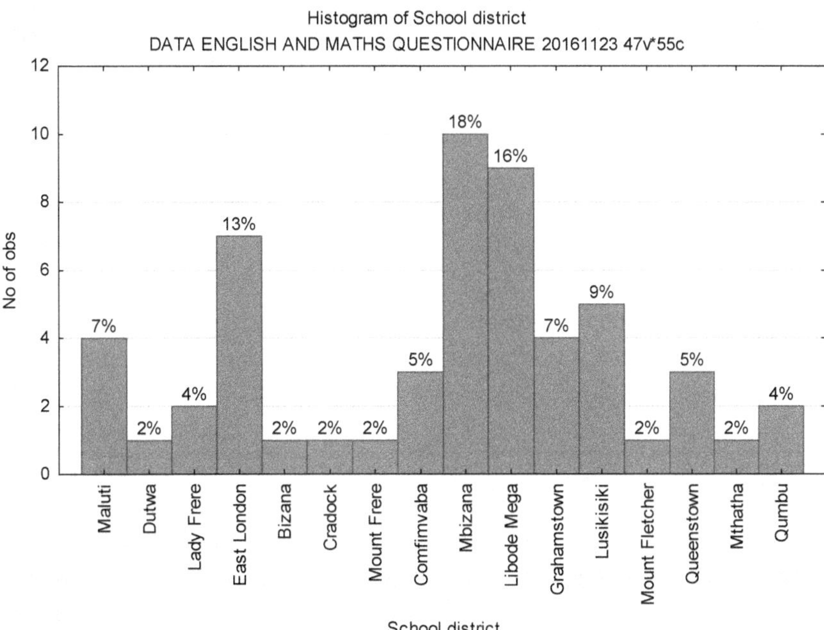

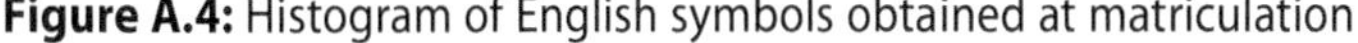

Figure A.4: Histogram of English symbols obtained at matriculation

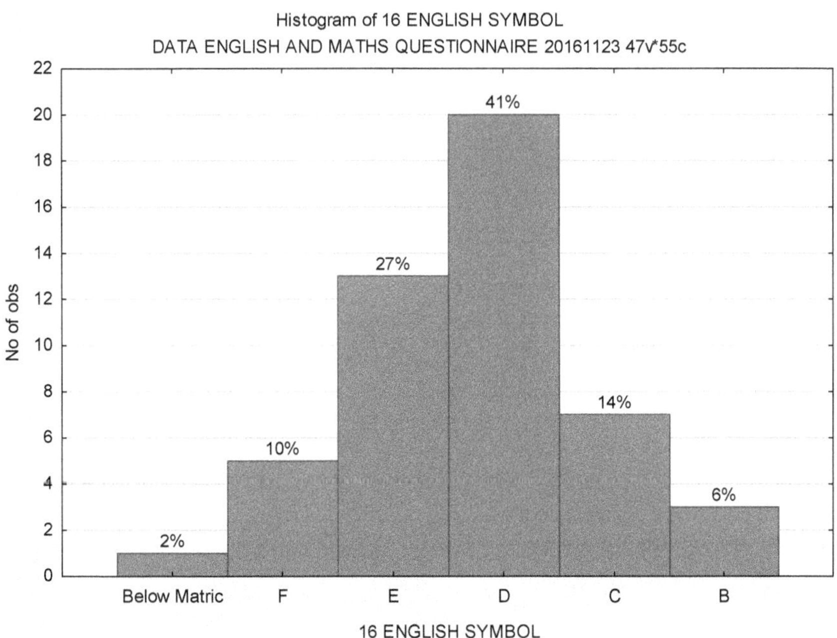

capture the symbols the teachers had obtained in English at matric[2]/school-leaving level. The matric English symbols are indicated in Figure A.4.

None of the teachers obtained symbol **A**, 6% obtained symbol **B**, 14% obtained symbol **C**, 41% obtained symbol **D**, 27 % obtained symbol **E**, 10% obtained symbol **F**, and 2% failed matric English. It is noteworthy that the majority of the teachers obtained symbol **D** and below. However, as established in previous research, matriculants who settle for teacher education qualifications are usually not the *crème de la crème* (Tshuma, 2017), and more than half of South African teachers indicated that teaching was not their first profession of choice (Ainley & Casterns, 2018). Figure A.4 provides information to confirm these attestations. Be that the case as it may, the fundamental question is: if a university (regardless of international ranking) claims to be a top educational institution, shouldn't it admit students with the lowest matric scores and turn them into star graduates?

To ascertain the participants' teaching experience, the number of years they have been practising as IP mathematics teachers was captured. 19% of the participants had been teaching IP mathematics for two years and less; this could be attributed to the fact that some teachers could have been teaching other phases and/or subjects such as Technology and were moved to teaching IP mathematics when their subject specialisations were removed from the curriculum. 32% of the participants had been teaching IP mathematics for an average of 5 years, while 27% of the participants had been teaching IP mathematics for an average of 12 years and these could be considered as seasoned IP mathematics teachers.

FINDINGS

The research question was: **To what extent are IP mathematics teachers in ECDoE schools proficient in English, the prescribed LoLT?** and the key finding was that sampled teachers ***indeed*** exhibited a lower proficiency than what is expected of a grade 7 learner. Based on JET's performance scales, a consolidation of these performance scales is presented in Table A.1:

Table A.1: English Proficiency assessments scores

	ENGLISH LANGUAGE PROFICIENCY SCORES
70% and above	(0) 0%
50% - 69%	(6) 11%
Below 50%	(49) 89%

Key:

- **70% and above:** satisfied requirements
- **50% - 69%:** cause for some concern
- **Below 50%:** cause for serious concern

2 A matric certificate is offered after completing the 12th grade of the South African ordinary school system. The issuing of certificates is standardised and quality assured by Umalusi Council in conjunction with the Department of Basic Education (DBE).

None of the teachers met the satisfactory requirements in the English language proficiency assessment. While 11% of the teachers scored between 50% - 69%, and thus are causes for some concern, 89% of the teachers scored below 50% and are causes for ***serious concern***. Thus, the majority of the teachers are causes for serious concern in this language proficiency assessment. These consolidated scores reiterate that much more still needs to be done to improve the linguistic competencies of IP mathematics teachers.

The teacher English language proficiency scores are presented in a histogram in Figure A.5 in order to see the nature of the distribution of the scores as well as to identify possible outliers:

Figure A.5: Histogram of English Language Assessment %

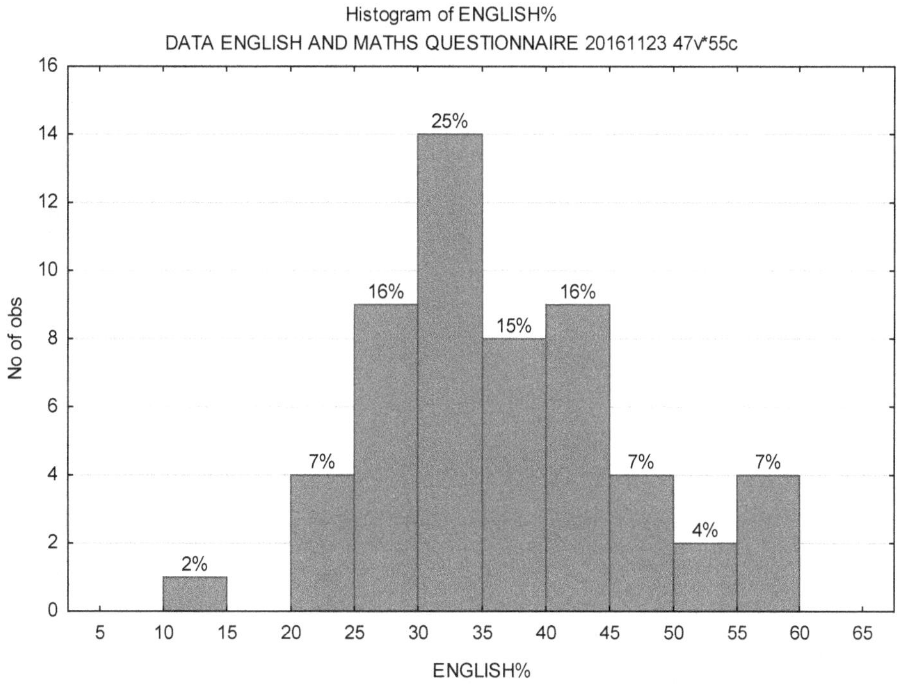

Out of the 55 teachers who wrote the test, the minimum score obtained was $\frac{9}{80}$ = 11%; while the maximum score was $\frac{48}{80}$ = 60%. It is troublesome to note that on average, teachers exhibited a lower proficiency than what is expected of a Grade 7 learner.

Research analysing initial teacher education across five South African universities reveals that "... the situation with respect to the language of learning and teaching, predominantly English, is of particular concern" (Taylor, 2015: 23). The participants' English language proficiency as exhibited in Figure 5 confirms Taylor's findings. However, merely establishing and/or confirming this low proficiency level is just the beginning. More needs to be done to change the status quo and ultimately contribute to the improvement of teacher classroom practice.

Teacher Proficiency in the LoLT and English Symbol in Matriculation

Demographic data presented the teacher participants' English language symbols obtained at matriculation. The data on teacher English language matriculation symbols were correlated to the teacher English language proficiency scores. Figure A.6 illustrates the correlation between the teachers' English language proficiency level versus English symbols obtained at matriculation.

Figure A.6: English symbol at matriculation vs English Language proficiency assessment means

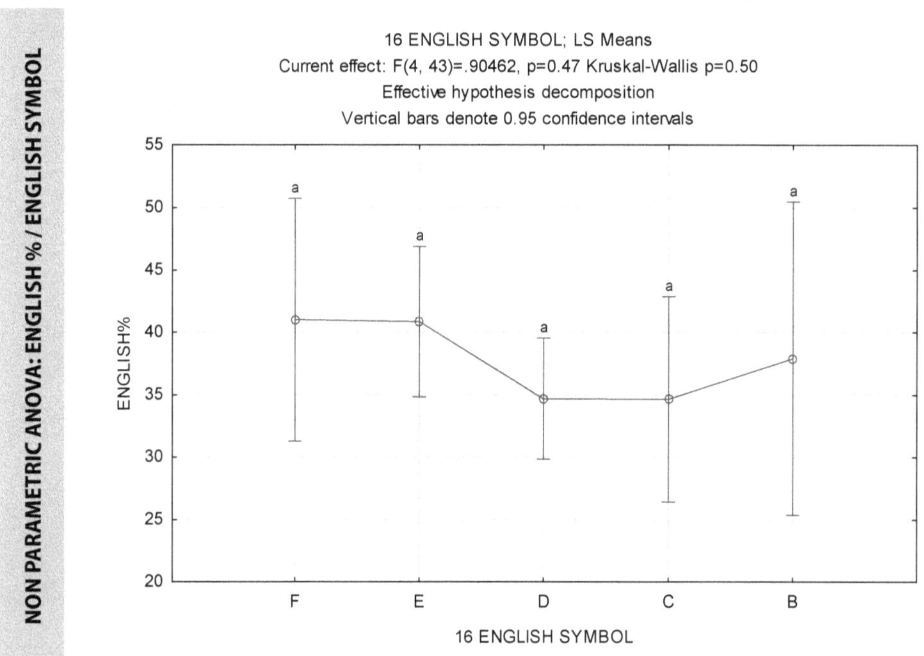

In the demographic data it was established that the majority of the teachers obtained symbol **D** and below; and the group that obtained the D symbol had the lowest LS mean; thus there is a correlation between the matric English symbol and the English language proficiency level.

Teacher Proficiency in the LoLT and Teacher Initial Education Qualification

Closely related to the matric English symbol obtained is the teacher qualification as established in participant demographic data. According to the demographic information, all the participant teachers are qualified; however, the learner performance at GET level as detailed in literature is significantly low, thus a gap exists. The fact that all participants are qualified teachers is a good starting point; however, if these qualified teachers are exhibiting language deficiencies as indicated in the teacher language proficiency assessment, it means there is still more that needs to be done. Thus, the qualifications provide a good starting point to further develop the teachers to become better classroom practitioners in order to bridge the gap between the qualified teachers and the significantly low learner performance at GET level. The next steps should be a review of the teacher education curriculum during and after ITE in order to provide modules that will linguistically upskill teachers for the purposes of changing the status quo.

With regard to the two outliers identified in the demographic section, the ABET trained teacher who ended up in mainstream classroom teaching is recommended for furthering their studies and alleviating the shortage of mathematics teachers. This research envisages that such teachers need special upskilling in order to cope with the demands of mainstream classrooms. The requirements of an ABET course are not the same as those of a formal curriculum and therefore, their upgrading into the mainstream curriculum teaching posts needs to be sufficiently supported. On the other hand, one of the participants was in the process of completing a Masters Degree in Education. By virtue of being in the process of completing such a high-level qualification, it would have been expected of this participant to obtain relatively higher marks in a basic English language proficiency assessment; however, this was not the case and arguably a possible cause of concern.

Teacher Proficiency in the LoLT and IP Mathematics Teaching Experience

Demographic data statistically analysed the teacher participants' experience in teaching IP mathematics. The data on experience in teaching IP mathematics were correlated to the teacher English language proficiency assessment scores. Figure A.7 illustrates the correlation between English language proficiency levels versus experience in teaching IP mathematics.

As established in the demographic data section, about 60% of the participants are experienced teachers who have been teaching IP mathematics for between 5 and 12 years. As illustrated in Figure A.7, with a correlation coefficient of $r = 0.1159$, there is a positive correlation between English language proficiency and experience in teaching IP mathematics. This correlation implies that as teachers gain more experience in teaching mathematics, their proficiency in English language somewhat improves and this could be attributed to the exposure the teachers gain as they engage with the teaching and learning content. Further research is needed to establish and confirm to what extent prolonged exposure to teaching and learning content alone improves teacher language proficiency in the LoLT.

The sections above provided information gathered through the teacher English Language Proficiency assessments by initially analysing the raw scores and then relating the scores to data collected from the teacher demographic data questionnaires. The section below presents samples of teacher responses from sub-Section E of the English Language Proficiency Assessment with a total achievable mark of [4]. Although forming part of the English language proficiency assessment, the

Figure A.7: Correlation between experience (Years) vs English %

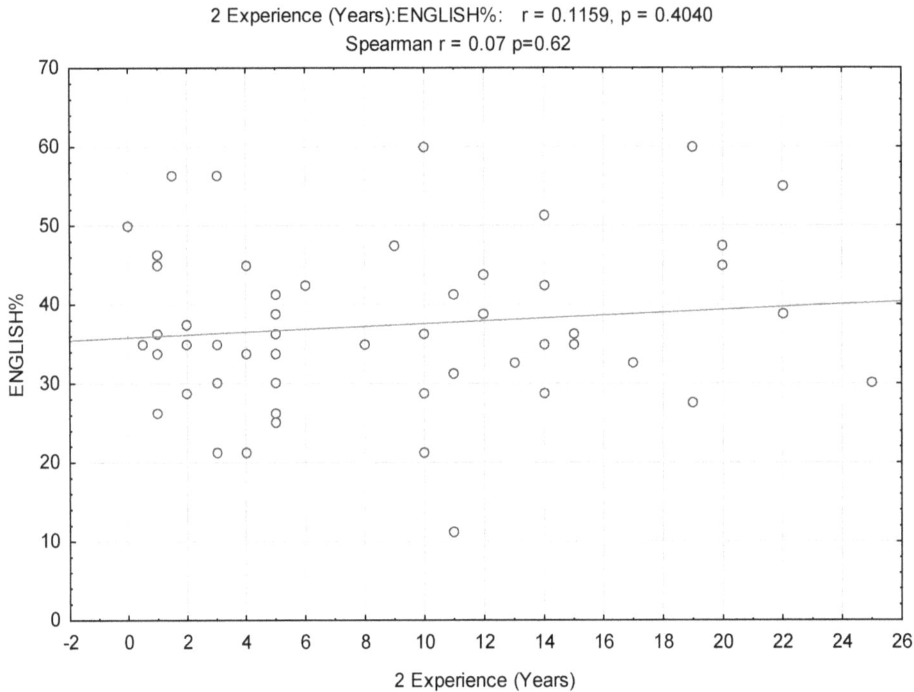

question was based on a typical lower cognitive demand *mathematical word problem* which requires simple recall of facts and interpretation of tabulated information as illustrated in Table A.2:

The sample responses illustrated in Table A.2 exhibit a variety of linguistic challenges ranging from spelling mistakes; punctuation (*omission of the question mark*); grammatical mistakes; repetition and duplication of information; provision of information that does not answer the question; incomplete answers and unattempt to answer questions. Thus, the teachers' language competency is bordering around mediocrity as portrayed in the language competency assessment. However, the assessment component particularly illustrated above (*sub-Section E*) had the highest scores; thus, teachers are relatively comfortable with answering questions with lower-order linguistic demands; and, such questions constitute only a minority of the teaching and learning content at IP level. Inasmuch as the relatively high performance in lower-order questions is highly commended, IP teachers need to broaden their command of the English language to include even higher-order questions for the purposes of improving the quality of instruction using English as LoLT.

Table A.2: Sample Teacher Responses to a Typical English Language Proficiency Assessment Question

QUESTION

SECTION E.

Read the table of results from a survey which asked children about their favourite foods. Note that they could choose more than one food item and many did so.

Table 1: Children's favourite foods

FOOD ITEM	PERCENTAGE OF CHILDREN WHO CHOSE THE ITEM
Chips	36%
Fresh vegetables	15%
Toasted cheese	25%
Chicken	65%
Hamburgers	27%

37) Write two questions that you could ask Grade 4 to 6 learners about the information in the table.

a) ____________________ (2)

b) ____________________ (2)

Total marks for Section E = 4 marks

T10

37. **Write two questions that you could ask Grade 4 to 6 learners about the information in the table.**

a. Which is most favourite food children likes?

b. Which is the least of favourite food

T40

37. **Write two questions that you could ask Grade 4 to 6 learners about the information in the table.**

a. ____________________

b. ____________________

Adapted from JET, 2015

RECOMMENDATIONS

The section above presented the findings from the current research, and the section below presents the recommendations based on these findings. These recommendations focus on the different target audiences in initial teacher education, namely: prospective IP mathematics teachers, pre-service IP mathematics teachers, in-service IP mathematics teachers as well as decision makers responsible for deploying newly qualified teachers.

Recommendations: Prospective IP Mathematics Teachers at Registration Level

English second-language speakers' academic success is dependent on their proficiency and competency in academic language (Brown, 2007). Against the backdrop of the South African education system, providing essential skills that will ensure language competency among aspirant IP mathematics teachers would be positive way of modelling good practice for emulation by the teachers in their own classroom settings where the majority of the learners are second-language speakers of the English language. Low levels of academic performance and unsatisfactory overall rates in some South African schools and higher institutions of learning suggest that responses to the teaching and learning needs of the underprepared learners are not sufficient (SAUVCA, 2003; Parades, 2010; Eid, 2012). At university registration, prospective IP mathematics teachers should have good passes in matric English. If, due to quota systems, enrolment includes prospective teachers with below-average passes in the language of instruction, language bridging courses should be introduced to ensure that prospective teachers attain the required minimum level.

Recommendations: Pre-service IP Mathematics Teachers

With a maximum of 60% and a minimum of 11% obtained in the teacher English language competency assessment, teachers revealed a lower proficiency in English than what is expected of a Grade 7 learner. Even though English is a foreign language to the majority of the teachers who use it as the language of instruction in South Africa, the mere fact that the current LiEP requires teachers to use English as the medium of instruction calls for high levels of proficiency in that language. The Minimum Requirements for Teacher Education Qualifications (MRTEQ)[3] policy documents require IP teachers to at least have Additional Language proficiency in English and prescribes that one of the competencies of a newly qualified teacher is to "know how to communicate effectively in general, as well as in relation to their subject(s) in order to mediate learning" (MRTEQ, 2015: 20). However, the answers provided by the teachers in the English Language Proficiency Assessment revealed grammatical and idiomatic errors and this research consequently infers that

3 The MRTEQ provides "a basis for the construction of core curricula for Initial Teacher Education (ITE), as well as for Continuing Professional Development (CPD) Programmes that accredited institutions must use in order to develop programmes leading to teacher education qualifications" [p 8]. The policy is based on the 2011 MRTEQ policy, with revisions in place in order to align it with the Higher Education Qualifications Sub-Framework (HEQSF).

Additional Language proficiency in the medium of instruction is not good enough for an IP mathematics teacher. Therefore, pre-service teachers currently enrolled for IP mathematics education courses need to be upskilled in the prescribed LoLT.

Recommendations: In-service IP Mathematics Teachers

Since South Africa's democratisation, English has become the lingua franca in public life and the language is prevailing as the LoLT at primary and secondary school levels as well as in higher education (Cuvelier, Du Plessis & Teck, 2003). With many educational institutions in a state of change, Cuvier et al. (2003) further contend that School Governing Bodies (SGB)s are mandated the power to decide on the LoLT for their schools. A common practice that has been adopted by most SGBs is additive bilingualism, where FP is taught in mother tongue, while English is taught as a first additional language in preparation for a transition that takes place in the IP where English becomes the LoLT. While this is a plausible attempt to balance the attention given to marginalised languages, SBGs and School Management Teams (SMT)s that have adopted this policy need to further support it by mandating stakeholders in teacher education to ensure the transition is smooth enough and does not compromise the quality of IP mathematics teaching and learning.

Recommendations: Decision Makers Responsible for Newly Qualified IP Mathematics Teachers' Deployment and Class Allocation

Proficiency in English is necessary for teachers to engage in high-quality mathematics instruction in English. The fact that the majority of South African teachers (*and learners*) in ordinary schools are not first-language speakers of English, but are prescribed to use English from the IP level onwards without much preparation for the phase change presents a gap in policy, practice and research. Addressing this gap needs to be prioritised for meaningful mathematics education to take place in multilingual IP classrooms. At school level, SMTs must use proficiency in English as one of the selection criteria used to allocate Grade 4 (*the entry level of IP*) mathematics teachers. Education officials responsible for deploying newly qualified IP mathematics teachers in under-resourced schools should ensure that these teachers have a firm grasp of the language of instruction. If not, these teachers should be enrolled for continuous teacher development courses that prioritise proficiency in the language of instruction.

The specific contribution made by this research is supplementing the current body of knowledge regarding language in mathematics education in general, and specifically on teacher language competencies in the prescribed LoLT. Without promoting hegemony or dominance of the English language, immediate measures need to be put in place to capacitate pre-service as well as in-service IP mathematics teachers to be confident and proficient in the use of English as LoLT, especially at IP level where the transition from mother-tongue to second-language instruction occurs. Failure to prioritise these eminent issues will continue to see South Africa occupying the lowest position in its mathematics and science education (TIMSS,

2013) compared to other sub-Saharan countries. Without drastic changes, under-resourced schools, staffed by under-qualified teachers will continue producing half-baked school leavers, who cannot cope with the demands of tertiary education.

CONCLUSION

Teacher language competency in the prescribed language of instruction is a critical, but almost tragically forgotten aspect of teaching and learning IP mathematics. To avoid settling for mediocrity, trivialising mediocrity as if it is the norm and/or guessing what the LoLT requires of them, teachers need to be proficient enough in the language of instruction. Teachers cannot do this alone without the support of other stakeholders in education. Therefore, didacticians, teacher educators, educational linguists and teacher education curriculum developers are hereby challenged to provide adequate linguistic preparation and support to pre-service as well as in-service mathematics teachers for the attainment of language skills which can be used to improve mathematics instruction. Teachers who are confident and proficient in the LoLT are better able to deal with the linguistic needs of IP learners within the South African education system whose language of instruction has just changed from home language to second language, and these teachers are most likely to produce better results.

CHAPTER 4

FACTORS INFLUENCING WORD PROBLEM SOLVING

INTRODUCTION

Why are word problems viewed as difficult?

Mathematics is a subject that deals with problems which involve a process of analysis, computation and other mental skills, and word problems are frequently used to teach mathematics because they have been used as the primary means of instruction for thousands of years. Furthermore, they have always been an important part of mathematics education, in which the linguistic component is a fundamental part because problems are embedded within a text and difficulties in solving the problem are related to terminology (Swetz, 2012). In word problems, relevant information is presented in the form of a short narrative rather than in mathematical notation. Learners from foundation phase through to tertiary education face difficulties in solving word problems. Both linguistic and numerical complexity contribute to the difficulty in solving word problems.

To establish factors contributing to the difficulties faced by learners in solving word problems, Daroczy, Wolska, Meurers and Nuerk (2015) reviewed studies conducted in the past three decades. This chapter draws on Darocz et al.'s (2015) systematic overview of linguistic and numerical aspects relevant to solving word problems as well as their interaction, and concludes with a ten-step instructional model that can be adapted in mathematics education to improve learner performance in solving mathematical word problems.

INTERACTION BETWEEN LANGUAGE AND CONTENT IN WORD PROBLEMS

One of the most challenging tasks of a mathematics teacher for English Language Learners (ELLs) is balancing the linguistic and cognitive demands of learners' tasks. Teachers always have to ask themselves: What is the language my learners need to succeed in this mathematics task? Cummins (2000) states that language and content are acquired most successfully when learners are challenged cognitively and provided with the contextual and linguistic scaffolding required for successful task completion. Mathematics, like every other academic discipline, has a complex and specialised language that is different from everyday conversation (Sigley & Wilkinson, 2015). Therefore, proficiency in language may affect mathematical achievement on instructional and assessment levels (Kempert, Saalbach, & Hardy 2011; Pavón, & Cabezuelo, 2019).

The link between language and mathematics is especially evident in the case of word problems. It is known that learners' performance on arithmetic word problems is a reliable predictor of their subsequent mathematical competence. Solving arithmetic problems is a cognitive task that relies on language processing (Van Rinsveld, Brunner, Landerl, Schiltz, & Ugen, 2015; Pavón, & Cabezuelo, 2019). Thus, the learning of mathematics is strongly related to language processes. According to Bialystok (2001), language and mathematics share common features such as abstract mental representation, conventional notations, and interpretive function, making mathematics a subject that has cognitive effects on ELLs.

LANGUAGE PROFICIENCY AND MATHEMATICAL COMPLEXITY

A high linguistic proficiency has a positive impact on mathematics and literature reveals a significant relationship between language and mathematics achievement. A cognitive modelling process is required to solve mathematics word problems, in which learners identify and extract relevant pieces of information and, at the same time, suppress any misleading or irrelevant linguistic or numerical information (Kempert, Saalbach & Hardy, 2011; Pavón, & Cabezuelo, 2019) that is embedded in the problem context. Variations in the linguistic aspect of word problems affect problem representation and problem-solving performance. According to Abedi and Lord (2001), the following features can be revised in order to adapt the linguistic register of mathematical word problems to the linguistic level of learners:

- Familiarity or frequency of non-math vocabulary: change of unfamiliar or infrequent words
- Voice of verb phrases: change passive to active verb forms
- Length of nominals: shorten long nominals
- Conditional clauses: separate conditional sentences or change the order of conditional and main clause
- Relative clauses: remove or recast
- Question phrases: change complex question phrases to simple question words
- Abstract or impersonal presentations: make information more concrete

Some changes involve more than one feature. However, linguistic adaptations must be done cautiously since simplification and elaboration processes may result in undesired outcomes. Proficiency in the language of learning and teaching and arithmetic skills can strongly predict learners' ability to solve mathematical word problems. In order to classify learners by linguistic proficiency, reading comprehension could be used as a suitable instrument (Kempert, Saalbach, & Hardy, 2011; Pavón, & Cabezuelo, 2019).

INDIVIDUAL DIFFERENCES AND SOCIAL FACTORS

Unsuccessful word problem solvers can experience negative social health and life outcome. Beyond social consequences, some research has focused on individual differences and group differences, such as learners with and without learning disabilities and learners with and without developmental disabilities. Different learners – such as those with calculation difficulty, or word problem difficulty – may struggle with different types of word problems. Besides domain-general capabilities like intelligence quotient (IQ), the roles of domain-specific knowledge and processes were investigated to get a complete account of problem solving – basic cognitive abilities, visual, reading skills, mathematical skills, and metacognitive abilities involved in the solution process. Boonen et al. (2014) and Oostermeijer et al. (2014) explored the role of spatial ability and reading comprehension in word problem solving, since good word problem solvers do not select numbers and relational keywords but create a visual representation (Darocz et al., 2015; Stacey, 2012; Schley & Fujita, 2014; Powell & Fuchs, 2014; Darocz et al., 2015; Kingsdorf & Krawec, 2014).

Social factors like schooling, teachers and peers are critical because the way of responding, the assessment criteria, the presence of illustrations next to the text, or solution models used by the teachers influence word problem performance. Learners and teachers usually exclude real-world knowledge from their word problem solution, which is consistent with the observation that the problem solving process is influenced by social, cognitive and epistemic behaviour settings. Linguistic and pedagogical factors affect learners' understanding of arithmetic word problems. Both individual differences and social factors contribute to word problem performance (Berends & Van Lieshout, 2009; Jimenez & Verschaffel, 2014; Gasco & Villarroel, 2014).

SUBCATEGORIES OF WORD PROBLEMS

Different types of word problems are presented for various learner levels. Swanson et al. (2013) investigated the role of strategy instruction and cognitive abilities on word problem solving accuracy. The mathematical word problems they used were: addition, subtraction and multiplication without any further description of the problem type. Different categories of word problems may lead to different solution strategies and different error types. Different semantic problem types result in different errors and have a different difficulty level (Vicente et al., 2007).

The *mathematical content* of word problems is also a significant factor. Algebra word problems require translation into a mathematical formula, whereas arithmetic word problems are solvable with simple arithmetic or even mental calculation. In contrast to arithmetic word problems, algebraic reasoning word problems share the same numerals and signs (Powell & Fuchs, 2014) and the manipulation of those numbers and signals differs based on the question or expected outcome.

Standardised phrases and the *idea that every problem is solvable* are important attributes of many word problems. Textbooks suggest that every word problem is solvable and that every item of numerical information is relevant (Pape, 2003). There are non-standard word problems (Jimenez & Verschaffel, 2014) which can be non-solvable word problems or, if they are solvable, some have multiple solutions and may contain irrelevant data (Csikos et al., 2011). Learners give incorrect answers to non-standard word problems because these seem to contradict their mathematics-related beliefs learned in the classroom.

The dominant factor in determining the learners' solution strategy is semantic structure. Solution strategies have systematically been the focus of word problem research and have addressed the following questions:

- How do learners and teachers solve word problems?
- Why do they make different errors and at which level of the solution process do they do so?
- Which kind of semantic representation do they create of the word problem?
- Which skills are necessary for the solution process?
- Are teachers sufficiently skilled to teach word problems?

Adapted from Darocz et al (2015)

Theories on word problem solution processes draw on text comprehension theories. When solving problems, the learner first integrates the textual information into an appropriate situation model/mental representation of the situation being described in the problem, which then forms the basis for a solution strategy. Solving word problems is not a simple translation of problem sentences into equations. Often both word problems and the corresponding numerical problems are done without language translation (Schley & Fujita, 2014). Word problem solving is an exercise in text processing required for understanding the problem which is highly dependent upon language comprehension skills (Jimenez & Verschaffel, 2014; Kingsdorf & Krawec, 2014). Successfully solving word problems requires at least three distinct processes:

i. understanding and constructing the relation between text and arithmetic task,
ii. linguistic understanding of the word problem itself, and
iii. solving the arithmetic tasks.

Many learners can successfully solve calculation tasks and possess good text comprehension skills, but fail to solve word problems correctly. This suggests that other factors, such as solution strategies and building up a mental model of the task, also play a major role in the word problem performance.

LINGUISTIC COMPLEXITY

In linguistics, complexity is grounded in research on language evolution and language acquisition (Bulté & Housen, 2012; Ellis, 2003). Linguistic complexity involves the words and their lexical and morphological aspects, the way these words can be combined syntactically to form sentences, the text structure, and overall discourse. The implication of linguistic encoding differences is twofold: first, the difficulty of word problems is language-specific, thus linguistic manipulation leading to increased word problem complexity in one language may not have an effect on another, more complex language. Second, the performance of ELLs on word problems presented in English may be affected by the differences between the learner's mother tongue and English, the language of the problem presentation. In the following two sections, the main findings on aspects of linguistic complexity that affect performance are briefly summarised.

STRUCTURAL FACTORS

The relationship between linguistic structure and learner performance on word problems depends on the number of letters, words, sentences, mean words, and sentence length, or the number of complex and long words. Research interrogating comprehension difficulties in word problems for English language learners (ELLs) reveals the following. Regarding vocabulary, comprehension difficulties which result in problem-solving difficulties for ELLS emanate from the presence of unfamiliar words, polysemous words, idiomatic or culturally specific lexical references. Regarding sentence structure, factors that play a role include noun phrase length, the number of prepositional phrases and participial modifiers, the presence of passive voice and complex clause structure such as relative, subordinate, complement, adverbial, or conditional clauses (Abedi & Lord, 2001; Shaftel et al., 2006; Thevenot et al., 2007; Martiniello, 2008; Darocz et al., 2015).

SEMANTIC FACTORS

A factor that is related to word problem difficulty is the presence or absence of explicit verbal cues whose semantics hint at the expected operation and directly lead toward the solution. Verbal cues include words and phrases of different categories: conjunctions ("and" for addition), adverbs ("left," "more than," "less than" for subtraction), or determiners ("each" for multiplication). Reserch reveals that learners focus on linguistic verbal cues and perform translation directly to the mathematical operation (Van der Schoot et al., 2009). Verbal cues lead to mathematical interpretation; even small differences in phrasing in cue words can cause significant changes in performance. This is especially relevant for young learners who connect words such as "join," "add," "get," "find," or "take away" with concepts such as *putting together*, *separating*, *giving away*, or *losing*. A problem can be reworded by adding verbal clues so that the underlying mathematical relation is more explicit. For example, the word problem:

"There are five marbles. Two of them belong to Mary. How many belong to John?"

can be reworded as

"There are five marbles. Two of them belong to Mary. The rest belong to John. How many belong to John?"

Adapted from Cummins (1991)

This kind of conceptual rewording is useful to improve learners' performance on word problems, and thus changes in wording can influence representation (Vicente et al., 2007).

Semantic or object relations between the objects described in the problem also relate to difficulty. The correlation between object relations and mathematical operations reflects a structural correspondence between semantic and mathematical relations. For this reason, the semantic structure properties of a word problem are an important factor contributing to difficulty rather than the syntactic structure (Yeap & Kaur, 2001). Information load and the presence of content irrelevant to the core solution (numerical or linguistic distractors) result in higher error rates. Semantic structure of word problems influences learners' choice of mathematical solution strategy.

NUMERICAL COMPLEXITY

Arithmetic word problems have to be transformed mentally into an arithmetic problem and usually require an arithmetic solution. This means transforming words and numbers into the appropriate operation. Since the arithmetic problem has to be solved in the end, numerical representations and arithmetic processes will also play an important role in the solution process (Nuerk et al., 2011). Numerical processes in current word problem research are categorised into five categories:

i. the property of numbers;
ii. required operation;
iii. mathematical solution strategies;
iv. relevance of the information; and
v. other numerical processes and representations.

NUMBER PROPERTIES

Hines (2013) suggests that parity influences the difficulty of addition and subtraction, but not multiplication, and tasks containing odd numbers are more difficult than those with even ones. Furthermore, most word problems, especially for learners, contain single-digit numbers (Powell & Fuchs, 2014; Haghverdi et al., 2012).

REQUIRED OPERATION

Most errors stem from learners' failure to understand the language used in word problems, thus, the linguistic embedding of the calculation problem (Schumacher & Fuchs, 2012), and arithmetic computation errors themselves (Raduan, 2010; Kingsdorf & Krawec, 2014). Some errors may result from correct calculations performed on incorrect problem representation and different operations may lead to different solution strategies. The most common operations used in word problems are addition and subtraction (Schumacher & Fuchs, 2012). Choosing the correct operation depends on the type of the numbers given in the problem. Carry operations (*such as, 28 + 47; the decade value 1 from the unit sum **15** has to be carried over to the decade sum*) make multi-digit addition more difficult for learners and adults (Nuerk et al., 2015). However, solution strategies differ between learners and adults – primary school learners compute and search for the correct results, while adults also decide based on the rejection of the incorrect result. Three different cognitive processes are identified for adults: unit sum calculation, carry detection, and carry execution (Moeller et al., 2011). Inability to execute one of these processes may lead to poor performance in carry problems. Carry addition problems require larger working memory resources (Furst & Hitch, 2000). If cognitive load/working memory demand is high, because both the linguistic and the numerical complexity of the word problem are large, this may lead to over additive problems in the domain-general processing stages involved in word problem solving (Figure 4.1).

Figure 4.1: Factors Affecting Mathematical Word Problem Solving, Darocz et al., (2015)

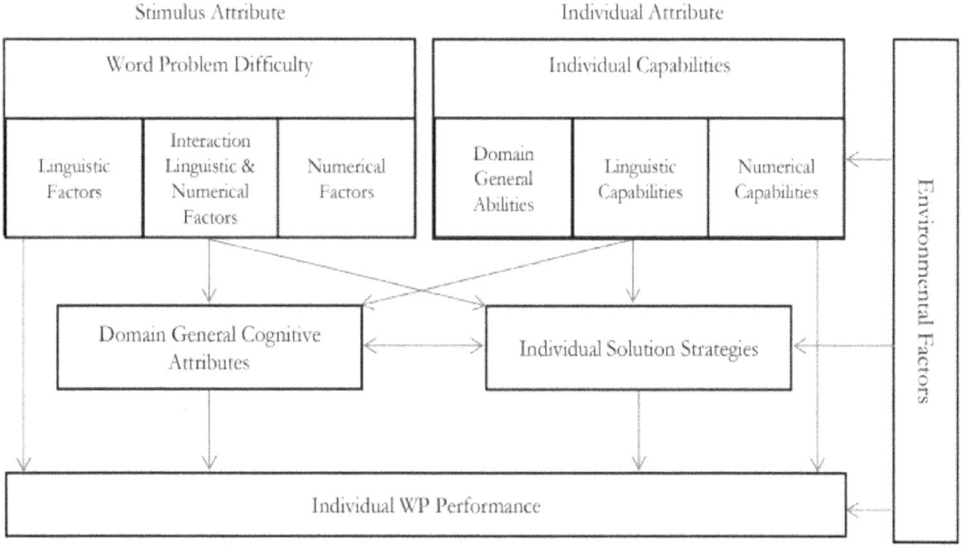

Stimulus Attribute
Individual Attribute
Word Problem Difficulty
Individual Capabilities
Linguistic Factors
Interaction Linguistic & Numerical Factors
Numerical Factors
Domain General Abilities
Linguistic Capabilities
Numerical Capabilities
Environmental Factors
Domain General Cognitive Attributes
Individual Solution Strategies
Individual WP Performance

MATHEMATICAL SOLUTION STRATEGIES

Different mathematical solution strategies are influenced by linguistic factors such as wording, semantic categories and propositions. However, how solvers come up with mathematical solution strategies can be influenced by number magnitude. Such variables make word problems harder and influence numerical representations. The position of the unknown variable has an effect on representation (Garcia et al., 2006). The strategy of counting on from larger is easier if the bigger number is represented first (Wilkins et al., 2001). Even for adults: 4 + 2 = 6, and 2 + 4 = 6, which are mathematically equivalent, may psychologically imply different meanings.

The sequence of the numbers: whether a problem starts with the smaller or with the larger number (Verschaffel & De Corte, 1990), the position of the numbers and particular words (Schumacher & Fuchs, 2012) influence learners' solution of primary school addition and subtraction problems.

INFORMATION RELEVANCE

The relevance of the information provided is a significant factor in word problems. Learners have to extract the relevant information from the text in order to carry out the correct solution. Secondary information distracts learners from recognising the underlying mathematical relations (Schley & Fujita, 2014). This extra information may also be presented in the form of an extra number or an extra operational step. Problem complexity increases with the addition of steps (Terao et al., 2004), as well as the addition of irrelevant information to the problem (Kingsdorf & Krawec, 2014; Darocz et al, 2015). Presence of irrelevant information and the need for an extra step reduces the accuracy of the solutions because learners believe that all of the numbers in a word problem should be used. Two-step word problems are much more difficult than one-step word problems. However, it cannot be concluded that the reason for two-step word problems being more difficult is arithmetic complexity, because in two-step problems, the word problem has also become more difficult linguistically as it usually contains more phrases and semantic distractors.

CONNECTING LINGUISTIC AND MATHEMATICAL FACTORS

There are so many linguistic influences on numerical cognition and arithmetic, for instance, number word structure plays an important role. Learners growing up with regular number word structure usually perform better in a variety of numerical tasks from basic verbal counting up to arithmetic (Dowker et al., 2008). In addition, reading direction influences numerical processes (Shaki et al., 2009; Fischer & Shaki, 2014). Finally, grammatical and syntactic properties of elementary number words influence early number acquisition (Sarnecka, 2013) and spatial-numerical representations (Roettger & Domahs, 2015). O´Neill (2013) states that language and mathematics might originate from the same roots, and that between language and mathematics in word problems there is deep connection: "the cognitive ability that drives symbol processing is the connection between language and maths". Mental arithmetic in word problems is influenced by language processing (Darocz et al., 2015; Zarnhofer et al., 2013).

Word problems require some connection between linguistic and mathematical understanding; therefore, it is common for learners to make more errors when solving word problems compared to number problems (Koedinger & Nathan, 2004). Learners are able to solve several types of addition and subtraction problems before they start formal schooling, and understand numerical concepts before seeing word problems in their curricula (Garcia et al., 2006). Therefore, learners always have the necessary basic arithmetic skills. This may lead to the misconception that numbers may play a lesser role than they actually do and factors other than computational skills are a major source of difficulty with word problems. It is noteworthy to state that there are difficulties in solving word problems that are not attributed to poor reading comprehension skills or to the lack of general mathematical skills. Nevertheless, there is evidence that linguistics and numerical factors are linked in word problem solving.

LEXICAL CONSISTENCY

Some word problems contain linguistic markers such as "less" or "more." In the direct translation strategy learners simply associate "less" with subtraction and "more" with addition. Learners search for linguistic markers and keywords, they construct a mental model of the problem and plan their solution on the basis of this model. Successful learners use the problem model strategy while unsuccessful learners rely on the direct translation strategy focusing on linguistics. This leads to wrong solutions in lexically inconsistent texts, where "more" is associated with subtraction and "less" with addition such as in the word problem adapted from Boonen et al. (2013) demonstrating lexical inconsistency:

> *"At the mall, a bottle of milk costs R7. That is R2 more than at the spaza shop. How much will a bottle of milk cost in the spaza shop?"*

The difficulty for learners in comprehension and finding the correct solution is due to the fact that the word "more" is associated with addition, but the correct solution is not 7 + 2 but 7 - 2, given the way the text is organised. Linguistic semantic consistency with respect to the required mathematical operation is an important determinant of task difficulty. Inconsistent language results in a high error rate and longer response time. Learners find it easier to convert the relation term "more than" into a subtraction operation than the relational term "less than" into an addition operation (Pape, 2003; Van der Schoot et al., 2009).

When reading comprehension and arithmetic skills are not the key factors to the difficulties in solving word problems, a possible explanation is that linguistic complexity and numerical complexity rely on the same resources: working memory. There are multiple linguistic and numerical factors which contribute to a problem's complexity. Generally, problem solving performance is related to the ability of reducing the accessibility of no target and irrelevant information in the memory (Passolunghi & Siegel, 2001). Working memory contributes to early arithmetic performance, including word problem solving (Lee et al., 2004) due to semantic memory representation "less than" which is more complex than "more than". Changes in the structure of the text place more demand on the working. This will probably also be influenced by instruction specifying how learners have to solve a word problem, the method of evaluation and the assessment system.

PEDAGOGICAL IMPLICATIONS: MATHEMATICAL WORD PROBLEM SOLVING

Below is a ten-stage process of systematically solving word problems adapted from Ilany & Margolin, (2010). Although the process involves a number of steps, teachers can adapt the process to the needs and levels of their learners.

Reading the Problem

Stage 1 involves reading the problem *from the bottom up*, as a way of collecting the details. The action of reading at this stage is an exposure to the meaning, where the location of the meaning is in the text. The process of reading is an accumulative process from the smallest units (the words) to the largest units (the whole text).

Understanding the Linguistic Situation

Stage 2 involves reading the problem for a second time. Reading at this stage is referred to as *warming up*: a multi-directional search as a way of *brainstorming*. Relevant questions at this stage:

1. Are all the words clear?
2. Are all the sentences clear?
3. What are the keywords?
4. Do I understand the keywords?
5. What is the question?
6. Do I understand the question?
7. How can I describe the problem in my own words?

Understanding the Mathematical Situation

The mathematical situation is the mathematical context of the problem which relates to two types of information: the data and the question. Understanding the mathematical situation requires examination of the data and the question and to have a good grasp of the subject at hand. In **Stage 3**, the learner needs to identify the known facts and the logical-mathematical conditions of the problem – the connections and relationships between the mathematical data and the logical analysis of the problem (Hershkovitz & Nesher, 2003). Relevant questions at this stage:

1. What is my relationship with the mathematical subject of the problem?
2. Is there a difficulty in the problem?
3. Are all the data clear?
4. Are there implicit data in the problem?
5. Are there superfluous data?
6. Do I understand the connection between the data in the question?
7. Is it possible to demonstrate the problem in particular instances?

Matching the Mathematical Situation to the Linguistic Situation

Stage 4 involves rereading the problem *from the top down*. Reading at this stage is an accumulative process from the combining of mathematical knowledge schemas with the schemas in the text. Processing the information necessary for solving the word problem is one of the main difficulties in the solution of mathematical word problems. Processing the literal information to change it into a mathematical exercise or an algebraic equation is done by understanding the *literal clues,* the words that support (helpful clues) or the words that deceive (misleading clues) as clues for choosing the arithmetic operations needed to solve the problem such as the use of the words *more* or *less* (Nesher, Greeno & Reilly, 1982). Relevant questions at this stage:

1. Do the nouns in the question appear again in a more general unit? (given *apples* and later *fruits* – the problem learner needs to understands that apple is a fruit).
2. Do the connectives that appear in the question relate to different mathematical sizes? (For example: if some number *is* 7 and the product *is* x what *is* the multiplying number?)
3. Are there literal clues in the problem, that is, certain words that help as a clue for choosing the arithmetic operation required for solving the problem?
4. Is it possible to demonstrate the problem by means of a picture, a table, a diagram, or a graph?

Bringing up Ideas for a Solution

The meaning of solving a problem is to find an arrangement of steps starting from the given situation (in the problem) until the desired goal is reached. Usually, learners are given problems similar to those they solved in the past. **Stage 5** involves asking the question: *Do you know a problem close to this one?* (Polya, 1945). In most cases there is no difficulty in bringing up problems that are already solved and are close to the problem at hand. Relevant questions at this stage:

1. Is the problem unique?
2. Have I encountered similar problems?
3. Is it possible to construct a schema for solving the problem on the basis of past experience?

Screening the Ideas

Stage 6 involves checking whether suggested ideas for solving the problem really help to solve the problem. It is necessary to screen the ideas, and to retain only relevant ideas. Relevant questions at this stage:

1. Does the idea help to solve the question?
2. How does the idea help to solve the question?

Building a Mathematical Model

Stage 7 involves constructing a schema that represents the network of connections between their previous knowledge and the schemas in the mathematical text defining the problem and comprehending the situation it describes; building a mathematical model of the mathematical principles relevant to the problem; understanding the relationships and the conditions pertaining to the problem; and using of the mathematical model. Mathematical model building is constructing representations in mathematical language such as an equation. Relevant questions at this stage:

1. What will I do as a first stage to solving the problem?
2. Do I know how to solve the problem and to build an appropriate mathematical model?
3. What mathematical model should I use to solve the problem?

Finding the Solution

Stage 8 involves solving the word problem to reach the expected solution. It is important to check if this is a unique solution or if there is another possible solution; all possible solutions must be found. Relevant questions at this stage:

1. Is the solution unique?
2. What are all the possible solutions to the problem?

Control

Stage 9 involves checking that the solution to the problem is suitable to the problem and checking if there is a need to return to the original problem, to reread it, and double-check. Relevant questions at this stage:

1. Does the solution make sense?
2. Is the solution appropriate to the linguistic situation?
3. Is the solution appropriate to the mathematical situation?
4. Does the mathematical model that I used fit the problem?

This stage is the most important. Often it seems as if the learner has found the solution, but the solution does not make sense, and so the learner needs to redo the process from the beginning: testing the solution, and checking all the steps that lead from the data to the solution.

Efficiency

Stage 10 involves checking whether the solution is efficient, whether it is possible to solve in a different way, and whether there exists a shorter method of solution. Figure 4.2 presents a teaching and learning model for solving mathematical word problems.

Figure 4.2: Teaching and Learning Model for Mathematical Problem Solving

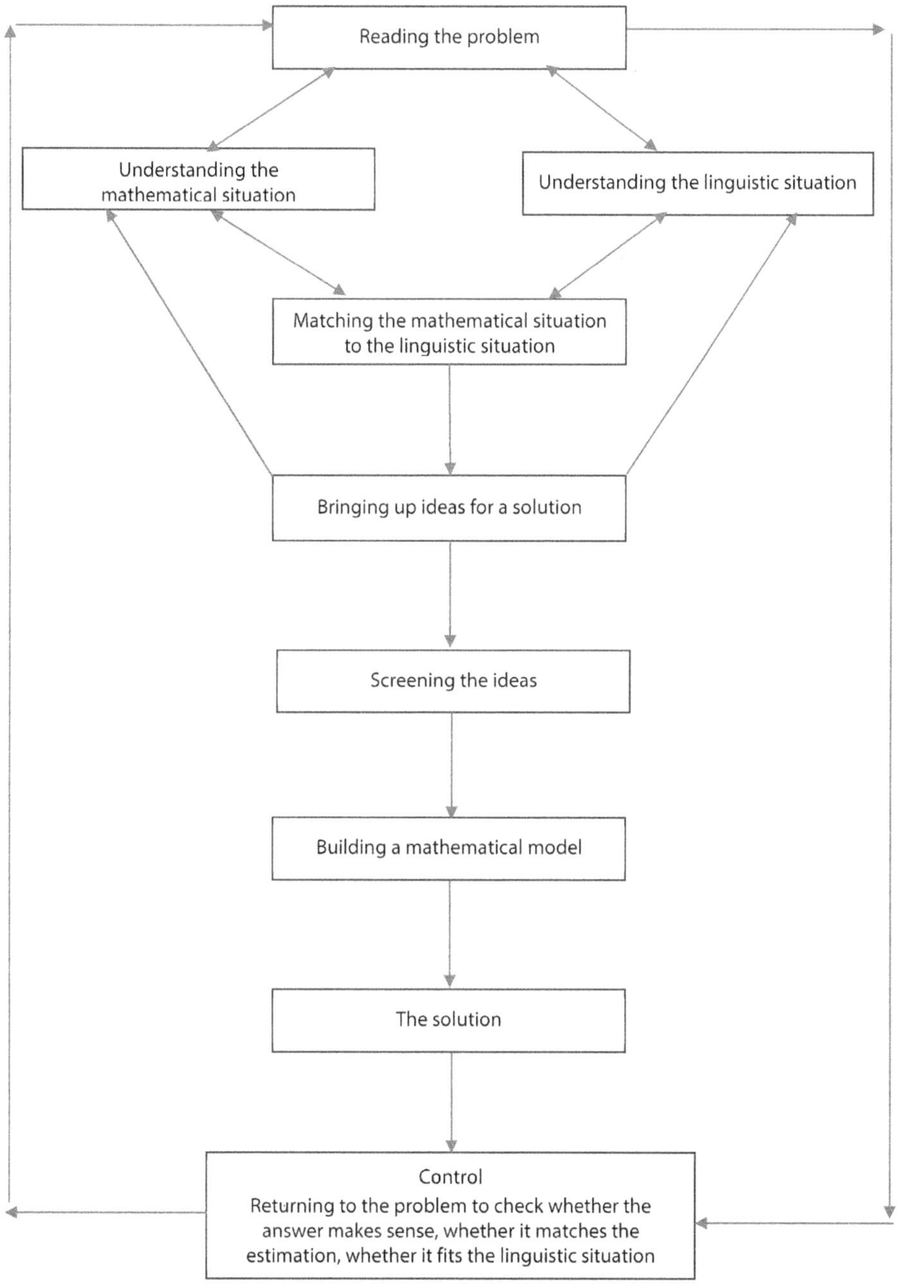

Reading the problem

Understanding the mathematical situation

Understanding the linguistic situation

Matching the mathematical situation to the linguistic situation

Bringing up ideas for a solution

Screening the ideas

Building a mathematical model

The solution

Control
Returning to the problem to check whether the answer makes sense, whether it matches the estimation, whether it fits the linguistic situation

Adapted from Ilany & Margolin (2010)

It is important to note that in every word problem it is necessary to pass through all the stages. However, different learners need to focus on different stages (since some of the stages are already *automatic*). During instruction, it is necessary to go over a different stage each time, and to locate the stages with specific difficulties for different learners and to focus on them.

CONCLUSION

Word problem difficulty is influenced by the complexity of linguistic factors, numerical factors, and their interrelation. To better understand the difficulty of word problems, it would be desirable to manipulate such variables and their interaction following the principle of isolated variation. To support a systematic investigation, the variables to be manipulated also need to be discussed against the backdrop of linguistics, and numerical cognition. A differential-psychological approach to word problem solving acknowledges that different learners are challenged by different types of word problems. Weak learners may have problems with linguistically complex word problems, and arithmetically weak learners with arithmetically complex problems. This understanding is crucial in promoting strategic learning of word problems – a challenge for a majority of learners.

Dealing with the text of a mathematical problem is a multi-staged process which requires a number of cognitive actions: interpreting symbols and graphs, understanding the linguistic situation, finding a mathematical model, and matching between the linguistic situation and the appropriate mathematical model. The ten-stage instruction and learning model becomes a complex thought process when fully understood and internalised. It is a meta-cognitive process that contributes to learners' conceptualisation (Natstasi & Clements, 1990). Knowledge of meta-cognitive processes assists in problem solving and improves the ability to achieve goals.

Teachers need to explain to the learners the differences between natural and mathematical languages and the possibility of combining them in word problem solving. Teachers need to use the instruction and learning model innovatively and adapt the model and its various stages both to learners in different levels/grades and to the nature of the problem and its complexity. The instruction and learning model is suitable for IP, SP and FET learners, and in university departments of teacher education.

Scaffolding the methods of word problem solving enables learners to cope with more complex problems. Adapting the model for different learners will enable teachers to help every learner according to the learner's needs and to pinpoint the focus of difficulty for each learner. Innovative use of the suggested model of instruction and learning will also help both pre- and in-service teachers, their teacher preparation, teaching practice and the schools in which they are deployed. Understanding the model will enable the newly qualified teacher to understand that meta-linguistic awareness, syntactic and semantic awareness, and awareness of mathematical schemas are necessary for solving problems in mathematics. Furthermore, the way the problem is worded and its correspondence to reality can significantly affect learners' ability to solve the problem.

CHAPTER 5

MATHEMATICS TEACHING AND LEARNING IN A SECOND LANGUAGE

INTRODUCTION

Does learning in a second language play any significant role in mathematics instruction?

For many learners worldwide access to higher education and lucrative labour markets depends on becoming fluent in a second language. Being fluent in a second language is essential for gaining meaningful access to education, the labour market and broader social functioning. In sub-Saharan Africa the question of how best to develop second-language fluency amongst large parts of the citizenry remains a critical component of education planning. South Africa is one such country facing the challenge of how to most effectively equip the majority of its population with a second language, in this case English. This presents a challenge to Language in Education Policies (LiEP) regarding when and how in the school programme the transition to the second language should occur.

To understand the challenges faced by Intermediate Phase (IP) teachers while presenting mathematics in a second language such as English, in turn necessitates an understanding of second-language acquisition. This chapter explores diverse issues related to second-language acquisition and learning from multiple perspectives to expose the complexity of second-language learning. The chapter also describes bilingual transition models and the application of mathematics teaching and learning in a second language. The critical and often mishandled teacher language competency to sustain effective mathematics instruction in English (a second language for both the teachers and the learners) as a LoLT at IP level necessitates the understanding of second-language literacy acquisition and learning.

Second-language acquisition is a sub-discipline of applied linguistics. It is easier to acquire and learn the other basic interpersonal communication skills, particularly by individuals who spend huge amounts of time interacting with native speakers (Cummins, 2000: 2). Second-language speakers who have attained a high degree of fluency and accuracy in everyday spoken English have the corresponding academic language proficiency, which dispels the myth that speakers who exhibit language disparity have special educational needs, when they actually need more time to learn the new language. Cummins further notes that the common underlying proficiency in language learning provides the basis for the development of both the first language and the second or additional languages.

THE NEXUS BETWEEN FIRST ADDITIONAL LANGUAGE AND SECOND LANGUAGE

The Curriculum Assessment and Policy Statements (CAPS) document describes a first additional language as a language which is not a mother tongue, but which is used for certain communicative functions in society. It is also called a second language and assumes that learners do not necessarily have any knowledge of the language when they begin schooling (DBE, 2011a: 12). In South Africa, there is a distinction between home language, referred to as a first language, and the first additional language i.e. the second language.

This second language is taught in addition to one's home language and it may be taught formally in kindergarten, reception year 'Grade R', and at school. The second language may be introduced earlier by parents at home, and/or as learners interact with others in their immediate environment (Mohlabi-Hlaka, 2016: 11; Manyike, 2007: 14; National Council for Curriculum and Assessment, 2006: 04).

A clarification of the use of the term 'additional language' within the South African context is that the term 'additional language' is preferred to 'second language', because the language will exist alongside the first language and be of equal but not necessarily of greater importance to the learner. This clarification is important for its relevance to South African constitutional stipulations and the (LiEP) (DAC, 2003: 8).

It is of paramount importance that the teachers who are entrusted to manage the transition period from learning to read at Foundation Phase to reading to learn at Intermediate and higher phases are adequately trained and master the first additional language beyond mediocre levels in order to facilitate a smooth transition later (Tshuma, 2016, 2017). Therefore, this chapter is particularly interested in the second-language acquisition models as reviewed in the following section, because the LoLT at IP level is a second language to 90% of the learners and teachers.

MODELS OF SECOND-LANGUAGE ACQUISITION

This section gives an overview of the models of second-language acquisition, which attempt to explain how a second language is acquired. The seven models are input hypothesis, output hypothesis, interaction hypothesis, acculturation model, automaticity model, conditions outcomes model and the associative model (Sibanda, 2014). Each of these models highlights different but crucial features of the language acquisition process.

INPUT HYPOTHESIS

The input hypothesis model contends that the second language is acquired in the same way as the first language: through comprehensible input. Input is necessary for the acquisition of a second language provided it is comprehensible to the language learner and that the learner has a low affective filter (attitudinal and mental dispositions that would not interfere with the acquisition of the input).

OUTPUT HYPOTHESIS

The output hypothesis acknowledges the role of input in second-language learning; however, it deems input as inadequate to account for second-language acquisition. Despite the vast input to which second-language learners are exposed, and which positively impacts on their oral proficiency, their productive language skills of speaking and writing are limited compared to those of native speakers (Swain, 2005).

INTERACTION HYPOTHESIS

Within the interaction process, interlocutors negotiate meaning and facilitate comprehension, which leads to language learning. The input is not just consumed; it is queried, recycled and paraphrased, which increases its comprehensibility (Mayo & Soler, 2002). Within the interaction pattern, there is modified input and modified interaction.

From the three major hypotheses, it is clear that second-language learning is based on the interplay between input and output within a context of interaction (Sibanda, 2014). The acculturation model, automaticity model, conditions outcomes model, and the associative model (Bialystok, 2009, 1982; Van Lier, 1996) also explain second-language acquisition but their application in classroom teaching is limited.

ACCULTURATION MODEL

The acculturation model posits that the extent of second-language learning is associated with the social and psychological distance between the first and second language (Barjesteh & Vaseghi, 2012). The culture of the second-language system and how the learners view and are viewed by the new target language group impacts on their acquisition of the language. The model accounts for natural rather than taught second-language acquisition.

AUTOMATICITY MODEL

The model distinguishes between explicit and implicit language knowledge and the degree to which the individual analyses, monitors and uses two languages with ease, using language with automaticity or control (Bialystok, 2009). The model has limited application to second-language instruction, but provides a framework for cognitive learning in the second language.

CONDITIONS OUTCOMES MODEL

The conditions outcomes model combines input and output factors in second-language acquisition (Van Lier, 1990). It identifies the critical conditions required for language acquisition such as learner receptivity, attention and practised intake. The model places emphasis on optimal learner output or production within authentic and meaningful interaction. Awareness, autonomy and authenticity are key principles in second-language acquisition.

ASSOCIATIVE COGNITIVE MODEL

Ellis (2006: 100) posits, "Second language learning is governed by the same principles of associative and cognitive learning that underpin the rest of human knowledge." This perspective is based on the fact that high-frequency constructions are more readily processed than low-frequency ones. Such associative learning of a language from frequency of usage applies to words, letters, morphemes or syntactic patterns.

A broader picture of second-language learning views all the models outlined above as interconnected perspectives that enrich the understanding of second-language acquisition, rather than viewing them as independent and mutually exclusive. Although there is diversity in the second-language acquisition models, there is consensus on the stages learners pass through in their acquisition of a second language (Sibanda, 2014). Different stages of second-language acquisition are outlined below.

STAGES IN SECOND-LANGUAGE ACQUISITION

One similarity between first-language acquisition and second-language acquisition is that both proceed by stages and are characterised by developmental orders. Language is acquired through predictable and sequential stages of language development (Orosco & Hoover, 2009; Baca & Cervantes, 2004; Collier & Combs, 2003). Thus, second-language learners go through the same stages; however, these stages are sometimes named differently.

In the early stages of learning a second-language, learners pass though developmental stages similar to those of learning a first language. Initially, learners may err in their use of the grammar and vocabulary, just as first-language learners do. Although the process of second-language acquisition varies with each learner, depending on various factors, all second-language learners pass through the general developmental stages. With increased exposure to the English language, second-language learners progress from acquiring social language to the more complex academic language. Social language is considered conversational, contextualised language and can be developed within two to three years. Academic language is defined as utilising the combination of cognitive skills and content knowledge necessary for successful academic performance at secondary and university level. Achievement of academic proficiency can take between 7 to 10 years, if all of the schooling takes place in the second language (Thomas & Collier, 2002).

Language acquisition is a process that involves five main stages. The first three stages lay a foundation for language learning while the last two stages focus on using the language for academic discovery as they are characterised by reading grade-level texts and compiling organised essays. Table 5.1 summarises general behaviours of second-language learners at each stage of second-language acquisition.

Table 5.1: Behaviour of second-language learners at each stage of language acquisition

	STAGE OF SECOND-LANGUAGE ACQUISITION	GENERAL BEHAVIOUR OF SECOND-LANGUAGE LEARNERS
Language Foundation	**Silent / Receptive Stage** ▪ 10 hours to 6 months ▪ 500 receptive words	▪ Point to objects, act, nod or use gestures ▪ Respond using Yes/No ▪ Speak hesitantly
	Early Production Stage ▪ 6 months to 1 year ▪ 1000 receptive/active words	▪ Produce one or two word phrases ▪ Use short repetitive language ▪ Focus on key words and context clues
	Speech Emergence Stage ▪ 1 to 2 years ▪ 3000 active words	▪ Engage in basic dialogue ▪ Respond using simple sentences
Academic Discovery	**Intermediate Fluency Stage** ▪ 2 to 3 years ▪ 6000 active words	▪ Use complex sentence ▪ State opinions and original thoughts ▪ Ask questions
	Advanced Fluency Stage ▪ 7 – 10 years ▪ Content area vocabulary	▪ Converse fluently ▪ Understand grade-level classroom activities ▪ Argue and defend academic points ▪ Read grade-level text books ▪ Write organised and fluent essays

Collier (1995)

The five stages of language acquisition described above provide a general framework for understanding how second-language learners progress; however, learning is an on-going, dynamic and fluid process that varies from one learner to another. Learners may move back and forth between stages, depending on the academic demands of a lesson and the amount of participation required.

KEY ELEMENTS OF AN EFFECTIVE LANGUAGE-LEARNING ENVIRONMENT

For language acquisition to occur, learners must:

- receive understandable and meaningful messages and
- learn in an environment where there is minimal anxiety.

Collier (1995)

An awareness of these two principles can assist teachers to create a natural language-learning environment in their classrooms. Table 5.2 outlines five key elements of an effective language-learning environment.

Table 5.2: Elements of an effective language-learning environment

ELEMENT	DESCRIPTION
Comprehensible input	Teachers can make their language more comprehensible by modifying their speech, avoiding colloquialisms, speaking clearly, adjusting teaching material, adding redundancy and context, and scaffolding information within lessons.
Reduced anxiety level	A learner's emotions play a pivotal role in assisting second-language learning. Teachers can assist learners by creating a comfortable environment that encourages participation without fear of feeling embarrassed.
Contextual clues	Visual support such as tangible objects, facial expressions or models of abstract concepts makes language more comprehensible.
Verbal interaction	Learners need opportunities to work together to solve problems and use English for meaningful purposes. They need to give and receive information and complete authentic tasks.
Active participation	Lessons that encourage active involvement motivate second-language learners, engage them in the learning process and help them to remember content more easily.

Adapted from Collier (1995)

Use of these strategies can assist learners to better access the content material, in particular mathematical content, which has its own specific register. Thus, the IP mathematics teachers should draw on these strategies when assisting learners.

FACTORS THAT CAN ENABLE OR CONSTRAIN SECOND-LANGUAGE DEVELOPMENT

The integrated model for literacy development illustrated in Figure 5.1 incorporates both home and school factors that influence language development.

Figure 5.1: Integrated Model for Literacy Development

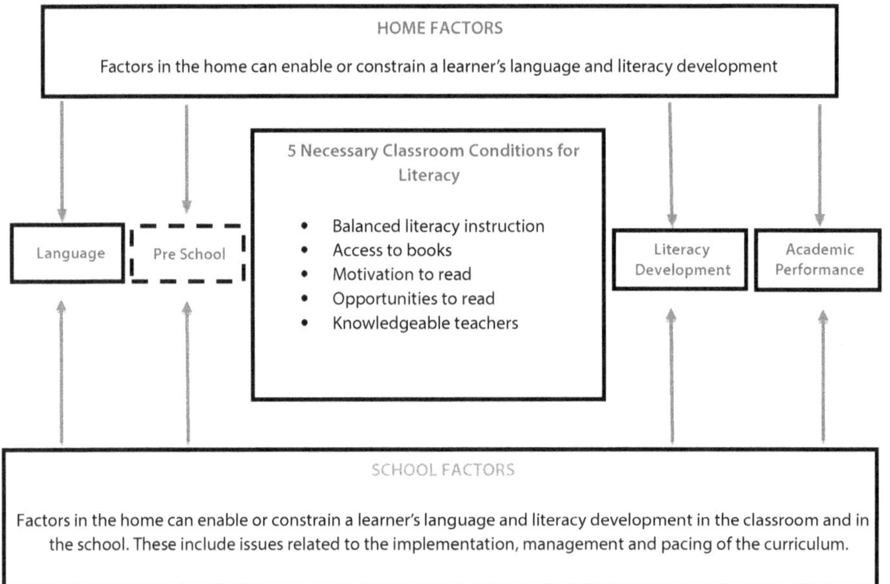

Adapted from Pretorius, (2014).

The model also emphasises language development being initiated in the preschool, followed by formal learning in the classroom where certain conditions such as balanced instruction, access to reading material, motivating learners to read and provision of opportunities to read under the supervision of a competent teacher should be met. Once literacy is developed, academic performance improves.

Learners in under-resourced communities have limited educational opportunities outside the classroom especially if the LoLT is not aligned with the home language; and literacy practices in their homes may be influenced by the parents'/guardians' limited education or resources.

FACTORS INFLUENCING SECOND-LANGUAGE LEARNING

The pace at which second-language learners progress through the five stages of language acquisition and develop conversational and academic fluency depends on learner variables (internal) and environmental factors (external), (Sibanda, 2014). Learner variables include age, aptitude, motivation and attitude, personality, cognitive style, hemisphere specialisation and learning strategies. Environmental factors include the type and quality of instruction and input, the environment, the material of instruction (graded, sequencing, ungraded, skill-oriented materials), among others. The section below elaborates on some of these factors namely age, aptitude, personality, motivation and attitude as well as the role of the first language.

AGE OF THE LEARNER

Age affects second-language learning in a number of ways: older learners acquire the second language the fastest, followed by adults and then by young learners – a finding that favours acquisition of a second language during adolescence. The exception is in pronunciation where younger learners fare better and can eventually acquire native accents and pronunciation more easily than older learners. Older learners often do not speak in front of peers because they are usually self-conscious and feel vulnerable about taking risks and making mistakes. Class discussions and textbook-reading levels are more academically demanding for older second-language learners than for younger learners; therefore, it may take longer for older learners to achieve grade-level performance in content classes (such as mathematics classes).

APTITUDE

Aptitude relates to innate attributes learners are individually endowed with which facilitate or constrain their acquisition of a second language. Carroll (1991) identifies four sub-components of aptitude, namely phonetic coding ability (capacity for sound discrimination), associative memory (capacity to connect native- and second-language equivalents), grammatical sensitivity (appreciating the function of words in sentences), and inductive language analysis (ability to identify patterns, especially in verbal material, which may involve implicit or explicit rule representation).

Aptitude on its own is not adequate unless it is coupled with motivation and a positive attitude.

LEARNER MOTIVATION AND POSITIVE ATTITUDE

Learner motivation may be integrative, instrumental or resultative. In second-language acquisition, parental and peer attitudes may also shape learner attitudes. Learner attitude towards the learning context and the language impacts on their language learning. Their attitude towards the native culture and people will affect the motivation to acquire the language.

PERSONALITY FACTORS

Self-esteem correlates with second-language learning. Introverts usually perform better on reading and grammar, while extroverts usually persevere, ultimately resulting in the promotion of second-language acquisition. Anxiety can facilitate or constrain, and therefore can interfere with language acquisition positively or negatively depending on the circumstances. Willingness to take risks and use the language facilitates language acquisition whereas being sensitive to rejection constrains language acquisition.

LITERACY IN A FIRST LANGUAGE

The learner's proficiency in the first language interferes with their acquisition of the second language as they can transfer some skills to second-language acquisition. The complementary nature of the first language to the second language is expressed in the observation that "the mother tongue is the launch pad for the second language" (Morgan & Rinvolucri, 2004: 8), and that the most effective way to begin to learn the meaning of a word is by translation into the first language (Nation, 2003). The degree to which the first language would positively impact on second-language learning is dependent on the level of learner proficiency in the first language. The higher the proficiency, the more the first language eases the acquisition of the second-language forms.

The factors influencing second-language learning are presented from a learner's perspective. These factors are deemed necessary for equipping teachers of ELLs on the linguistic characteristics that impact on their use of the second language as a medium of instruction. However, there is a gap in the literature regarding factors influencing teachers' second-language learning for the purposes of using the language to deliver subject knowledge. The section below explores the implications of second-language acquisition on the teaching and learning process using a second language.

PEDAGOGICAL IMPLICATIONS: SECOND-LANGUAGE ACQUISITION

The major limitation in second-language acquisition research is the number of variables regarding the comparability of groups who underwent different second-language learning experiences. In South Africa, for example, it is invalid to simply compare schools that adopt a straight-for-English approach with schools that transition from first language to English in Grade 4, because these two kinds of schools possess different characteristics, operate under different settings and are usually located in different environments (Hulstjin, 1997; Sibanda 2014). In addition, studies must be conducted over a period of several years. Furthermore, the desired outcome should not be English proficiency at the end of the treatment period, but at a later stage once those in a bilingual programme have transitioned to English as language of instruction. The focus should be on achieving long-term second-language acquisition. The majority of studies have not used data with a long enough time span to address this fundamental research question.

Some immigrant learners in South Africa may enter the country with limited or interrupted schooling. In some cases, learners may originate from rural communities where literacy and schooling were not emphasised, while others may originate from countries where political turmoil prevented them from attending school regularly. Such learners face the additional challenge of learning appropriate school behaviour and expectations at the same time as they are learning content and concepts, and the second language used as the language of learning and teaching. Learners with such backgrounds and with limited or no first-language support may spend 7 to 10 years striving to achieve academic parity with their peers (Thomas & Collier, 2002). Teachers can assist these learners by explicitly modelling appropriate school behaviour (as discussed under 'Semiotic approaches' in Chapter 3). It is also essential for the teacher to assess such learners' background knowledge before introducing a new topic in order to identify gaps and create opportunities to build and strengthen the missing background knowledge. Assessing circumstances around second-language learners' decisions to move or relocate implies that the teacher will be better able to understand the learners' emotional and psychological wellbeing, as well as factors affecting the learners' motivation and academic achievement.

Understanding second-language acquisition is essential in determining how learners become bilingual: "*Who learns how much of what language under what conditions*?" (Spolsky, 1989: 3). The "who learns" is concerned with individual differences and is constantly changing. The 'how much of what language' part is concerned with the specific language skills that are being developed, the measures used to assess this skill development, as well as cultural influences. The element "*under what conditions*" focuses on situation and context, thus the learning environment and the strategies influencing language learning. The section below presents a review of bilingualism and its application in contexts where a second language is used as a medium of instruction.

BILINGUALISM, BICS AND CALP

The theories on bilingualism referred to in the section below are largely drawn from Sibanda (2014). At this juncture, it is noteworthy to state that the majority of teachers and learners in under-resourced schools are multilingual (competent in more than two languages)[1] instead of being bilingual as they can speak more than two languages. This section therefore focuses on bilingualism for the purposes of understanding the transition models (determination of language of instruction) adopted by different schools in South Africa when learners move from Foundation Phase (FP) to the Intermediate Phase (IP).

TYPES OF BILINGUAL EDUCATION

Bilingual education may be interpreted as merely teaching and learning through the medium of two languages. However, bilingual education is complex and some questions need to be addressed in order to assess the learning context. These questions include:

- Are both languages used in the classroom?
- For how long are the languages being used in school?
- Are two languages used by all or some learners?
- Are two languages used by the teachers or just by the learners?
- Is the aim to teach a second language or to teach through a second language?
- Is the aim to support the home language or to move to an alternative majority language?

Adapted from Baker and Prys Jones (1998: 464)

There are several different types of bilingual education programmes and within these types there are further subdivisions.

IMMERSION VERSUS BILINGUAL TRANSITIONAL MODELS

English language proficiency impacts on academic opportunities through its association with educational success. However, Casale and Posel (2011) demonstrate that English proficiency also improves labour market returns directly. As such, South African curriculum developers are challenged by the question of when and how the teaching of English should be introduced in schools, and when and how a transition to English as the Language of Learning and Teaching (LoLT) across the curriculum should take place. Research reveals several theoretical models, each with numerous variations that have been tried in different parts of the world.

In Figure 5.2 Taylor and Coetzee describe one end of the spectrum as the immersion model with variations occurring in between, and the other end of the spectrum as the transition model. In the immersion model learners are taught in the second language right from the beginning of their school life. The straight-for-English approach is a type of the immersion model. In contrast, there are various types of bilingual models. Transitional models prescribe that a learner's first language be used in the first years of schooling followed by a transition to

1 Many South Africans, whose main language is not English or Afrikaans, speak three or more languages (Adler, 2001; Essien, 2003).

Figure 5.2: Immersion versus bilingual transitional models

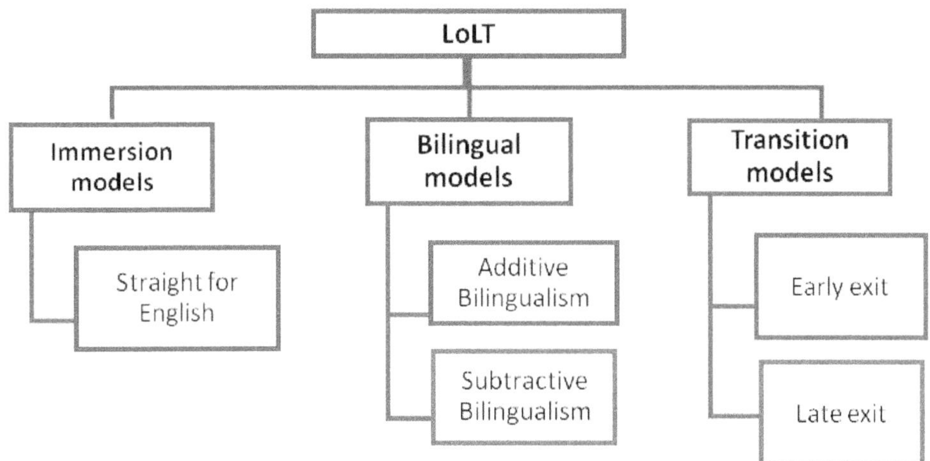

Adapted from Taylor and Coetzee (2013)

the second language as the LoLT. In early-exit transition models the transition to second-language instruction occurs after at least three years of schooling. In late-exit transition models the transition occurs after about six or eight years of schooling. There are also various models of additive or subtractive bilingualism in which the first and the second language are used alongside each other, with the balance of use changing through the school years.

It is important to note that these are general classifications and it is not assumed that all bilingual education contexts can be classified under one of the above. The primary concern lies with the **early-exit context** – learners transitioning from learning mathematics through mother-tongue instruction to learning mathematics through the medium of English.

THEORIES OF BILINGUALISM

Three theories of bilingualism are discussed in this section: the threshold hypothesis and the theory that differentiates between social and academic language proficiency.

CUMMINS (1976) – THRESHOLD HYPOTHESIS

This hypothesis states that the level of mother-tongue proficiency already reached by a learner determines if they will experience cognitive deficits or benefits from learning in a second language (Cummins, 1976). This implies that there is a certain 'threshold' that one must reach in their first language before the benefits of studying in a second language can develop. For those who begin learning in a second language before achieving this level, there will be serious learning difficulties (Cummins, 1976). Therefore, the threshold hypothesis implies that learners who have learnt through the medium of isiXhosa but have not developed their language to a sufficient level will experience difficulties when transferring to learning through the medium of English, and therefore have challenges when learning mathematics through the medium of English. Those who have reached the 'threshold' in both mother tongue and English are at an advantage. Given that both languages are interdependent and proficiency in both is important, the languages cannot be looked at in isolation. Baker (1996) developed a triple-storey house analogy in order to demonstrate the bilingual linguistic requirement at each level as illustrated in Figure 5.3. Ideally a bilingual learner needs to progress beyond the second level in order to attain cognitive benefits from learning in a second language.

Figure 5.3: Bilingual linguistic requirements

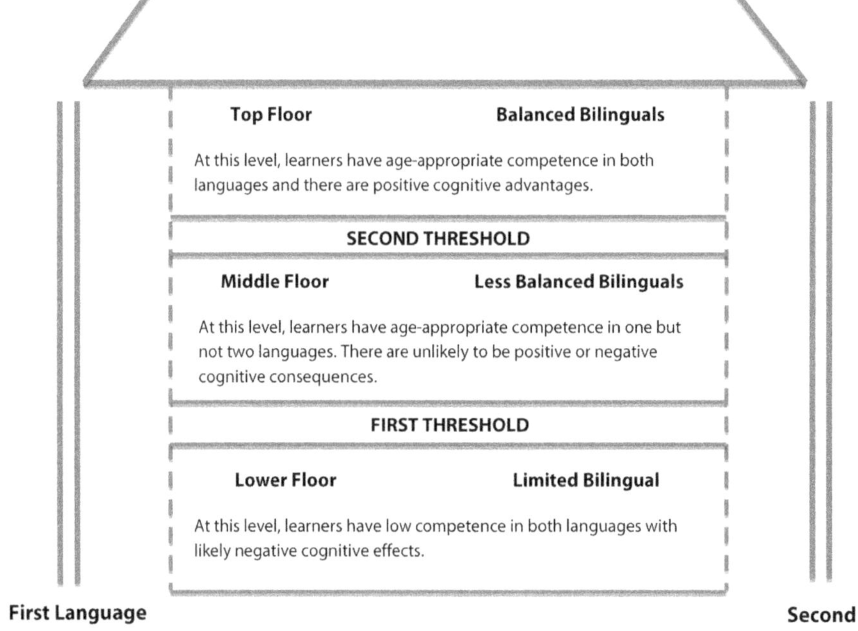

Baker (1996: 149)

In 1979 Cummins refined his Threshold Hypothesis and this led to the development of his **Developmental Interdependence Hypothesis**, which had a more in-depth focus on the relationship between a learner's two languages. The Interdependence Hypothesis proposed that the level of proficiency already achieved by a learner in their first language would have an influence on the development of the learner's proficiency in their second language (Baker, 2001). Therefore, the greater level of proficiency achieved in the first language will allow for a better transfer of skills to the second language.

BASIC INTERPERSONAL COMMUNICATIVE SKILLS VERSUS COGNITIVE ACADEMIC LANGUAGE PROFICIENCY

Another theory about language acquisition that can help teachers understand and better manage the challenges faced by second-language learners is the distinction between social and academic language proficiency. Cummins, (1981) suggests that there are two types of language proficiency, which can be classified as individual registers which bilingual learners have to develop and accomplish in their first and second languages:

- Basic interpersonal communication skills (BICS);
- Cognitive academic language proficiency (CALPS).

BICS relates to communication skills and conversational competence. It relies on the phonological, syntactic and lexical skills required to function in everyday contexts – most of the time these contexts are cognitively undemanding and contextually supported (May, Hill & Tiakiwai, 2004). Second-language learners generally develop conversational fluency (BICS) within two years of studying a second language (Cummins, 2000).

CALP relates to the fluency in the more technical, grade-appropriate academic language. Second-language learners develop cognitive academic language fluency (CALP) within five to seven years, depending on the learner's age and level of home-language literacy. Cummins's theory of second-language acquisition makes a distinction between two types of language proficiencies, namely BICS and CALP. The difference between two types of language proficiencies as the basis of language learning and development is illustrated in Table 5.3.

Table 5.3: Language proficiencies for second-language acquisition

LANGUAGE PROFICIENCY	FUNCTION
Basic Interpersonal Communication Skills (BICS)	▪ The surface skills of listening and speaking acquired quickly by language-learning individuals during the language-learning process.
Cognitive Academic Language Proficiency (CALP)	▪ The language learner's ability to cope with the academic demands placed upon their various subjects during the language-learning process.

Cummins (2000)

Failure to understand the distinction between these two types of language proficiency may lead to inappropriate assumptions about a learner's language ability; for instance, second-language learners may be exited from direct English instructional programmes because they appear to be fluent in conversational English; however, they may lack the necessary academic language, reading and writing skills essential for success in the content area (Cummins, 2000).

Baker (2001) supports Cummins's descriptions of BICS and CALP by noting that BICS occurs where there is contextual support of language delivery. Contextual support occurs in the form of face-to-face situations which provide non-verbal support to secure understanding. In addition, actions like eye contact, hand gestures, instant feedback, cues and clues support verbal communication. On the other hand, CALP supports academically oriented contexts. This involves higher-order thinking skills that are required by the curriculum.

CALP is required for context-reduced academic situations. CALP demands manipulation of the surface features of a language in impersonal contexts (May, Hill & Tiakiwai, 2004). The skills required are higher-order in nature such as analysis, synthesis and evaluation. Cummins (2000) argues that these skills are a prerequisite for CALP as they provide learners with the facility to use language as an instrument of thought in problem solving and this justifies the assertion that it takes five to seven years for learners to acquire academic-language proficiency in a second language. This is further complicated by the fact that learners not only have to develop proficiency in the academic register in English, but also learn new mathematical content in that language.

An appropriate mathematical example demonstrating the difference between BICS and CALP:

> A learner is given a mathematical question such as: '***You have 20 rands. You have 6 rands more than me. How many rands do I have?***' At the higher CALP level the learner will conceptualize the problem correctly as 20 minus 6 equals 14. At the BICS level the word 'more' may be taken to mean 'add-up' with the learner getting the wrong answer of 26. The BICS learner may think of 'more' as used in basic conversation. However, in the mathematics classroom this illustration requires that 'more' be understood by the mathematical phrasing of the question.
>
> Baker (1996)

THE RELATIONSHIP BETWEEN BICS AND CALP

The iceberg analogy can be used to demonstrate the underlying concepts of BICS and CALP. Language skills such as comprehension, speaking, pronunciation, vocabulary and grammar lie above the surface and are used in conversation (BICS). Below the surface lie the academic language skills such as analysis, synthesis and semantic meaning as illustrated in Figure 5.4.

According to Cummins (1979), in order for bilingual learners to master academic-language proficiency, their second language must be well developed. While second-language learners may pick up oral proficiency (BICS) in their new language of learning in as little as two years, it may take up to seven years to acquire the language skills (CALP) necessary to function successfully in a second-language classroom. Mathematics is located within this CALP and in order for learners to attain mathematical academic-language proficiency, their underlying proficiency must be well developed (Cummins, 1979). Language for academic purposes requires understanding and the use of classroom discourse, which includes the teacher's verbal instructions and lessons. Communication therefore develops as a result of exposure to this form of education.

Figure 5.4: The relationship between BICS and CALP

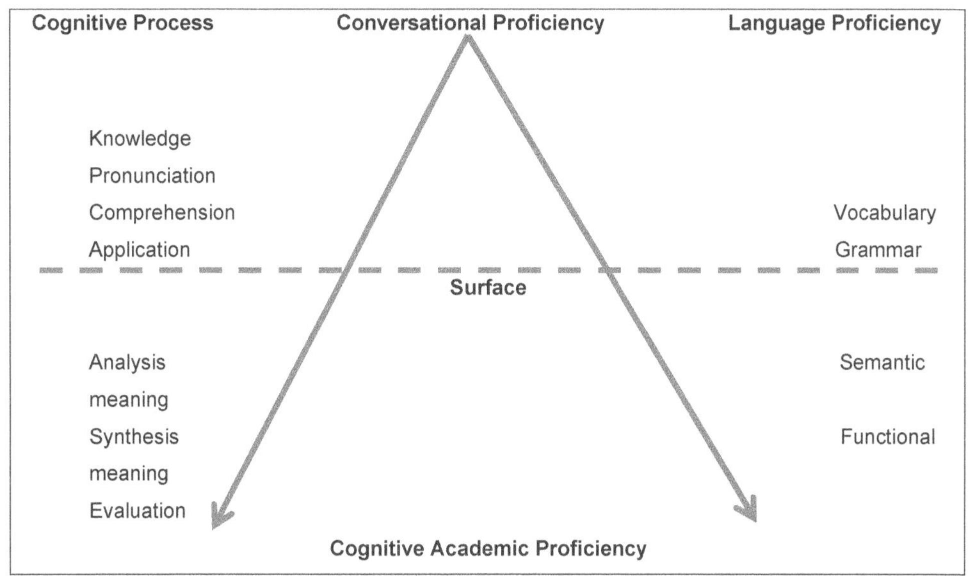

Baker (2001: 170)

PEDAGOGICAL IMPLICATIONS: BILINGUALISM, BICS AND CALP

Distinguishing between BICS and CALP has implications for teachers of second-language learners. Learners will succeed only when they have developed an appropriate level of language proficiency in the language of instruction so that they can cope with the context-reduced, cognitively demanding situations that arise in learning environments as illustrated in the fourth quadrant in Figure 5.5. On the other hand, learners working at a context-embedded level may be hindered in developing their understanding of the content of the lesson, while also failing to develop higher-order cognitive processes (May, Hill & Tiakiwai, 2004). Teachers of bilingual learners must avoid assuming that their learners are proficient in the language of instruction if the learners demonstrate conversational competence. Teachers must acknowledge that bilingual learners experience greater difficulties in acquiring the academic-language proficiency in the language of instruction and also that appropriate teaching strategies and interventions need to be implemented in order to facilitate bilingual learners' understanding of mathematics. Figure 5.5 illustrates the cognitive demands of language in mathematics learning and teaching.

Figure 5.5: The cognitive demands of language in mathematics learning and teaching

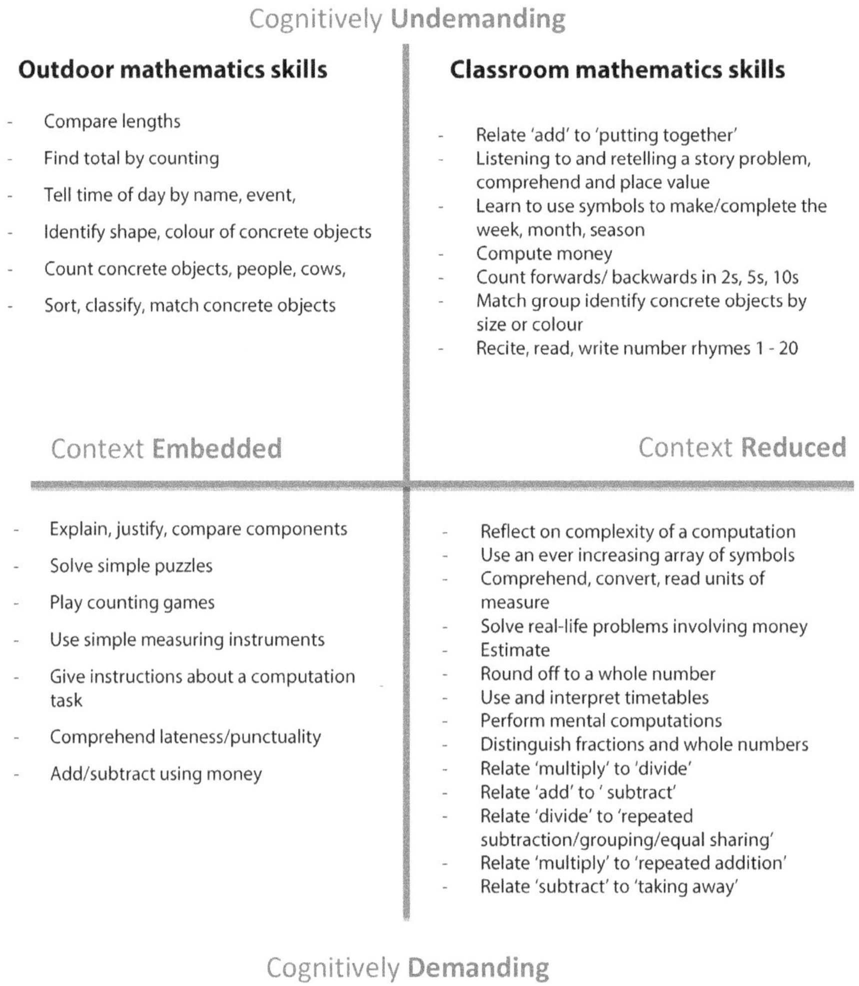

Kasule and Mapolelo (2005)

As illustrated in Figure 5.5, even when the context-embedded setting is maintained, the linguistic demands on the learner become increasingly complex. The range of outdoor mathematical skills relies on games learners play using their first language in very familiar and less rigid settings (Kasule & Mapolelo, 2005). However, the classroom mathematical skills appearing on the right side of the quadrant demand greater linguistic proficiency because they are learnt in a rigid classroom setting through a second language. Unfortunately, mathematics classroom learning becomes increasingly context-reduced, and more cognitively demanding from one grade to the next. The section below specifically explores mathematics teaching and learning in a second language.

MATHEMATICS TEACHING AND LEARNING IN A SECOND LANGUAGE

Given the increase in international migration, and the dominance of English as a language for learning and teaching mathematics, many learners face a transition to learning mathematics through the medium of English (Barwell, Barton & Setati, 2007). However, this section is specifically concerned with addressing the teachers' English-language competency in association with mathematics instruction and investigating the difficulties encountered with the English mathematics register when English is the learners' second language of learning. The contemporary theory regarding learning in a second language is translanguaging, which is an umbrella term that incorporates other dated theories such as code switching. The section below exposes the two widely used theories and how they are used in mathematics instruction.

PEDAGOGICAL TRANSLANGUAGING

The term 'translanguaging' was created by Cen Williams in 1984, drawing on the Welsh word ***trawsiethu*** to name a pedagogical practice which switches the language mode in bilingual classrooms. This is done by assigning different languages to the lesson's input (reading/listening) and output (speaking/writing), and this is systematically varied. Translanguaging is "the process of making meaning, shaping experiences, understanding and knowledge through the use of two languages" (Baker, 2011). Translanguaging is based on the observation that learners pragmatically use their two languages inside and outside the classroom in order to construct understanding (Probyn, 2015; Baker, 2011; García, 2009; 2012).

García (2009) propounded translanguaging as a hybrid language use that is systematic, strategic, affiliative and sense making. Thus, translanguaging uses both languages to purposefully complement one another in knowledge transfer. To connect translanguaging to Cummins's bilingual theoretical frameworks, Baker (2011) notes four advantages of translanguaging:

- Promoting a deeper and fuller understanding of the subject matter. In order to read and discuss a topic in one language, and then to write about it in a different language, the subject matter has to be processed and understood, since learners' prior knowledge of, as well as the interdependence of, their two languages is utilised;
- Helping bilingual learners develop oral communication and literacy in their new, weaker language, since they have to work in their two languages;
- Facilitating home-school cooperation – learners are able to ask their minority language parents for help in solving their assignments; and
- Developing English-language learners' competence and content learning at the same time.

Thus, translanguaging allows the integration of learners with different levels of second-language competency, and consequently their cognitive academic language proficiency (CALP) and basic interpersonal communication skills (BICS) can develop jointly. Thus, translanguaging can occur in any educational setting: bilingual, English Second Language or mainstream. It can serve as a scaffold for learning English (Baker, 2011). It can also be a powerful way for learners to use their languages as a resource for their learning.

CODE SWITCHING IN MATHEMATICS CLASSROOMS

Although the process of language switching appears to be an unconscious and unplanned behaviour, the mathematics teacher's role in the process is essential because the teacher decides which language to use in order to enhance the learners' mathematical ability. Code switching is often associated with terms such as 'code mixing' and 'borrowing' (Chikiwa, 2016), which are described below.

- **Code switching**

Cook (2001) defines code switching as a natural phenomenon in settings where speakers share two languages. It refers to the alternate use of two or more languages in the same utterance or conversation. It is a trait of bilingualism and a form of translanguaging in which there is juxtaposition within the same speech exchange of passages of speech belonging to two different grammatical systems or subsystems (Gumperz, 1982). With most of the world becoming multilingual (Jegede, 2012), as a result of globalisation and migration of people within and beyond countries and continents, the use of more than one language in daily speech is becoming more prevalent than it was. Code switching occurs when one substitutes a word or phrase in one language with a phrase or word in a second language (Heredia & Altarriba, 2001). It is the use of elements from two languages in the same utterance or in the same stretch of conversation (Genesee, Paradis & Crago, 2004). It is the use of two or more linguistic varieties or elements from other languages within the same utterance or conversation. This occurs naturally in day-to-day conversations especially in multilingual societies (Martin, 2007; Metila, 2009; Mahadhir & Then, 2007). Code switching is a diverse linguistic resource from which teachers can choose to draw in order for them to communicate mathematical concepts effectively and successfully (Mati, 2004).

- **Code mixing**

Bokamba (1989) defines code mixing as the embedding of various linguistic units such as affixes, words, phrases and clauses from a cooperative activity, where the participants must reconcile what they hear with what they understand. This resonates with Ncoko, Osman and Cockcroft's (2000) notion that code mixing involves the mixing of affixes, words, phrases and clauses from more than one language within the same sentence and speech situation. Thus code switching and code mixing are in some cases used interchangeably (Mahootian, 2006), even though code mixing often refers to intra-sentential code switching only. Code switching and code mixing are usually used complementarily (Chikiwa, 2016), with code switching for alternation between sentences and code mixing for alternation of two languages within a sentence (Winford, 2003) or the mixing of two or more languages or language varieties in speech. Speakers practise code-mixing when they are fluent in both languages.

- **Borrowing**

Borrowing is defined as the introduction of single words or short phrases from one variety of language into the other. It is the incorporation of features of one language into another. Such borrowing may include integrating and assimilating structural features (phonics, semantics), figurative language, metaphor, verbs, nouns, adjectives, connectives, similes and many others from one language into another, since any linguistic feature can be transferred from any language to any other language. Borrowing is a means of using one primary language, but mixing in words or ideas from another. In borrowing, code mixing and code switching the speaker is actively using both languages, alternating between them or incorporating features from one into another to achieve predetermined goals. (Chikiwa, 2016; Treffers-Daller, 2007; Hughes, Shaunessy & Brice, 2006).

TEACHER CODE SWITCHING IN THE CLASSROOM

Teacher code switching is common in many multilingual mathematics classes, where the LoLT is neither the teacher's nor the pupils' first language. In multilingual classrooms code switching is a reality and is used for various reasons. Even if there is an official policy about this, teachers make individual spontaneous decisions about language choices, and these are mostly determined by the need to communicate effectively. Being entrusted to implement the school curriculum, teachers do so through the best pedagogical options available to them (Chikiwa, 2016; Halai, 2009; Zevenbergen, 2001). This section seeks to highlight that these assumptions often lead to linguistic malpractices in IP mathematics classrooms.

Teachers often use code switching for different purposes. Some of these purposes include:

- Code switching for pedagogical and curriculum access reasons;
- Code switching for communicative purposes;
- Code switching for social reasons;
- Code switching for classroom management purposes; and
- Code switching for identity purposes.

Adapted from Chikiwa (2016)

The linguistic relevance of these teacher code-switching purposes is clarified further by Gumperz (1982), who describes them as metaphorical functions, thus, further clarifying the purposes of code switching when used in a conversation. The six metaphorical functions of code switching suggested by Gumperz (1982) are quotation, addressee specification, interjections, reiteration, message qualification, personification versus objectivisation, and situational code switching. These functions are briefly described in Table 5.4.

Table 5.4: Functions of code switching

FUNCTION	DESCRIPTION
Quotation	▪ A form of code switching used when quoting or reporting someone else's discourse.
Addressee Specification	▪ A code-switched message that aims at a particular/different addressee. ▪ Code switching that occurs when a teacher is directing a message to one of the learners or a specific group of learners.
Interjections	▪ Code switching that is used to express an emotion or sentiment.
Reiteration	▪ Teacher code switching that is used to emphasise, clarify or amplify the intended message. ▪ A code-switched message repeated from one code to another, either literally or in a modified way.
Message qualification	▪ Code switching used by a teacher to add more information in order to qualify the main message. ▪ A code-switched message that elaborates what has been said.
Personalisation vs. Objectivisation	▪ Code switching that relates to the distinction between talk about action and talk as action; the degree of speaker involvement in a message reflects personal opinion or knowledge whether it relates to specific instances or has the authority of generally known facts.
Situational code switching	▪ Code switching that results from a change in social setting: topic or participants.

Adapted from Gumperz (1982: 75 – 84)

Though the list is not exhaustive, the functions of code switching stated above capture the dynamic nature of teacher code-switching practices in an attempt to convey meaning in multilingual mathematics classrooms (Mahootian, 2006).

Since teaching and learning is a two-way process that involves both the teacher and the learner, teacher code switching is directly related to learners' code switching. Research has shown that learners' language practices in the classroom reflect those practices their teachers expose them to (Rollnick, 2000; Barton & Neville-Barton, 2003). Learners have the teacher as their 'more knowledgeable other' (Vygotsky, 1981) and model in the classroom. Thus, the teachers' use of code switching in turn interferes with the learners who may, as in many cases in South African classrooms, already be using it (Chikiwa, 2016).

Most textbooks and other teaching resources, particularly in mathematics, are still written mainly in English and most teachers are expected to teach in English. However, most teachers in Xhosa-dominant schools in the Eastern Cape Province use isiXhosa to teach or engage in code mixing and code-switching in order to be understood by their learners.

English is the dominant language of instruction and is regarded as the language of power in Africa, even though relatively few people have it as a first language. Setati and Adler (2000) undertook an extensive review of research that had taken place in South Africa. Their focus was on mathematics learning and bilingual education. They found a significant relationship between language proficiency and mathematical achievement. In particular, they noted that oral proficiency in English in the absence of mother-tongue instruction was negatively related to achievement in mathematics. However, an important aspect highlighted by the research is that the findings cannot be attributed to the learner's language ability alone. Factors including social, cultural and political issues also need to be considered, as they have a significant impact on the teaching and learning process and schooling in general.

PEDAGOGICAL IMPLICATIONS: MATHEMATICS TEACHING AND LEARNING IN A SECOND LANGUAGE

Research on multilingualism in mathematics classrooms and other learning areas has indicated that code switching does not necessarily imply a breakdown in communication, but can actually serve as a resource or a competence available for use by multilingual learners and teachers; (Adler, 2001; Chikiwa, 2016; Bullock & Toribo, 2009; Howie, 2003; Moschkovich, 2002; Setati, 2005). Therefore, teachers need to take advantage of the presence of multilingualism in their classrooms and use it to achieve meaningful learning and teaching. The perception that code switching is a teaching and learning resource has attracted much attention in a range of mathematics education studies in and outside of South Africa (Chikiwa, 2016; Chitera, 2009; Halai & Karuku, 2013; Jegede, 2012, 2005, 2008; Vorster, 2008).

This chapter highlights the need for mathematics teachers to be enlightened on different methods of code switching and their applications to promote an effective use of the teaching strategy in IP mathematics classrooms. Once teachers are confident in the practice of code switching, they will be able to easily and purposefully switch from one language to the other, compared to embarking on the practice without planning. Planned code switching would also enable teachers to provide coherent explanations in both the mother tongue and the prescribed LoLT, and not only provide a watered-down version of mathematical concepts. IP mathematics teachers' mastery of code switching will ensure that teachers leave the learners with the correct explanations of the mathematical concepts in the language of assessment. Thus, teachers will be able to explain concepts in their mother tongue and still manage to take the learners back to the English version of the concepts which learners are most likely to encounter during assessment.

Why are there so many theories in Second-Language Acquisition? Why isn't there just one theory? What is it about second-language acquisition that invites a diffusion of theoretical perspectives? one may ask. To understand this, one might consider a parable about four blind men and an elephant. When the men came across the elephant for the first time, the first one, holding its tail said: "The elephant is like a rope." The second one holding onto a giant leg said: "The elephant is like a tree." The third one, feeling the animal's massive body said: "The elephant is like a wall." The fourth one, holding the trunk said: "The elephant is like a snake." Thus, second-language acquisition is a complex set of processes. Researchers have investigated different parts of second-language acquisition to understand the complex phenomenon. The different theories complement each other.

CONCLUSION

This chapter contributes to the broad debates on the nature of second-language learning and acquisition among multilinguals and seeks to explain the intricacies of mathematics teaching and learning in a second language. The reviewed literature sheds more light on the nature of the classroom dynamics influencing the acquisition of a second language by learners. By investigating the presence and nature of these factors, it is apparent that the learners' knowledge of critical vocabulary is influenced by the teacher's competence in the language of instruction. An IP mathematics teacher needs to understand the intricacies of second-language acquisition as (in most cases) the learners will be learning in a language which is not their mother tongue; and the teacher needs to understand the stages these learners go through in order to reach out to them effectively. On the other hand, (specifically in under-resourced schools), the prescribed language of instruction for IP mathematics is a second language to the teacher; and the teacher therefore needs to be prepared to present mathematics content in a language foreign to both the teacher and the learner.

Since the majority of the South African population is multilingual rather than bilingual, the pedagogical and non-pedagogical practices that may affect FP learners' successful transition from mother tongue education to using English as LoLT in the IP level are discussed. To improve learner performance in mathematics (and other subjects), there is a need to further investigate the following aspects at the time of the transition from FP to IP:

- FP learners' age at the time of the transition;
- FP learners' mother-tongue proficiency;
- IP teachers' quality of teaching; and
- IP teachers' language competency in the target LoLT.

These investigations would provide pertinent information regarding ***both*** learners' and teachers' readiness for the transition and thereby determine the most appropriate transition model for meaningful instruction.

Being proficient in conversational English does not guarantee successful learning in mathematics. In communities using a medium of instruction which is a second language to both the learners and the teachers, it is important to equip teachers with the necessary linguistic skills to facilitate mathematics instruction (specifically skills for using language as a resource in mathematics instruction as discussed in Chapter 2 of this book). There should not be assumptions that pre-service teacher training alone, adequately equips teachers to facilitate a smooth transition from mother-tongue to English-medium mathematics instruction without the support of in-service professional development courses in language competency (Tshuma,

2017). The literature reviewed in this chapter concurs with other educationists (Adler, 2001; Barwell, 2009; Chikiwa, 2016; Jegede, 2012; Setati, 2008), who declare that **teaching mathematics in a second language (such as English) requires more teacher effort as it calls for the teaching of content, language of mathematics and the English language**.

Chapters 4 and 5 of this book specifically focus on the intricacies involved in solving mathematical word problems as well as the presentation of mathematical concepts in a second language. **Case Study B** focuses on teacher mathematics word problem solving; and the purpose of the case is to analyse the Eastern Cape Department of Education (ECDoE) Intermediate Phase (IP) mathematics teachers' Pedagogical Content Knowledge regarding word problems. This case study draws on the Error Analysis approach, which analyses errors on written mathematical tasks focusing on the written aspects of mathematics and the steps used in solving a word problem. A standardised teacher Mathematics Word Problem assessment piloted in five South African universities was administered on 55 Intermediate Phase (IP) mathematics teachers purposefully selected from 16 education districts in the ECDoE. Data were quantitatively and qualitatively analysed. Results show that teachers' aptitude in solving mathematical word problems is significantly low. Whether the weak aptitude in solving mathematics word problems is attributed to a low proficiency in the language of instruction (as revealed in Case A), or general innumeracy or specifically to a poor mastery of word problem solving strategies calls for further research.

CASE STUDY: B

INTERMEDIATE PHASE MATHEMATICS TEACHERS' PEDAGOGICAL CONTENT KNOWLEDGE CONCERNING WORD PROBLEMS: THE EASTERN CAPE PROVINCE

INTRODUCTION

Pollak (1969: 393) states that learners are involved in the application of mathematics "mainly through ... 'word' problems". Although this observation is focused mainly on learners; its applicability to Intermediate Phase (IP) mathematics teachers is illustrated in the example below:

Picture the scene: While presenting a continuous professional teacher development course in mathematical thinking and problem solving, a lecturer assigned a group of Intermediate Phase (IP) mathematics teachers from the Eastern Cape Province of South Africa the following word problem to solve: "***While sharing one full pizza, Enhle eats ¼ of the pizza and Mandisa eats ⅓ of the pizza. Which of the two girls ate more pizza***?"

Figure B.1: A symbolic comparison of ¼ and ⅓

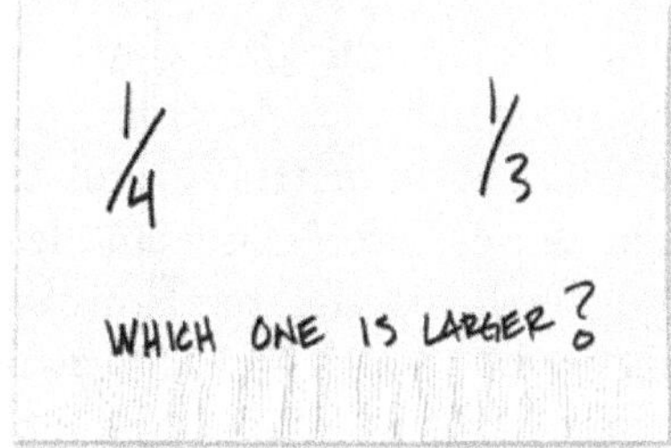

Most of the teachers responded that Enhle ate more pizza slices since 4 is bigger than 3. On realising that the majority of the teachers were missing the mathematics behind the problem, teachers were requested to explain their answers. After fruitless deliberations among the teachers, the lecturer provided a diagrammatic representation of the problem as illustrated in Figure B.2:

Figure B.2: A diagrammatic comparison of ¼ and ⅓

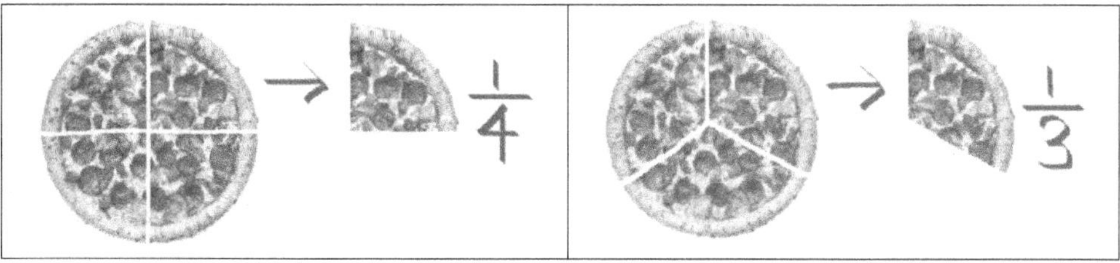

Adapted from Barwell, Leung, Morgan and Street (2002)

The added clues as presented in the diagram seemed to help the teachers and ALL agreed to say ⅓ is bigger than ¼ as the diagrammatic presentation clearly illustrated the solution. The lecturer confirmed that ⅓ is indeed greater than ¼ and provided other realistic examples to support the answer. It was intriguing for the teachers to discover that in a similar survey conducted by a fast-food outlet in the USA the company noticed that the sales for the *Third Pounder Hamburger* were very low compared to those of the *Quarter Pounder Hamburger.* In a focus group discussion customers believed (and indicated) that they were getting less meat because the "3" in ⅓ is smaller than "4" in ¼, hence "customers believed they were being overcharged". Although the general populace may be excused for such innumeracy, the same level of innumeracy among mathematics teachers (*even after factoring in the language barrier*) is a serious cause of concern.

The explicit teaching of the solution steps of a mathematical problem usually starts as early as in grade 1 in primary school. To ventilate how and why word problems are taught, it is of paramount importance to investigate in-service teachers' pedagogical content knowledge concerning word problems. This case study briefly defines a word problem, strategies for solving word problems, the role of the teacher's pedagogical content knowledge in solving word problems, research questions, conceptual framing, research design and methodology, findings and discussion.

WHAT IS A WORD PROBLEM?

Word problems are often considered a separate genre of mathematical tasks – an add-on to the prescribed learning material. According to the nationwide survey of the Committee of Mathematics Education (Hungarian Academy of Sciences 2016), many teachers would eliminate word problems from the first and second grades of public schooling. Without generalising, the trend practised by Hungarian teachers resonates with what is practiced by South African teachers as depicted by Sepeng (2013). This view is based on an understanding of what a word problem is and how word problem solution steps are usually introduced in the schools. Solving word problems requires a relatively high level of reading skills and reading comprehension, and it is not in the mathematics class that such skills and abilities should be improved and assessed.

Peters (2011: 7) and Verschaffel, Van Dooren, Greer & Mukhopadhyay (2010: ix) define a word problem as a "verbal description of a problem situation wherein one or more questions are posed of which the answers can be obtained through the application of mathematical operations to information available in the text". This definition involves several types of word problems from mere routine tasks to more sophisticated realistic word problems.

Among the several possible roles word problems may play in the classroom, there seem to be two critical purposes. Word problems are often used in a way that Palm (2006: 42) described as "many of them are ... merely ordinary school mathematics tasks 'dressed up' with an out-of-school figurative context". Maybe learners' superficial strategies in word problem solving are rooted in both the teachers' teaching practice as well as the teachers' own capability to solve the mathematical word problem. Another usual aim in the mathematics classroom (especially in the foundation phase) is teaching an algorithm to be followed step by step. Over-reliance on either of these two purposes may overshadow other possible functions word problems should fulfil, such as the developing self-explanation or shaping a positive attitude towards mathematics (Fonseca & Chi, 2011; Csíkos & Szitányi, 2020).

WORD PROBLEM SOLVING STRATEGIES

The roots of mastering a mathematical word problem solving strategy originate in teachers' instructional strategy, a part of which is teaching word problem solving algorithms (Csíkos & Szitányi, 2020). Word problem solving algorithms are explicitly taught to Foundation Phase learners. At this phase level, teachers use situations expressed in words to represent an arithmetic operation. In this phase the basic purpose of introducing word problems is only understanding the structure of an operation, as illustrated in Figure B.3 demonstrating the idea of multiplication as repeated addition.

Figure B.3: A word problem illustration for visualising a simple arithmetic ***operation***

Adapted from Lehotanová, (2013).

Later, when learners are already acquainted with arithmetic operations, the main purpose of using word problems is to make learners recognise and understand what the text describes and select a suitable model for problem solving.

In South African teacher education modules/course packs/textbooks and learners' textbooks, several steps of word problem solving are listed and expected to be followed in the classroom. Most of the examples involve reading, understanding, planning (and estimation), drawing (if needed), calculations, verification, answering the question. These steps represent a kind of common understanding of what a word problem solution should look like in written form and how the solution steps are to be scored. While some of the steps are necessarily present in all kinds of word problems, others are more specific to certain types of word problems only. For example, sometimes the solution comes not from executing a single arithmetic operation, and often there is no need to estimate before making calculations. The last two steps may be combined to form the interpretation of the solution, but these steps usually prove to be just mechanical activities performed, and in cases where a learner fails to accomplish it, they are penalised.

Neményi and Szendrei (1997) describe four main steps of word problem solving, namely understanding the problem, seeking for a mathematical model and turning the original problem into a mathematical problem, solving the mathematical problem, and interpretation of the mathematical solution. This general four-step model echoes Pólya's (1945) problem solving strategy phases which involve: Understanding the text, preparing a plan for the solution, implementing the plan and checking the results. The difference between the seven-step and the four-step processes requires the consideration of instructional and assessment methods, dilemmas of equity and even deep philosophical questions of mathematics education (Csíkos & Szitányi, 2020).

THE ROLE OF TEACHERS' PEDAGOGICAL CONTENT KNOWLEDGE IN TEACHING WORD PROBLEM SOLVING STRATEGIES

Blomeke and Delaney (2012) distinguish between different aspects of knowledge of mathematics for teaching: namely mathematics content knowledge (MCK) and pedagogical content knowledge (PCK). MCK includes the substantial elements of mathematics which encompasses knowledge of procedures and concepts and ways in which this knowledge is organised (Grossman, Wilson and Shulman, 1989). PCK, which is a subset of MCK, involves knowledge that is related to the practices of teaching (Beswick, Callingham & Watson, 2012; Bowie, Venkat & Askew, 2019). This case particularly focused on PCK because of the nature of the assessment items which required the teacher to explain how they would teach a particular concept – practices of teaching.

Teachers' pedagogical content knowledge (PCK) was coined by Shulman (1986). Among the knowledge components necessary for anyone to be a successful teacher, one form of teachers' knowledge is of utmost importance, namely the knowledge form that is built on the content knowledge but goes beyond that by including powerful analogies, illustrations or demonstrations, in order to make the content to be learnt comprehensible to others. Therefore, mathematics teachers must possess a very high level of mathematical common content knowledge as well as additional knowledge components (Ball et al., 2008).

According to Hill et al. (2008), teachers' knowledge about what makes a problem difficult for a student highly depends on the teacher's level of mathematical subject-matter knowledge. As for word problems, it is clear that teachers must not only have the necessary knowledge to solve the tasks, but they must have further knowledge components, a repertoire of word problem solving strategies that are somehow comprehensible to the learners. Shulman (1986: 9) denotes that "there are no single powerful forms of representations", and mathematics teachers' pedagogical content knowledge regarding word problems should adopt different solution strategies and techniques (and the corresponding specialised content knowledge).

Teachers' classroom practices for teaching word problems were studied by Chapman (2006), who claimed that there had been only a few studies conducted on pre-service teachers' difficulties with word problems, and there is a lack of research on how pre-service teachers handle the contextual factors of word problems. There is an even more profound lack of research conducted on in-service teachers' difficulties in with word problems (Tshuma, 2017).

According to Shulman (1986), finding out why learning a topic is easy or difficult is an integral part of the PCK. Chapman (2006) identifies two types regarding how teachers view word problems. Teachers who adopt a paradigmatic-oriented approach usually blame the realistic context of word problems as a main barrier to providing a mathematical solution. In contrast, teachers who follow the narrative-oriented approach focus on the psychological factors of the solution. They focus on learners' attitudes, motivation and potentially different critical interpretations of the text. According to them, failures should not be attributed mainly to a lack of mathematical skills but rather to a lack of meaningfulness.

Finally, another important aspect of teachers' pedagogical content knowledge is how they assess learners' word problem solving activities. Verschaffel et al. (2010) emphasises that the way learners' mathematical performance is assessed has a strong influence on teaching practice. They revealed that current summative evaluation practices have several characteristics that are not in line with what researchers would find fruitful for assessing higher-order thinking skills. It is highly probable that teachers' formative assessment techniques applied in classroom settings are in line with how their learners will be assessed in high-stakes tests. Independently of the current high-stakes testing practices, it is worth finding out how teachers in the classroom assess the learners' solutions provided for word problems (Csíkos & Szitányi, 2020).

OBJECTIVES

Interrogating teachers' pedagogical content knowledge regarding mathematics word problems seeks to:

i. Determine IP mathematics teachers' aptitude in solving mathematical word problems.
ii. Determine the extent to which teacher aptitude in solving mathematical word problems relates to English symbol obtained at matriculation.
iii. Determine the extent to which teacher aptitude in solving mathematical word problems relates to experience in teaching IP mathematics.

iv. Determine the steps IP mathematics teachers follow to solve a given word problem.
v. Identify strategies commonly used by IP mathematics teachers in solving word problems.

CONCEPTUAL FRAMING

The conceptual framework grounding this case is Newman's (1983) approach to analysing errors on written mathematical tasks. Newman's approach to analysing errors on written mathematical tasks focuses on the written aspects of mathematics and the steps used in solving a word problem (Newman, 1977, 1983). Her framework for interrogating mathematical tasks is of great significance, considering that teachers and their learners continuously encounter written mathematical tasks when dealing with textbooks, exams and other forms of external assessments. She states that when confronted with a written mathematical word problem, a person needs to go through a fixed sequence of events, as illustrated in Figure B.4, in responding to the problem.

Figure B.4: The Newman hierarchy of causes of errors (Newman, 1977, 1983)

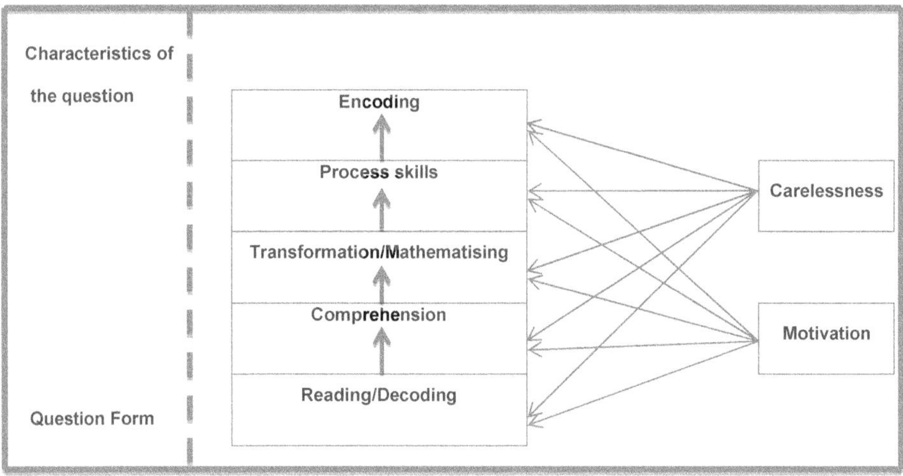

Newman's approach draws attention to the importance of language factors in mathematics teaching and learning. It is based on her research conducted with Grade 6 learners and was further developed by Casey (1978), Clements (1980), Clarkson (1980) and Watson (1980), who depicted the hierarchy as a non-prescriptive but useful tool in teaching mathematics problem solving. The approach reveals that 50% of errors made in mathematics tasks occur before the application of process skills. With this in mind, mathematics teachers need to pay particular attention to whether the learners are able to comprehend the mathematics word problems they are tasked to solve, and so should mathematics teacher educators in teacher educator institutions.

Newman also described an additional category classified as *carelessness* or *imprecision* of errors resulting from unknown factors. In addition to carelessness, a learner who has read, comprehended and worked out an appropriate strategy for solving a mathematics problem might be discouraged from proceeding further because of a lack of *motivation*. To identify errors made in written mathematical tasks and to reduce the occurrence of careless mistakes and lack of motivation, the Newman approach recommends that the following steps be implemented:

- Read the question (*Reading/Decoding*);
- What is the question asking you to do? *(Comprehension)*;
- Suggest a method that can be used to find an answer to the question. *(Transformation/Mathematising)*;
- Explain how you worked out the answer, or what you are doing to work out the answer *(Processing)*; and
- Write down the answer to the question *(Encoding)*.

Thus the Newman procedure involves an oral question-and-answer session after failing to solve mathematical questions, so as to determine sources of difficulty encountered while answering. Application of Newman's approach in this case involves interviewing the respondents, asking them questions with the intention of establishing whether they can carry out the following steps: read the question, comprehend what they have read, conduct an appropriate mental transformation from the words of the problem to the choosing of a mathematical strategy, employ the process skills required by the selected strategy, and encode the answer in an appropriate written form. This framework provides justification for employing mathematics word problems as a means of assessing mathematics performance, while addressing potential language issues encountered when confronted with mathematics instruction using English as the LoLT. In particular, analyses of data based on the Newman procedure have drawn special attention to:

- the influence of language factors on mathematics learning; and
- the inappropriateness of many 'remedial' mathematics programmes in schools in which there is an over-emphasis on the revision of standard algorithms.

(Clarke, 1989)

According to Newman (1977, 1983) a person wishing to obtain a correct solution to a one-step word problem such as: "***The marked price of a book was R20. However, at a sale, 20% discount was given. How much discount was this?***" must ultimately proceed according to Newman's hierarchy.

Newman used the word "hierarchy" based on the reasoning that failure at any level of the above sequence prevents problem solvers from obtaining satisfactory solutions, unless they arrive at correct solutions by chance through faulty reasoning. Usually, problem solvers often return to lower stages of the hierarchy when attempting to solve problems (Casey, 1978). (For example, in the middle of a complicated calculation someone might decide to reread the question to check whether all relevant information has been taken into account.) However, even if some of the steps are revisited during the problem-solving process, the Newman hierarchy provides a fundamental framework for the sequencing of essential steps.

According to Clements (1980: 4), errors due to the form of the question are essentially different from those in the other categories, because the source of the difficulty resides fundamentally in the question itself rather than in the interaction between the problem solver and the question. This distinction is represented in Figure B.4 by the category labelled "Question Form" being placed beside the five-stage hierarchy. Two other categories – *carelessness* and *motivation* – have also been shown separately on the hierarchy. Although, as indicated, these types of errors can occur at any stage of the problem-solving process. A careless error, for example, could be a reading error or a comprehension error. Similarly, someone who had read, comprehended and worked out an appropriate strategy for solving a problem might decline to proceed further in the hierarchy because of a lack of motivation. For example, a problem-solver might exclaim: "What a trivial problem. It's not worth going any further."

METHODOLOGY

The research followed a mixed methods approach, incorporating quantitative and qualitative strands. Ethical clearance to collect data was obtained from University of Stellenbosch Research Ethics Committee and ECDoE Strategic Planning Policy Research and Secretariat Services. The data collection techniques included a teacher mathematics word problem assessment and an error analysis interview.

The mathematics word problem assessment was provided by JET Educational Services and is a standardised teacher assessment, conforming to the South African context and piloted in five different university teacher education departments in the country. A mathematics word problem assessment (quantitative) was administered to 55 IP mathematics teachers. The teacher assessment (developed using already existing tests, including Wits Maths Connect Project, Mathematics Knowledge for Teaching Measures, Annual National Assessments (ANA)[1] question papers 2009 – 2013 and Southern and Eastern African Consortium for Monitoring Educational Quality (SACMEQ) Mathematics test for teachers 2007) incorporated broad mathematics content area, the domains of knowledge, level of difficulty and school phase. The assessment is primarily based on the National Curriculum Statement (NCS) home language level assessment standards for Grade 7 level. The rationale used is that teachers, at the absolute minimum, need to show a proficiency in a language that is two years beyond that of their learners (JET, 2015). The assessment was made up of 45 items and covered a range of topic areas. The level of the assessment is quite low, as it is trying to establish the absolute minimum standards for teachers.

1 ANAs were phased out in 2015.

Considering that the research participants are qualified and practising teachers, the pass rate was thus set at 70%.

The researcher was the sole assessment administrator; however, this was not a challenge as the research is triangulated. Assessments were administered anonymously through the use of alpha numerical codes. Teachers were given 90 minutes in which to complete the test. On average, teachers spent 75 minutes on completing the proficiency test. No teacher required the full amount of time to finish the mathematics word problem assessment.

The Newman hierarchy of error causes for written mathematical tasks was incorporated to interview (qualitative) 10% of the sampled IP teachers after the administration of the mathematics word problems. This type of interview was used instead of an ordinary interview, because it solicited responses better suited to answer the research questions based on its likelihood of establishing IP mathematics teachers' pedagogical content knowledge concerning word problems.

The data collected were analysed using STATISTICA version 13 (Dell Inc., 2016), Dell STATISTICA (*data analysis software system*), version 13 software.dell.com) at University of Stellenbosch's Centre for Statistical Consultation, while JET Education Services conducted descriptive analyses of the quantitative data. Trends emerging from the two independent analyses of the quantitative data were compared and analysed qualitatively.

SAMPLING

The unit of analysis was language proficiency of 55 IP mathematics teachers from the ECDoE. For the purposes of this research, the teachers were studied specifically for their mathematics pedagogical content knowledge regarding word problems. The section below details the IP mathematics teacher participants' demographic information.

PARTICIPANT DEMOGRAPHIC DATA

The participants' ages ranged from early twenties to over 35. Two participants (3%) were below the age of 25, only one participant (2%) was between 26 – 29, another one (2%) was between 30 – 34, and the majority (93%) of the participants were above 35 as indicated in Figure B.5.

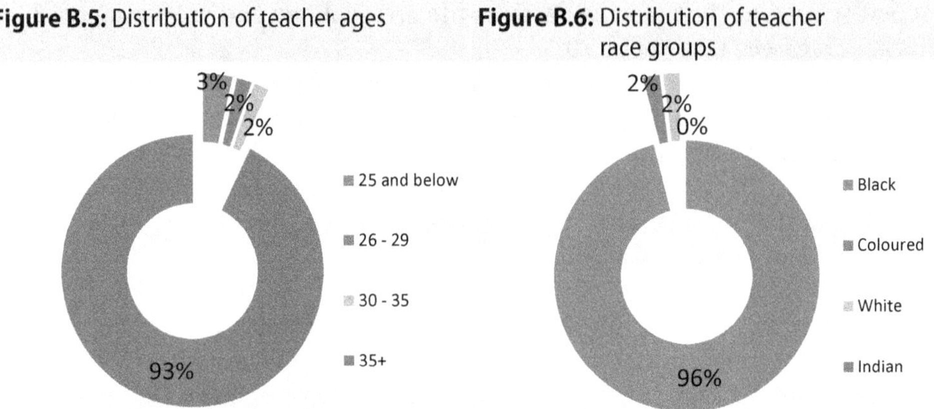

Figure B.5: Distribution of teacher ages

Figure B.6: Distribution of teacher race groups

The majority (96%) of the participants belonged to the black race group, 2% were coloured, 2% white and none were Indian, as indicated in Figure B.6. The distribution of participants across race groups is closely related to the participant teachers' home languages. IsiXhosa was home language to 91% of the participants, seSotho to 7% of the participants and Afrikaans to 2% of the participants, as indicated in Figure B.7.

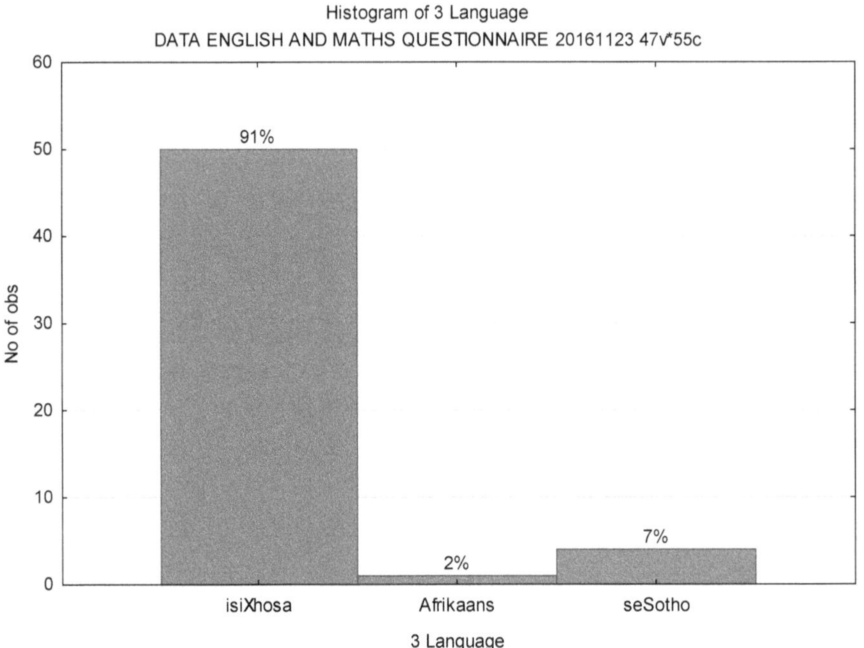

Figure B.7: Histogram of teachers' home languages

It can be inferred that teachers of the black race group constituted those whose home language is isiXhosa and seSotho, while the participants of the coloured and white race groups constituted those speaking Afrikaans as their home language.

Since the current research is based on interrogating the teachers' pedagogical content knowledge regarding word problems, it was necessary to capture the symbols the teachers had obtained in English at matric[2]/school-leaving level. The majority of the teachers obtained symbol D in matric English; and it is noteworthy that word problems at the IP level are presented (and supposed to be taught in English). To ascertain the participants' teaching experience, the number of years they had been practising as IP mathematics teachers was captured. 19% of the of the participants had been teaching IP mathematics for 2 years and less; this could be attributed to the fact that some teachers could have been teaching other phases and/or subjects such as Technology and were moved to teaching IP mathematics when their subject specialisations were removed from the curriculum. 32% of the participants had been teaching IP mathematics for an average of 5 years, while 27% of the participants had been teaching IP mathematics for an average of 12 years; and these could be considered as seasoned IP mathematics teachers. The section below outlines the case's limitations and delimitations.

According to the policy stipulation, the IP phase is comprised of three specific grades: 4, 5 and 6. However, changes in the DBE stipulations resulted in the IP phase including Grade 7 as well to accommodate primary schools that had enrolments ranging from grade R – 7. (As from the beginning of 2015, the DBE began the process of moving Grade 7, 8 and 9 classes housed in primary schools to neighbouring high schools – a process with good intentions but resulting in redeployment of teachers). As such, the information obtained from the teachers about the grades taught ranged from Grades 4 to 7 as indicated in Figure B.8.

Figure B.8: Distribution of IP mathematics grades taught

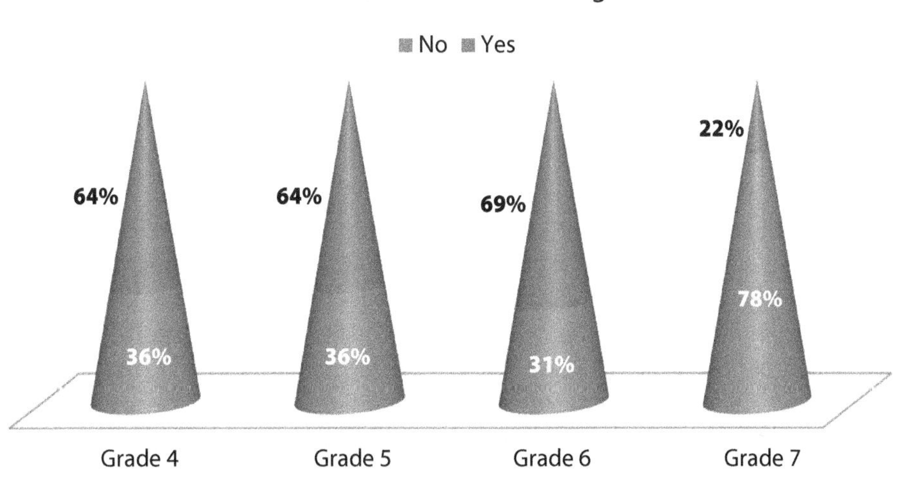

2 A matric certificate is offered after completing the 12th grade of the South African ordinary school system. The issuing of certificates is standardised and quality assured by Umalusi Council in conjunction with the Department of Basic Education, (DBE).

As indicated in Figure B.8, an average of 66 % of the teachers taught Grade 4, 5 and 6, and only 22% of the teachers taught Grade 7 classes.

LIMITATIONS

The research's limitation was that incorporating the error analysis schedule into the interview schedule focused on only three questions (pre-selected by the researcher) out of the **35** word problem questions making up the JET Mathematics Word Problem Assessment. The delimitation was that the pre-selected questions were all of relatively the same level of difficulty. More information would have been garnered if questions with different levels of difficulty had been selected and compared. Thus, more balanced results would have been obtained.

FINDINGS

The research question was: **To what extent are IP mathematics teachers in ECDoE schools competent in solving word problems?** and the key finding was that sampled teachers ***indeed*** exhibited a lower competence in solving word problems than what is expected of a grade 7 learner. This low competence suggests that learners are facing difficulties in solving word problems due to teachers' ignorance and lack of pedagogical content knowledge in teaching word problems. Based on JET's performance scales, a consolidation of these performance scales is presented in Table B.1:

Table B.1: Mathematics Word Problem Scores

	MATHEMATICS WORD PROBLEM SCORES
70% and above	(1) 2%
50% - 69%	(18) 33%
Below 50%	(36) 65%

Key:

- **70% and above:** satisfied requirements
- **50% - 69%:** cause for some concern
- **Below 50%:** cause for serious concern

Only one teacher met the requirement in the mathematics word problem assessment satisfactorily by scoring 70%; 18 (33%) of the teachers scored between 50% – 69%, which is a *cause for some concern*, while 36 (65%) of the teachers scored below 50% and this is *cause for serious concern*. Thus, the performance of the majority of the teachers is a cause for *serious concern* in mathematics word problem assessment. These consolidated scores emphasise that much still needs to be done to improve the pedagogical content knowledge of IP mathematics teachers.

The teacher **Mathematics word problem** scores are presented as a histogram in Figure B.9 in order to see the nature of the distribution of the scores as well as to identify possible outliers:

Figure B.9: Histogram of Mathematics Word Problem Score

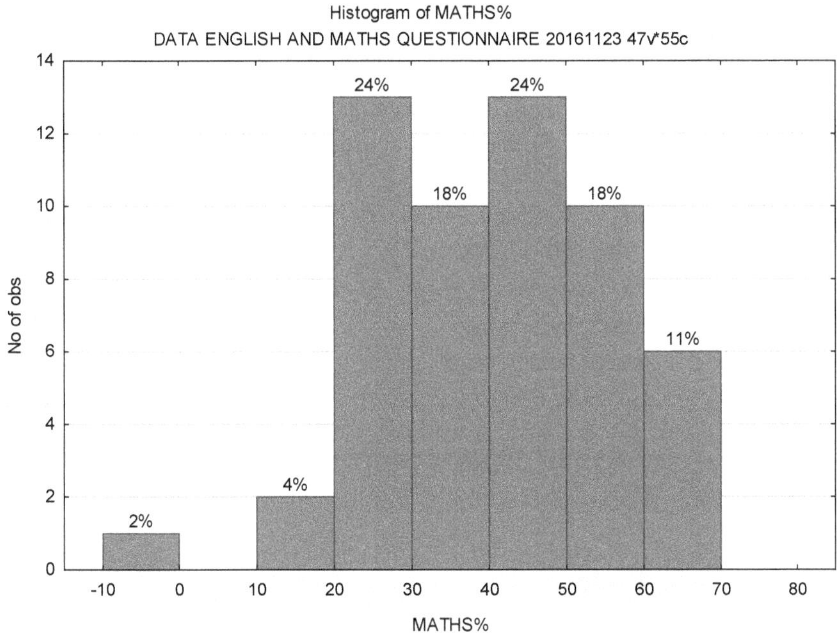

Figure B.10: English symbol at matriculation vs Mathematics Word Problem Assessment Means

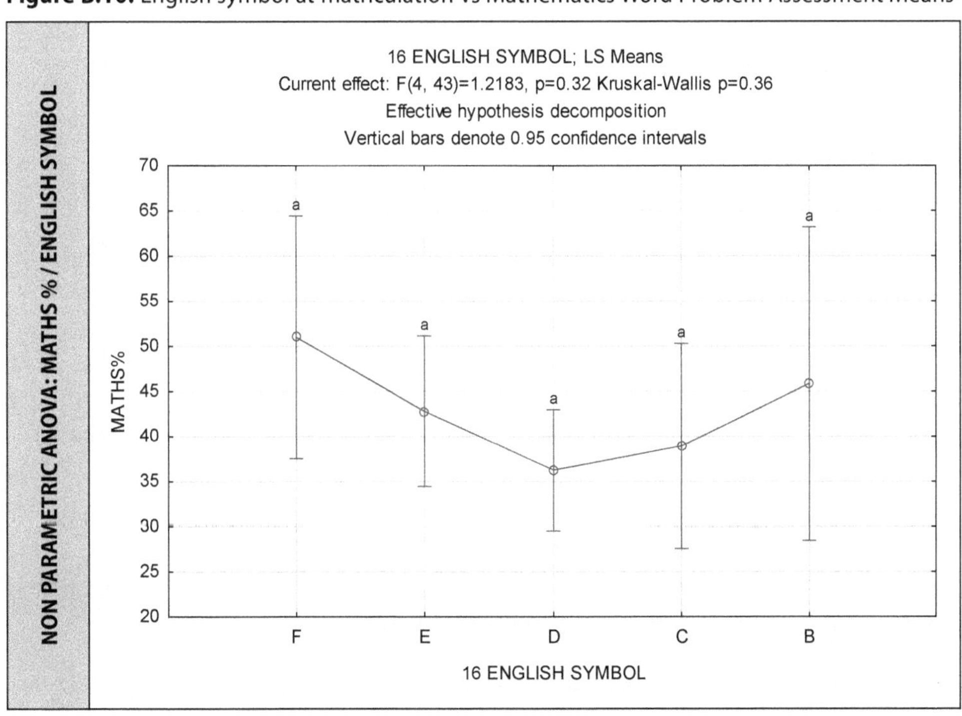

Out of the 55 scripts, one was spoilt; the lowest mark obtained was $^{13}/_{80}$ = 16 %, while the highest mark was $^{56}/_{80}$ = 70%. The modal mark was $^{34}/_{80}$ = 42%. It is worrying to note that on average, teachers revealed a lower performance than what is expected of a Grade 7 learner.

The data on matriculation English-language symbols were correlated to the teacher scores on mathematics word problem assessments. The two variables were compared using descriptive statistics as well as with an ANOVA test. Figure B.10 illustrates the correlation between teacher scores in teacher Mathematics word problem scores compared to the English symbol attained at matriculation.

Teacher Mathematics Pedagogical Content Knowledge Concerning Word Problems and Mathematics Teaching Experience

Demographic data statistically analysed the teacher participants' experience in teaching IP mathematics. The data on experience in teaching IP mathematics were correlated to the teacher scores on mathematics word problem assessments. Figure B.11 illustrates the correlation between teacher scores in Mathematics word problem assessment compared to experience in teaching IP mathematics.

Figure B.11: Correlation between Experience (Years) in teaching Mathematics vs Mathematics %

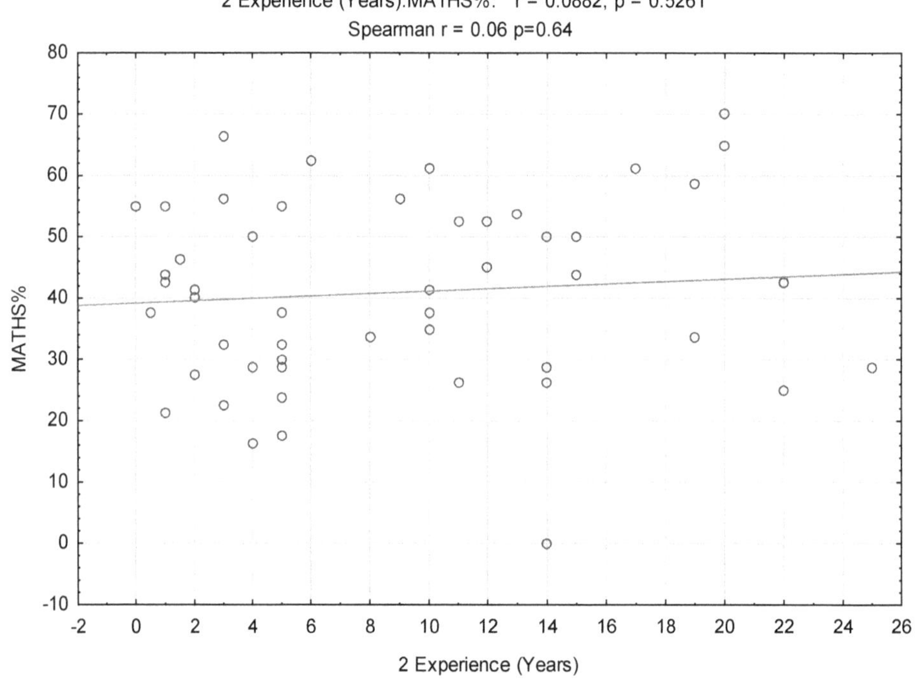

As established in the demographic data, about 60% of the participants are experienced teachers who have been teaching IP mathematics for between 5 and

12 years. As illustrated in Figure B.11, with a correlation coefficient of r=0.0882, there is a positive correlation between mathematics word problem assessment scores and experience in teaching IP mathematics. This correlation implies that as teachers gain more experience in teaching mathematics, aptitude in solving mathematics word problems improves, and this could be attributed to the exposure the teachers gain as they engage with the teaching and learning content.

Be this as it may, 27% of the participants were newly graduated teachers who had been teaching IP mathematics for between 0 and 4 years and they will still need to be upskilled in solving mathematics word problems. In addition, if the policy still stipulates that IP teachers continue to be generalists, it means more teachers who could have specialised to teach other subjects may find themselves teaching IP mathematics and they will be adding on to the number of teachers that need upskilling in solving mathematics word problems. Therefore, additional CPTD courses in word problem solving strategies will always be in demand as suggested by the majority (99%) of the teachers.

The sections above provided information gathered through the teacher Mathematics Word Problem Assessment by initially analysing the raw scores and then relating the scores to data collected from the teacher demographic data. The section below presents samples of teacher responses from the Mathematics Word Problem Assessment with a total achievable mark of [6].

As shown in Table B.2, sample responses exhibit a variety of linguistic challenges ranging from complete omissions – leaving the question unanswered, partial omissions – concentrating on the calculations (computation of digits), and not explaining the procedures as requested by the question. This could be attributed to the fact the teachers "are ensnared by the language rather than by the mathematics hidden therein" as suggested by Kambule (1984). Sample responses also included the use of one-word *'yes'/'no'* answers without supporting these answers, as well as the use of diagrammatic representations to compensate for the *missing written* explanations as an attempt to fulfil the requirements of the question. As a result, out of the six questions sampled in Table B.2 three questions had more than 50% of the teachers scoring zero.

Some responses were composed of detailed answers, though some of the detail included irrelevant information not answering the question. Other responses (such as **T45** answering **Question 13bi**) were composed of diagrammatic illustrations without any explanations. These illustrations confirm several empirical studies which indicate that drawings (both the type of drawing and the process of making drawings in the classroom) play a very important role in shaping learners' beliefs about and solution strategies for word problems (Willis & Fuson, 1988; Depaepe et al., 2010). In the two main teaching approaches defined by Chapman (2006), the main challenge with the pragmatic-oriented approach is finding the appropriate drawing that should definitely display some kind of mathematical content, whereas teachers following the narrative-oriented approach may seek a variety of possible

Table B.2: Sample Teacher Responses: Mathematics Word Problem Assessment

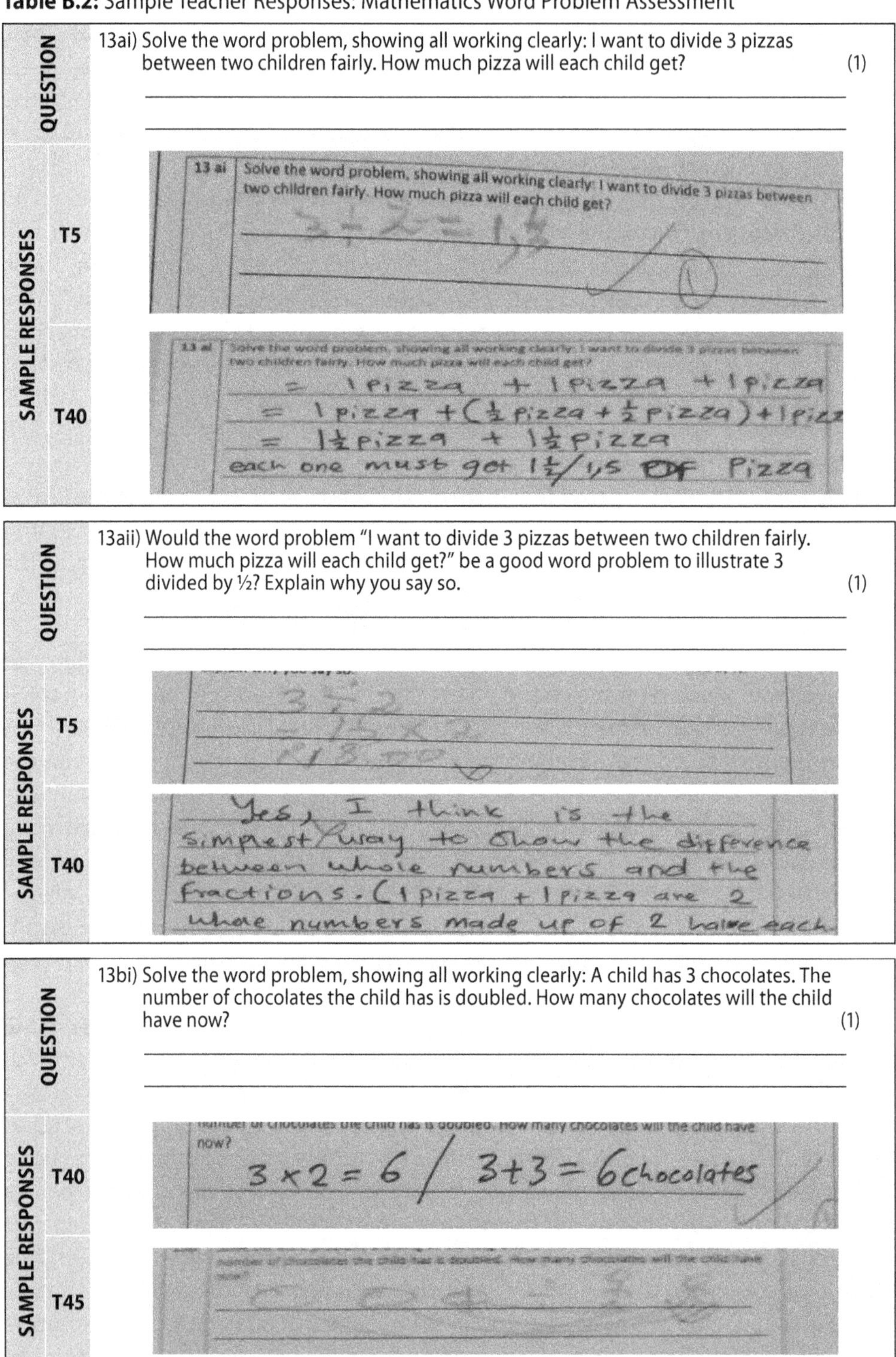

QUESTION		13ai) Solve the word problem, showing all working clearly: I want to divide 3 pizzas between two children fairly. How much pizza will each child get? (1)
SAMPLE RESPONSES	T5	13 ai Solve the word problem, showing all working clearly: I want to divide 3 pizzas between two children fairly. How much pizza will each child get? 3÷2=1,5
	T40	13 ai Solve the word problem, showing all working clearly: I want to divide 3 pizzas between two children fairly. How much pizza will each child get? = 1 pizza + 1 pizza + 1 pizza = 1 pizza + (½ pizza + ½ pizza) + 1 pizz = 1½ pizza + 1½ pizza each one must get 1½/1,5 OF Pizza
QUESTION		13aii) Would the word problem "I want to divide 3 pizzas between two children fairly. How much pizza will each child get?" be a good word problem to illustrate 3 divided by ½? Explain why you say so. (1)
SAMPLE RESPONSES	T5	3÷2 = 1,5 × 2 R18.00
	T40	Yes, I think is the simplest way to show the difference between whole numbers and the fractions. (1 pizza + 1 pizza are 2 whole numbers made up of 2 halve each.
QUESTION		13bi) Solve the word problem, showing all working clearly: A child has 3 chocolates. The number of chocolates the child has is doubled. How many chocolates will the child have now? (1)
SAMPLE RESPONSES	T40	number of chocolates the child has is doubled. How many chocolates will the child have now? 3×2 = 6 / 3+3 = 6 chocolates
	T45	number of chocolates the child has is doubled. How many chocolates will the child have now? [illegible]

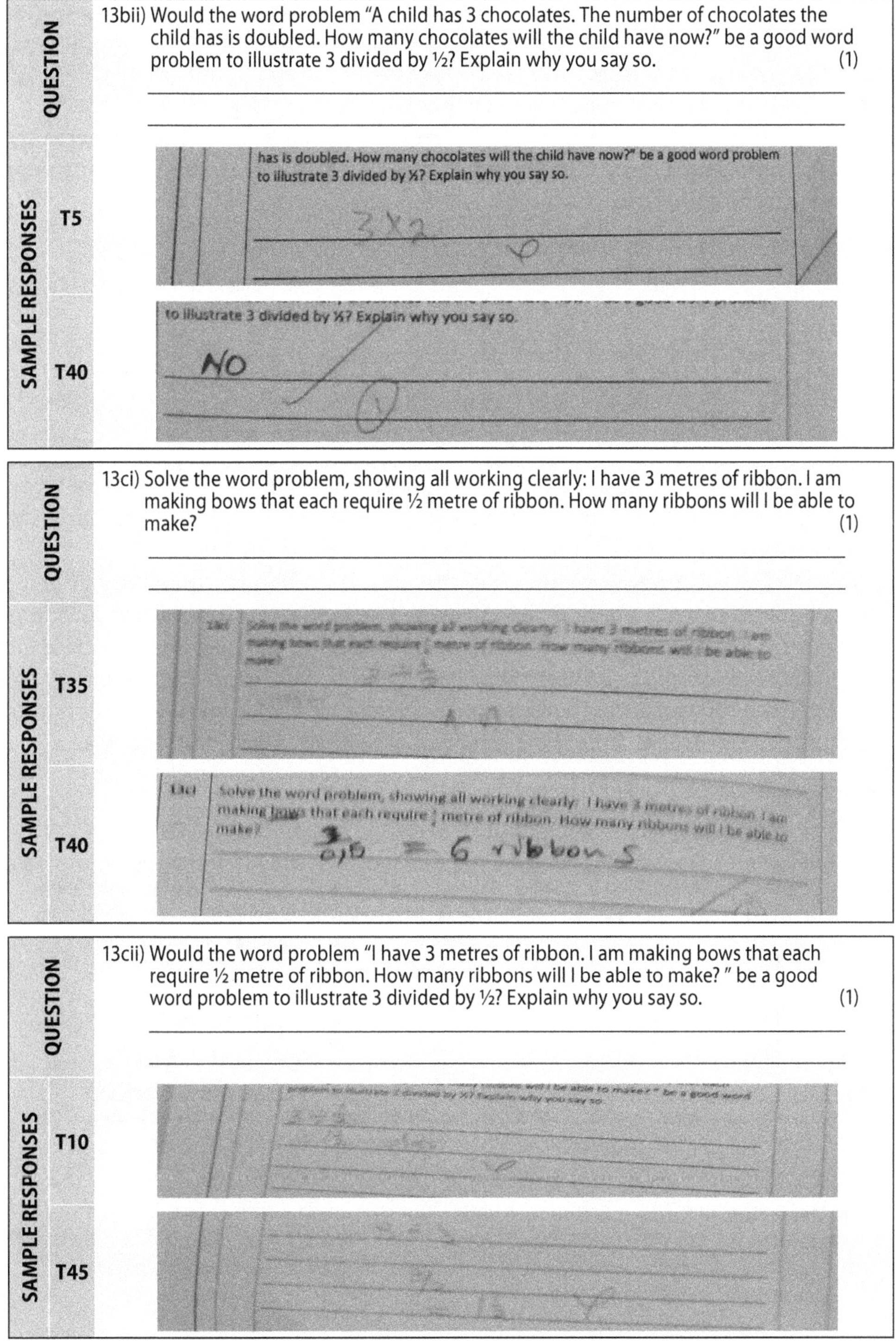

QUESTION

13bii) Would the word problem "A child has 3 chocolates. The number of chocolates the child has is doubled. How many chocolates will the child have now?" be a good word problem to illustrate 3 divided by ½? Explain why you say so. (1)

SAMPLE RESPONSES

T5

has is doubled. How many chocolates will the child have now?" be a good word problem to illustrate 3 divided by ½? Explain why you say so.

3×2

T40

to illustrate 3 divided by ½? Explain why you say so.

NO

QUESTION

13ci) Solve the word problem, showing all working clearly: I have 3 metres of ribbon. I am making bows that each require ½ metre of ribbon. How many ribbons will I be able to make? (1)

SAMPLE RESPONSES

T35

13ci) Solve the word problem, showing all working clearly: I have 3 metres of ribbon. I am making bows that each require ½ metre of ribbon. How many ribbons will I be able to make?

T40

13ci) Solve the word problem, showing all working clearly: I have 3 metres of ribbon. I am making bows that each require ½ metre of ribbon. How many ribbons will I be able to make?

3/0,5 = 6 ribbons

QUESTION

13cii) Would the word problem "I have 3 metres of ribbon. I am making bows that each require ½ metre of ribbon. How many ribbons will I be able to make?" be a good word problem to illustrate 3 divided by ½? Explain why you say so. (1)

SAMPLE RESPONSES

T10

problem to illustrate 3 divided by ½? Explain why you say so.

T45

Adapted from JET (2015b)

drawings each with their own possible advantages in the solution process. In a design experiment among grade 3 learners, Csíkos et al. (2012) found the latter approach beneficial in terms of both yielding better performance and inducing possible long-term effects on learners' metacognitive knowledge (Csíkos & Szitányi, 2020).

A descriptive analysis of the teacher scores on **Question 13** of the Mathematics Word Problem Assessment is indicated in Table B.3, with the items exhibiting poor performance[3] highlighted:

Table B.3: Itemised Analyses: Mathematics Word Problem Assessment

Q13AI			Q13AII			Q13BI				Q13BII			Q13CI			Q13CII		
0	1	Sum	0	1	Sum	0	1	11	Sum	0	1	Sum	0	1	Sum	0	1	Sum
8	46	54	41	13	54	3	50	1	54	33	21	54	20	34	54	35	19	54

Total Number of Assessment Items	6 (100%)
Total Number of Assessment Items in which ≥50% of the teachers scored 0	3 (50%)

An overview of the itemised analyses of **Question 13** of the Mathematics Word Problem Assessment indicated that at least 50% of the teachers scored 0 in 50% of the assessment items.

The teacher interview schedule incorporated the Newman hierarchy of error causes for written mathematical tasks (Newman, 1983). The error analysis questions were incorporated into the interview schedule, because these questions solicited for responses better suited to answer the research questions because of their likelihood to establish IP mathematics teachers' pedagogical content knowledge concerning word problems. Table B.4 presents the profiles of the 5 teacher interviewees.

Table B.4: Interviewee Profiles: Teachers

S	P	G	RACE GROUP	AGE	Q
S1	T1	Female	Black	4	▪ Primary Teacher's Diploma ▪ Advanced Certificate in Education (Mathematics)
S2	T2	Female	Black	4	▪ Primary Teachers' Diploma ▪ Advanced Certificate in Education (Technology)
S3	T3	Female	Black	4	▪ Primary Teachers' Diploma
S4	T4	Female	Black	4	▪ Primary Teachers' Diploma
S5	T5	Male	Black	4	▪ Primary Teachers' Diploma ▪ Advanced Certificate in Education (Management)

Key:

S – School
P – Participant
G – Gender
AGE – Age Group {**1:** 25 years and below; **2:** 26 – 29 years; **3:** 30 – 34 years; **4:** 35 years and above}
Q – Qualifications

3 The assessment items highlighted are those where more than half of the respondents obtained zero.

As illustrated in Table B.4, all the interviewees were middle-aged black teachers and four of the five were female, further confirming demographic data collected earlier in this case as well as the general patterns in the teaching profession in South Africa.

Error Analysis

Interviewees had to answer the error analysis question after completing the mathematical word problem assessment. The interviewees were informed that the error analysis question was based on three of the mathematics word problem assessment questions which they had answered previously. In addition, interviewees had access to their unmarked scripts on the three sampled questions and the researcher briefly explained that this particular question sought to identify error causes in written mathematical tasks. Table B.5 presents the responses to the fifth (error analysis) question on the teacher interview schedule.

Table B.5: Error Analysis on selected Mathematics Word Problems

Question 5: The following questions are based on the **JET mathematics word problems** you solved earlier: **(Question 13 or 14 or 15).**	
SAMPLE MATHEMATICS WORD PROBLEM	13ai) Solve the word problem, showing all working clearly: I want to divide 3 pizzas between two children fairly. How much pizza will each child get? (1) __________ __________ 13aii) Would the word problem "I want to divide 3 pizzas between two children fairly. How much pizza will each child get?" be a good word problem to illustrate 3 divided by ½? Explain why you say so. (1) __________ __________ 13bi) Solve the word problem, showing all working clearly: A child has 3 chocolates. The number of chocolates the child has is doubled. How many chocolates will the child have now? (1) __________ __________ 13bii) Would the word problem "A child has 3 chocolates. The number of chocolates the child has is doubled. How many chocolates will the child have now?" be a good word problem to illustrate 3 divided by ½? Explain why you say so. (1) __________ __________ 13ci) Solve the word problem, showing all working clearly: I have 3 metres of ribbon. I am making bows that each require ½ metre of ribbon. How many ribbons will I be able to make? (1) __________ __________ 13cii) Would the word problem "I have 3 metres of ribbon. I am making bows that each require ½ metre of ribbon. How many ribbons will I be able to make? " be a good word problem to illustrate 3 divided by ½? Explain why you say so. (1) __________ __________ Extracted from JET, (2015)

		Question Selected	Rating				
			5	4	3	2	1
T1	a) Read the question. (Reading / Decoding)	13	✓				
	b) What is the question asking you to do? (Comprehension)			✓			
	c) Suggest a method that can be used to find an answer to the question. (Transformation / Mathematising)				✓		
	d) Explain how you worked out the answer (Processing)			✓			
	e) Write down the answer to the question. (Encoding)			✓			
	Comments	Question 13 was selected. The interviewee managed to read the question with ease, the comprehension was rated as above average as the interviewee provided contextual clues that could be used to interpret the question when asked what the question required them to do. The method suggested for solving the question was rated as average as the interviewee did not provide detailed answers for all the questions. The explanations of how the interviewee worked out the answer were rated as above average as the interviewee provided sufficient vocabulary to explain the computations they had undertaken. In the final question, the answer given was exactly the same as the one that had been written in the initial word problem assessment. The interviewee gave a practical example of how the problem could be simulated in class so that learners understand how to get to the final answer; however, the simulation did not indicate some steps that would make the problem better understood by the learners.					

		Question Selected	Rating				
			5	4	3	2	1
T2	a) Read the question. (Reading / Decoding)	13	✓				
	b) What is the question asking you to do? (Comprehension)			✓			
	c) Suggest a method that can be used to find an answer to the question. (Transformation / Mathematising)					✓	
	d) Explain how you worked out the answer (Processing)					✓	
	e) Write down the answer to the question. (Encoding)						✓

		Question Selected	Rating 5	4	3	2	1
	Comments	Question 13 was selected. The interviewee managed to read the question with ease, the comprehension was rated as good as the interviewee managed to rephrase the question in their own words when asked what the question required them to do. The method suggested for solving the question was rated as below average as the interviewee did not answer all the questions in full. The explanations of how the interviewee worked out the answer were also rated as below average as the interviewee lacked sufficient vocabulary to explain the computations they had undertaken. In the final question, the answer given was a mere repetition of what had been written in the initial word problem assessment. The interviewee literally read through the process undertaken in the initial word problem assessment without reflecting on each step undertaken, or why it was undertaken.					
T3	a) Read the question. (Reading / Decoding)	13	✓				
	b) What is the question asking you to do? (Comprehension)			✓			
	c) Suggest a method that can be used to find an answer to the question. (Transformation / Mathematising)					✓	
	d) Explain how you worked out the answer (Processing)					✓	
	e) Write down the answer to the question. (Encoding)						✓
	Comments	Question 13 was selected. The interviewee managed to read the question with ease, the comprehension was rated as above average. The method suggested for solving the question was rated as below moderate as the interviewee did not answer all the questions in full and did not provide appropriate methods for solving the question. The explanations of how the interviewee worked out the answer were also rated as below average as the interviewee lacked sufficient vocabulary to explain the computations they had undertaken, or to provide reasons why they had undertaken the specific computations. In the final question, the answer given was a mere repetition of what had been written in the initial word problem assessment. The interviewee simply read through the process undertaken in the initial word problem assessment without reflecting on each step undertaken, or why it was undertaken.					

		Question Selected	Rating				
			5	4	3	2	1
T4	a) Read the question. (Reading / Decoding)	13	✓				
	b) What is the question asking you to do? (Comprehension)				✓		
	c) Suggest a method that can be used to find an answer to the question. (Transformation / Mathematising)					✓	
	d) Explain how you worked out the answer (Processing)					✓	
	e) Write down the answer to the question. (Encoding)						✓
	Comments	Question 13 was selected. The interviewee managed to read the question with ease, the comprehension was average as the interviewee merely reread the question when asked what the question required them to do. The method suggested for solving the question was rated as below average as the interviewee did not answer all the questions in full. The explanations of how the interviewee worked out the answer were also rated as below average as the interviewee lacked sufficient vocabulary to explain the computations undertaken. In the final question, the answer given was a mere repetition of what had been written in the initial word problem assessment. The interviewee merely read through the process undertaken in the initial word problem assessment without reflecting on each step undertaken, or why it was undertaken.					
T5	a) Read the question. (Reading / Decoding)	–					
	b) What is the question asking you to do? (Comprehension)						
	c) Suggest a method that can be used to find an answer to the question. (Transformation / Mathematising)						
	d) Explain how you worked out the answer (Processing)						
	e) Write down the answer to the question. (Encoding)						
	Comments	The interviewee requested not to respond to this section of the interview schedule; hence it is remains unestablished how he would have fared in the ratings.					

Key: 5 – Excellent
4 – Good / above average
3 – Average
2 – Below average
1 – Poor

Discussion of Findings: Error Analysis

As indicated in Table B.5, one interviewee wished to answer all the other interview questions except question 5; as such the corresponding section is cancelled out in Table B.5 Just before the interview session, another interviewee was overheard by the researcher muttering the following comment to a fellow interviewee:

> '... *kuzothiwa abantwana bazobakwazi njani ukuthetha isilungu, nabo otitshala bengakwazi ukuthetha isilungu leso*'

translated as:

> '... *they will say how can learners be fluent in English when their teachers cannot speak English fluently*'.

Although this comment was uttered only in passing and not in response to any of the questions on the interview schedule, nevertheless, the comment provides a snapshot of the linguistic reality in some of the under-resourced primary schools in the ECDoE.

Questions 13, 14 or 15 which the error analysis questions focused on are typical word problems. They are descriptions of problems which required the interviewees to systematically explain the mathematical operations required to obtain the solution to the problem. The error analysis schedule intended to assess the interviewees' ability to (a) read/decode, (b) comprehend, (c) transform/mathematise, (d) explain/process, and (e) write/encode the solution. By systematically going through the above-mentioned steps, it would have been easy to identify the particular stage in problem solving that the interviewee faced challenges, thus identifying the exact stage where the error originated from.

For the purposes of processing and presenting the interviewee responses, a Likert scale ranging from 1 – 5 (with 5 representing excellent and 1 representing poor) was used to rate the responses. As illustrated in Table B.5 all the interviewees who participated in the error analysis schedule selected to base their responses on question 13 of the JET Mathematics Word Assessment Problems. This could have been because the interviewees deemed the first as the easier choice out of the three suggested questions; however, the interviewees fared differently in their answering of the error analysis schedule.

All the interviewees were able to satisfactorily answer section (a) of the question, which only required them to read the question. Section (b) assessed how well the interviewees understood (comprehended) the selected questions. The responses provided in answering section (b) were bordering on inadequate. The researcher was specifically looking out for the interviewees' ability to demonstrate that they had understood the question by identifying key words in the word problem and relating these key words to the problem to be solved. None of the interviewees identified any key words in their response to this section of the interview schedule. In the responses given in answering section (c), the transformation and mathematisation section, the researcher was looking out for the specific mathematical register used by the interviewees. Overall, this was done with average success. To score better marks, the researcher was expecting the interviewees to suggest at least one other different method of transforming the problem to indicate that the interviewees really understood what the question requested of them. In section (d), the processing and explanation stage, the researcher was looking out for systematic explanations which covered what needs to be done first before moving onto the next step. Overall, this section was also rated as below average. In section (e), the decoding of the problem, the researcher was looking out for solutions that indicated that further engaging with the word problem by talking and explaining through the different processes undertaken earlier would have helped the interviewees to identify their initial error in working out the word problem. However, none of the interviewees came up with a different solution to what they had initially obtained. Most importantly, none of the interviewees seemed to be able to identify the error they had made initially; however, they simply repeated what they had done in the first attempt at answering the question.

The analysis of the 2014 ANA reveals that solving word problems was challenging to learners (DBE, 2014). Problem solving is a complex process in which learners need to be coached (Klingler, 2012). It is a process by which an unfamiliar situation is resolved (Killen, 2007). Word problems in mathematics involve real-life questions where information is provided to perform computations to solve them (Verschaffel, Greer & De Corte, 2010).

Phonapichat, Wongwanich and Suliva (2014) identify some key difficulties that learners experience in mathematical problem solving, namely: understanding keywords appearing in problems; interpreting keywords in mathematical sentences; and figuring out what information to assume and what information from the word problem is necessary to solve the problems. Schmacher and Fuchs (2012) found that word problem solving becomes difficult for learners as the problem is presented linguistically. Word problems challenge learners to read and interpret the problem, represent the semantic structure of the problem and then choose a strategy to resolve it (Schmacher & Fuchs, 2012).

In as much as research indicates that learners experience numerous challenges related to understanding key words, interpreting these key words and figuring out what information is vital in solving a mathematics word problem, the responses indicated in Table B.5 indicate that some teachers also experience the same challenges as exhibited by the error analysis interview responses. As such, some interventions need to focus on equipping IP mathematics teachers with word problem-solving strategies which they will in turn pass on to their learners. Responses obtained in the section above indicate that teachers can read the word problem fairly well; however, systematically explaining the process that needs to be done to solve the word problem seems to be a huge task for them. In addition, this confirms the low marks that the teachers obtained in the actual word problem assessment (Tshuma, 2017).

The error analysis helped the researcher to identify the word problem solving stages that the interviewees were comfortable with and those that they were not comfortable with. However, the process only diagnosed a problem and a possible solution to this problem would be to familiarise teachers with the error analysis schedule for the purposes of applying the same model to assist the weaker learners in their classrooms. Familiarising teachers with such techniques would empower teachers to coach learners on how to tackle word problems within the IP mathematics curriculum.

Quantitative data from teacher mathematical word problem assessment indicated that the majority of IP mathematics teachers struggle with solving word problems. Data from the error analysis interview which meant to systematically assist teachers to identify errors in their written mathematics word problem tasks, based on Newman's hierarchy of error analysis method, were generally poor and not as detailed as they could have been. The teachers simply narrated what they had done without reflecting on why they had performed specific mathematical procedures, and thus obtained less than the expected results. One of the participants requested the interview to go ahead on condition that the error analysis section was omitted and this in a way reflected the teacher's perceptions regarding mathematical word problems. If the teacher was avoiding answering questions based on mathematical word problems, it may be assumed that the teacher does not teach that section explicitly to the learners.

CONCLUSION

According to Brodie (2017), mathematics pedagogy is a thronged relationship between learner, teacher and the **content knowledge** (*own emphasis*) situated within contexts. In this relationship, the role of the teacher is to bring the learner into a relationship with the content knowledge. Debates on mathematical pedagogy are about the nature of the learner's relationship with the content knowledge and what the teacher should know and **be able to do** *(own emphasis)* in order to influence that relationship. In this case, the role of the IP mathematics teacher in teaching IP mathematical word problems is to bring the learner into a relationship that results in an increased aptitude in solving word problems. But, how is this chain of reaction supposed to take off when the teacher's own aptitude in solving word problems (the catalyst to the reaction) is weak? Whether the weak aptitude in solving mathematics word problems is attributed to a low proficiency in the Language of instruction (as revealed in Case A), or general innumeracy or specifically to a poor mastery of word problem solving strategies calls for further research. However, before we even comment on teachers' pedagogical content knowledge, a lot more still needs to be done regarding teachers' content knowledge in mathematical word problems.

CHAPTER 6

LANGUAGE COMPETENCY FOR MATHEMATICS INSTRUCTION

INTRODUCTION

What linguistic preparedness is necessary for Intermediate Phase (IP) teachers to teach mathematics in a second language (specifically English)?

To understand the linguistic preparedness of (IP) teachers to teach mathematics in a second language such as English, in turn necessitates an understanding of language development. Language development is one of the major and crucial teacher competencies. The following section outlines various language theories in order to explore the needs of IP mathematics teachers in terms of language development, acquisition and usage, and attempts to determine the effect of these theories on the quality of their instruction.

LANGUAGE DEVELOPMENT AND ACQUISITION

One of the main focuses of this text is to interrogate the role played by teachers' language competency in mathematics instruction. It is therefore of paramount importance to highlight the theories of language development and acquisition that are associated with language competency.

MULTIPLE INTELLIGENCES OF LANGUAGE DEVELOPMENT

Gardner's theory of multiple Intelligences provides a clear picture of the development and acquisition of language among individuals in various specialities. In Table 6.1 Armstrong (2009) outlines the categories of Gardner's Multiple Intelligences as the basis of successful language development.

Table 6.1: Categories of Multiple Intelligences in Language Development

TYPE OF INTELLIGENCE	DESCRIPTION
Verbal – linguistic	▪ well-developed verbal skills and sensitivity to the sounds, meanings and rhythms of words
Logical – mathematical	▪ ability to think conceptually and abstractly, and capacity to discern logical and numerical patterns
Visual – spatial	▪ capacity to think in images, pictures colours and shapes, to visualise accurately and abstractly
Musical – rhythmic	▪ ability to produce and appreciate rhythm, pitch and timbre
Interpersonal	▪ capacity to detect and respond appropriately to moods, motivations and desires of others
Intrapersonal	▪ capacity to be self-reflective and in tune with inner feelings, values, beliefs and thinking processes

TYPE OF INTELLIGENCE	DESCRIPTION
Existential	▪ sensitivity and capacity to tackle deep questions about human existence
Naturalistic	▪ ability to recognise and categorise plants, animals and other objects in nature
Bodily – kinaesthetic	▪ ability to control one's body movements and to handle objects skilfully

Armstrong (2009)

Gardener emphasises that people possess intelligence in different degrees, which makes them unique in various contexts. It does not matter in which language the intelligences are expressed, they contribute strongly to one's language acquisition. Thus, it is important to establish what proficiencies are appropriate for each category. The Total Physical Response (TPR) tool, which draws on physical movement, provides an additional means of successful and positive language development. The tool considers gestures, movements and physical expressions as essential in learning vocabulary, expressions and simple commands which all contribute to the ability to use language effectively. Hence, for the development of a language, grammar rules, verbs and conjunctions (for example) are memorised for effective interpretation of written texts or passages that are related to certain physical movements (Richards & Rodgers, 2001).

Language acquisition occurs in first- and second-language settings: the former generally involves learner language development, and the latter generally involves adult language development (Armstrong, 2009). Thus, first-language acquisition usually occurs when an individual is young, while second-language acquisition usually occurs as the individual grows older. However, the language-acquisition settings are not mutually exclusive. For the purposes of this research, adult language development would refer to IP mathematics teachers' language development for the purposes of effective content delivery in the prescribed LoLT.

In first-language learning, language is intertwined with the development of cognition, while analysis of errors made during second-language learning reveals the development of an inter-language. The correction of the errors made in second-language learning concretises the grammar of the new language and ultimately helps to improve the language-acquisition process (Brown, 1973).

Language acquisition is related to the learning process, with an emphasis on understanding and communication. In this regard, language acquisition involves the interrelationships among the processes of listening, speaking, reading and writing, which are integral to communication (Cohn, 1990; Olivier & Olivier, 2013). In language acquisition, competency can be improved by implementing certain teaching strategies, and the outcomes depend upon the nature of the tasks and the ability of the learners. Performing a collection of activities with specified outcomes enables the learners of a language to master an identified part of the curriculum (Chamot & O'Malley, 1994; Brown, 2007).

COGNITIVE ACADEMIC LANGUAGE APPROACH

The Cognitive Academic Language Learning Approach (CALLA) promotes language development and acquisition by intensifying the practical implementation of English language learning (Chamot & O'Malley, 1994). CALLA emphasises the inclusion of listening, speaking, reading and writing activities in content subjects on a daily basis as the most vital mechanism for the development and acquisition of language. An increased frequency of the use of such activities would thus benefit individuals whose medium of instruction is a second language, and who usually show slow progress.

The three major theoretical foundations of CALLA are language development, content area instruction and explicit instruction in learning strategy (Chamot, 1994). These theoretical foundations develop cognitive abilities, linguistic abilities as well as metacognitive abilities. The fact that learners are viewed as mentally active participants in the teaching and learning process gives these theoretical foundations their dynamism. Other mental activities that are propounded by these theories for reinforcement at various levels include the application of prior knowledge to new problems, searching for meaning in new information, higher-level thinking and developing the ability to regulate one's own learning.

COMMUNICATIVE LANGUAGE THEORY

Griffiths (2003) describes the communicative approach to language learning as the techniques or devices which an individual may use to acquire knowledge. Griffiths identifies direct and indirect language-learning strategies that contribute positively to effective language learning. Direct and indirect learning strategies are characterised as follows:

- Direct language learning strategies involve clarification, verification, monitoring, memorisation, guessing (inductive inferencing) and deductive reasoning;
- Indirect language learning strategies involve creating opportunities for practising the language.

Communicative language theory describes the goal of language teaching as the act of developing communicative competences; a learner who acquires communicative competence acquires both knowledge and ability of language use. The communicative theory also incorporates the pragmatic competence and linguistic competence in which an individual learning a language communicates by addressing formal gatherings, handling examinations, dialogue, speech, report writing and academic arguments. General language features are taken into account in the process of language acquisition and learning. Thus, a learner learning a language consistently uses the language in various formats and therefore gains confidence (Richards & Rodgers, 2001).

General language features serve an essential role in allowing learners to participate actively in learning the language successfully and therefore becoming competent. Each general language feature consists of phonemics, morphics, syntax, vocabulary and speech. The three general features of language – namely linguistics, semantics and pragmatics – that focus largely on meaning derived from language expressions are outlined in Figure 6.1.

Figure 6.1: Communicative Competence Features

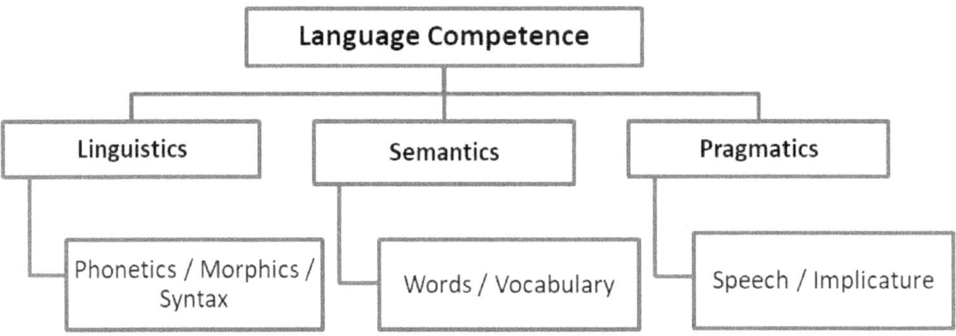

Adapted from Krashen and Brown (2007)

The communicative principles of language learning also include the abovementioned activities that promote language learning. The activities are used to carry out meaningful tasks that promote learning and also create the language that is meaningful to the learner. The activities are selected according to how well they engage the learner in meaningful and authentic language use, as well as promote second-language learning (Krashen & Brown, 2007).

Richards and Rodgers (2001) encourage language practice as a way of developing communicative skills in various communicative contexts. They suggest that the objective of the communicative approach is to provide the language speaker with opportunities to apply the language. The following pointers demonstrate that learners have acquired the required and appropriate communicative skills to use the language successfully:

- integrative and content levels that utilise language as a means of expression;
- linguistic and instrumental levels that utilise language as a semiotic system and an object of learning;
- an affective level of interpersonal relationship that utilises language as a means of expressing values and judgements about oneself and others;
- a level of individual learning that utilises language as a remedial process based on error analysis of language; and
- an extra-linguistic ability that prioritises language competency as the core of mastering subject-specific content.

Various linguistic researchers describe different conceptual perspectives that address conversational and academic language for communicative competence, as listed in Table 6.2.

Table 6.2: Theorists and concepts addressing conversational and academic language for communicative competence

LINGUISTIC THEORIST	LANGUAGE CONCEPT
Vygotsky 1962	Understanding scientific concepts
Bruner 1975	Demonstrating communicative and analytical competence
Donaldson 1978	Acquiring embedded and disembedded thought and language
Canale 1983	Developing communicative and autonomous proficiency
Mohan 1986	Application of practical and theoretical discourse
Snow 1991	Ability to contextualise and decontextualise
Cummins 2000	Developing the BICS and CALPS
Baker 2001	Appropriate cognitive process, conversational proficiency and language proficiency

These conceptual perspectives need to be acknowledged in the process of teaching and learning language for communicative purposes. The perspectives enable distinctions to be made between communicative characteristics. Richards and Rodgers (2001) describe complementary communicative competence features of language knowledge, such as knowing how to:

- use language for a range of different purposes and functions;
- vary the use of language according to the setting and participants;
- produce and understand different types of texts such as narratives, reports, interviews and conversations; and
- maintain communication despite having limitations in one's language knowledge.

It is therefore evident from the background provided on the poor literacy skills among teachers in South Africa that these aspects of language should form part of the curriculum as a means of raising the standards of language competency for the purposes of improving numeracy skills.

The characteristics of the communicative view of language are closely linked with acquiring the linguistic means to perform different kinds of communication as outlined below:

- language is a system of expressing meaning;
- language is used for interaction and communication;
- the structure of language reflects its functional and communicative uses; and
- the primary units of language are its grammatical and structural features.

As investigating IP mathematics teachers' language competency is one of the main objectives of this book, the characteristics mentioned above contribute towards successful communicative competence. The development of these communicative skills serves as an influential function that directs teaching and learning. The movement from one level to the other demonstrates the transition that occurs in terms of language acquisition and development within various contexts.

Evolution of the communicative language approach

The evolution of communicative competence implies that various linguistic experts present significant explanations of the movement from one concept to the other to explain the necessity for these conceptual changes. Figure 6.2 broadly illustrates the evolution of communicative language competence.

Figure 6.2: Communicative language competence

Chomsky (1975)
Linguistic competence

Hymes (1972)
Linguistic competence
Sociolinguistic competence

Canale and Swain (1980)
Grammatical competence
Strategic competence
Sociolinguistic competence

Canale (1983)
Grammatical competence
Strategic competence
Sociolinguistic competence
Discource competence

Celce-Murcia (1995)
Linguistic competence
Strategic competence
Sociolinguistic competence
Actional competence
Discource competence

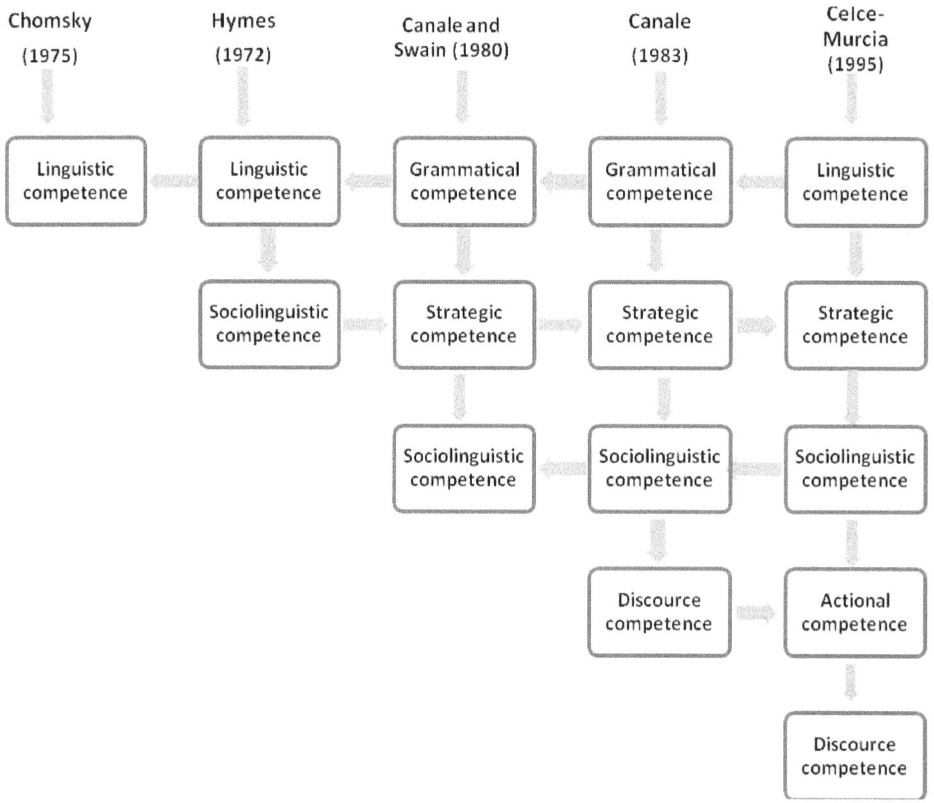

Adapted from Richards and Rodgers (2001)

The interconnectedness between the various language competences marks the successful progression and dynamic change, which contributes to the theories and understanding of language development. It is therefore important to consider that from the perspective of either the learner's or the teacher's language competency, all these competencies need to be demonstrated in various teaching and learning contexts.

Advantages of the communicative competence

The communicative language method allows the proficient teacher to provide a context that allows class interactions that are realistic and meaningful, but with the support needed to generate the target language. This occurs because language is a skill that needs to be practised in improvised settings. During the practise phase, there is a lot of preparation required to equip an individual learning a language with the necessary vocabulary, structures, functions and strategies to enable them to interact successfully (Ellis, 1997).

The communicative method focuses on language as a medium of communication and therefore recognises that all communication has a social purpose, as the individual learning a language has something to say or find out. In the same way, most of the communication activities that take place in the classroom maximise opportunities for learners to use their target language in a communicative way to promote meaningful activities. The communicative method is more learner-centred and is also directed by the needs and interests of learners. Learners are not limited to practising oral skills but are exposed to reading and writing skills, which are developed to promote their confidence in using and learning the language.

Communicative learners become independent, highly adaptable, flexible and responsive to facts, prefer social learning and a communicative approach, and enjoy taking decisions (Ellis, 1997). These learners like to learn by watching, listening to native speakers, talking to friends in English, watching television in English and learning new words by hearing them or learning though conversation.

Language approaches associated with language competency

The intervention approach, translanguaging and code switching are the three main language approaches associated with language competency in this context. The translanguaging and code switching approaches were covered in detail in Chapter 5, including how the two approaches can be used as resources in mathematics teaching and learning. The section below describes the intervention approach.

The intervention approach

Nathanson (2009) describes the intervention approach and emphasises the intervention process of cognitive processing to reading and writing ability as essential elements of language competency. These cognitive processes allow learners to abstract meaning from print by crosschecking and integrating multiple sources of information, such as syntactic, semantic, visual and phonological information.

According to Cazden and Clay (1992) and Nathanson (2009), learners should develop a rich cognitive network that allows them to read independently. By teaching reading and writing concurrently, the individual is enabled to make links between speaking, reading and writing. This also provides the teacher with valuable practical information to improve instruction. The literacy intervention approach matches the instruction to an individual's zone of proximal development, whereby the individual's level of potential development is determined through problem solving under the guidance of another capable individual or a mentor (Vygotsky, 1978; Ballantyre, 2008; Nathanson, 2008). Learning is therefore directed at the individual's zone of potential development.

Research findings confirm that South African schools are experiencing alarmingly low literacy and numeracy levels; as such, teachers should be equipped with necessary linguistic skills to reinforce successful mathematics teaching and learning as a way of promoting academic success. The literacy learning process depends on the types of opportunities that teachers provide for further development of cognitive networks in learners (Lyons, 1981; Nathanson, 2008). These early reading and writing experiences of an individual create a network of competencies which enable independent literacy learning. Similarly, if teachers are tasked with promoting literacy levels in schools, by extrapolation teacher educators should promote literacy levels among trainee teachers and teachers already in service.

Success in learning to read and write depends on different cognitive systems (Rumelhart & McClelland, 1986; Clay 2001; Nathanson, 2008). This suggests the need for constructive approaches to literacy learning by developing interacting competencies in reading and writing. The literature correlates with the book's objective of seeking to determine the relationship between language competency and effective instruction, specifically IP mathematics instruction.

Descriptive and prescriptive approaches to communicative language

Horne and Heinemann (2003: 37 – 40) classify communicative language in terms of descriptive and prescriptive approaches as illustrated in Table 6.3.

Table 6.3: Descriptive and prescriptive language approaches

DESCRIPTIVE LANGUAGE APPROACH	PRESCRIPTIVE LANGUAGE APPROACH
▪ Emphasises the description of actual language usage without making judgements about the right and wrong usage of the language	▪ Emphasizes the rules that govern language usage which are imposed on language users
▪ Describes and records language diversity as critical	▪ Prioritises standards of language usage as essential
▪ Positions the use of language as the first point of reference	▪ Positions rules of grammar for various texts as the point of reference
▪ Accepts non-standard varieties of language	▪ Does not accept or tolerate non-standard varieties of language and they are regarded as bad language

Horne and Heinemann (2003: 37 – 40)

The two approaches clearly depict the opposing views of the acceptable ways of using language. The fact that the views are opposing each other may create confusion for learners of a language and, most important, impact negatively on the ability and rate of achieving competence in language learning. However, Trask (1995) is of the opinion that effective communication should be guided by principles that allow educational institutions to teach a uniform standard of a language that promotes understanding. Thus, educational institutions should recognise the existence and value of other varieties of language.

RELATED APPROACHES TO AND THEORIES OF LANGUAGE COMPETENCY

English second-language speakers' academic success is dependent on their proficiency and competence in academic language (Brown, 2007). Therefore, teachers of ELLs should move beyond the functional syllabi stipulations and provide a content-rich and high-standard curriculum that prepares the second-language speakers to become academically successful in content learning. Against the backdrop of the South African education system, providing essential communicative skills that will ensure language competency among primary school teachers would be a positive way of modelling good practice for emulation by the teachers in their own classroom settings, where the majority of the learners are second-language speakers of the English language.

Low levels of academic performance and unsatisfactory overall rates in some South African schools and higher institutions of learning suggest that responses to the teaching and learning needs of the underprepared learners are not sufficient (SAUVCA, 2003; Paredes, 2010; Eid, 2012). Furthermore, teacher professional development initiatives like the ones proposed by this chapter, aimed at strengthening primary school teachers' language competency in the prescribed LoLT, need to be prioritised.

PEDAGOGICAL IMPLICATIONS: LANGUAGE COMPETENCY FOR MATHEMATICS INSTRUCTION

Failure to continue first-language development inhibits levels of language proficiency and cognitive academic growth. Thus, home-language proficiency is a strong indicator of second-language proficiency (Saville-Troike, 1988; Hakuta, 1990). Therefore, it will not be assumed that non-native speakers of English who have attained a high degree of fluency and accuracy in everyday spoken English have corresponding academic language proficiency. Some non-native speakers exhibit disparities in language usage and these disparities interfere with their ability to learn. Such disparities are deemed to be associated with mathematics teachers' ability to teach using English as LoLT.

In separate studies conducted by Maxwell (1998), Starkey (1998), and Stephen, Welman and Jordan (2004) the results revealed that more than 90% of black students in tertiary institutions lacked the comprehension skills required for successful completion of their studies. By extrapolation, pre-service mathematics teachers are a subset of this group. Poor comprehension skills among tertiary-level students results in high dropout rates, poor attendance of lectures and low examination pass rates.

Duffy (1993) and Stephen, Welman and Jordan (2004) state that teachers are faced with the challenge of helping learners to acquire the vocabulary and linguistic structures needed for successful academic performance. If the teachers are not competent in the language, they only expose learners to limited language functions deemed important to access academic content.

Brown (2007) is of the opinion that academic success depends on an individual's ability to understand the language of the textbooks and school discourse. Academic language is therefore inextricably tied to academic language functions and to higher-order thinking skills. These higher-order thinking skills need to be developed during the teaching and learning process.

Language competency assists an individual to construct and integrate acquired information into their own understanding, and linguistic incompetence therefore results from the lack of linguistic knowledge to fulfil various communicative needs that aid better performance. It is highly likely that challenges in understanding the LoLT spill over and affect other aspects of the academic environment ranging from the basic ones such as following directions to understanding instructions, reading, comprehension as well as the more complex ones such as interpretation of questions and discussions. Such challenges may result in an individual experiencing difficulty in expressing thoughts on paper as result of language limitations.

Gabela (2005) and Uys, Van der Walt, Van den Berg and Botha (2007) consider the use of English as a LoLT as a disadvantage for learners, considering that some teachers have a low proficiency in English as an additional language. This situation is compounded by the background of learners who need additional academic support in the area of language development. Dalvit, Murray, Mini, Terzoli and Zhao (2005) state that teachers' use of English is limited to the school environment, the concepts of a given subject and the knowledge taught. This predicament seems to intensify when the teachers have to improvise and use alternative strategies such as code switching to interact with the learners.

A diagnostic report compiled by chief examiners and markers of matric examination papers (DBE, 2013) identifies the following factors as responsible for low scores across the curriculum:

- lack of subject terminology;
- poor language usage;
- lack of vocabulary;
- misunderstanding of questions and instructions;
- difficulty in expressing oneself;
- misspelling of words; and
- poor reading and comprehension skills.

All the factors listed above are essential to language literacy and they are associated with performance as well as the ability to communicate in various school contexts. Although these literacy-related factors are diagnosed at matric level, they are usually carried over from lower grades and learners carry them right across the curriculum if they do not receive adequate corrective measures from their teachers.

Teachers who neglect early writing skills severely limit learners' opportunities to learn and thereby contribute to slower progress; furthermore, when most teachers respond to learners' inability to understand English, they resort to code switching to compensate for their own incompetence in oral language as well as low confidence levels in using English as a LoLT (Nathanson, 2008). This chapter contends that if teachers often resort to code switching in order to assist learners, they should be well versed in the practice and functions of code switching for meaningful learning to take place.

There is no consensus on what constitutes a sound second-language learning strategy or how this differs from other types of learner activities and their relationships (Rubin, 1980; O'Malley, Chamot, Stewner-Manzanares, Küpper & Russo 1985; Ellis, 1986; Griffiths, 2003). Communication strategies and language-learning strategies are therefore seen as two separate manifestations of language-learning behaviour. Thus, language-learning strategies compensate for inadequate resources.

Tarone (1981) and Griffiths (2003) believe that communication strategies expand language learning when individuals are assisted to express themselves. The communication may not be grammatically or lexically correct, but it exposes the individual to language input which may result in learning. Tarone and Griffiths reiterate that differentiating between communication strategies and language learning strategies on the grounds of motivation may create confusion where the learner has a dual motivation for both learning and communication. Learners may learn even though the basic motivation was to communicate; thus, it is difficult to establish a clear relationship between communication strategies and learning strategies.

There is widespread concern over South African learners' extremely low literacy levels and how this affects their ability to read (Townsend & Turner, 2000; Bloch, 2009; Ramphele, 2008; Spaull, 2016). These South African educationists also argue that poor reading skills impact on learners' academic performance and their overall development. A portion of these learners who end up choosing careers in teaching transmit their own inadequacies to their learners, sparking a chain reaction of mediocrity in the mastery of the LoLT and hence compromising the overall quality of learning and teaching in the country.

Rothwell (2010) proposes various constructionist models which describe the process of communication in various contexts as illustrated in Table 6.4.

Table 6.4: Constructionist models of communication

MODEL	DESCRIPTIONS
Linear	One-way model to communicate with others while assuming there is a clear beginning and end to communication: a message sent through an email, letter or a lecture presentation.
Interactive	Two-way communication which has added feedback such as the sender sending the message to the receiver, which in turn is channelled by the receiver back to the original sender.
Transactional	A model which assumes that people are connected through transactional communication and affects all parties involved such as talking or listening to a friend and providing feedback through facial expressions.

Rothwell (2010)

The social constructionist considers communication to be the product of sharing and creating meaning among speakers. The truth and ideas are constructed or invented through the social process of communication. This chapter intends to identify the most appropriate communicative model that can be adapted and used to improve IP mathematics teachers' quality of instruction using English as LoLT.

Cuvelier, Du Plessis and Teck (2003) highlight that parents usually dissociate themselves from deciding the LoLT used in schools, partly as a result of fear of disrupting the status quo in schools. According to the Language in Education Policy (LiEP) (DAC, 2003), the SBG is mandated with this task, and most often English is chosen as the LoLT. This makes the LoLT a critical aspect of the teaching and learning process. The South African government is well aware that the acquisition of initial literacy *and numeracy* in the home language for at least the first four years of schooling leads to improved literacy levels; however, the choice of the LoLT lies with the parents. Most parents choose to have their children taught in English because the language is perceived to be linked to better opportunities, which reflects the language of social and economic mobility, whether or not there is the appropriate supportive environment for teaching and learning in English. Oxford University Press, for example, published all core Grade 1-3 textbooks and workbooks in all 11 languages; but there was so little demand for African language editions after Grade 3 that virtually no publishers produced them, except for textbooks for learning African languages (McCallum, 2004: 180).

Studies conducted by the Organisation for Economic Co-operation and Development (OECD) showed that learners and teachers themselves prefer to use English as LoLT; however, prevailing practice in teacher education does not equip teachers to confidently use English as LoLT. The issues of LoLT and 'time on task' (TOT) are also relevant to learner achievement; the available data show that, where learners are taught in their home language by teachers whose first language is also the language in which they teach, learning outcomes are higher than for learners whose LoLT is not their (*or their teachers'*) first language (OECD, 2008: 188).

Without promoting the hegemony or dominance of the English language, immediate measures need to be put in place to capacitate teachers to be confident and proficient in the use of English as LoLT, especially at IP level, where the transition from mother-tongue to second-language instruction occurs. Failure to prioritise this will continue to see South Africa occupying the lowest position in its mathematics and science education compared to other sub-Saharan countries. Without drastic changes, under-resourced schools staffed by under-qualified teachers, will continue producing half-baked school leavers who cannot cope with the demands of tertiary education (Tshuma, 2016, 2017).

Since South Africa's democratisation, English has become the lingua franca in public life and the language is prevailing as the LoLT at primary and secondary school levels as well as in higher education (Cuvelier, Du Plessis & Teck, 2003). Since many educational institutions are in a state of change (Cuvier et al., 2003) further contend that School Governing Bodies (SGBs) are mandated with the power to decide on the LoLT for their schools. A common practice that has been adopted by most SGBs is additive bilingualism, where FP is taught in the mother tongue, while English is taught as a first additional language in preparation for a transition that takes place in the IP, when English becomes the LoLT. While this is a plausible attempt at balancing the attention given to marginalised languages, SBGs that have adopted this policy need to further support it by making sure the transition is smooth enough and does not compromise the quality of learning and teaching at IP level.

Language learning and teaching thus becomes the imperative for communicative competence (Bryant, Roberts, Bryant & DiAndreth-Elkins, 2011; Kock & Woods, 2004). Thus, the main objective of language teaching involves stimulation of an individual's oral communication to enable the individual to produce and receive communication in that language. For this reason, it becomes imperative to improve usage of language in various learning contexts, including the mathematics learning context.

CONCLUSION

This chapter reviewed literature on perspectives of language competency for mathematics instruction. The literature reveals language competence as a predictor of academic performance. Language competence is a complex skill that needs to be developed continuously and consciously among practising teachers. It can best be developed through practice when an individual reflects on how they acquire language and express themselves through communicative activities. It is therefore imperative to use the correct strategies that promote the use of language to yield positive results.

A large number of studies that have been carried out globally, in sub-Saharan Africa and in South Africa, continuously investigating learner language competence and its relationship to performance on the assumption that the teachers have adequate resources and are working in an environment conducive to supporting the learners. Of significant importance is the lack of international, regional and local studies conducted to specifically investigate the relationship between language competence and mathematics instruction among primary school teachers. This chapter points out that even fewer studies have been conducted on IP mathematics teachers deployed in the Eastern Cape Province, one of the most poorly resourced provinces in South Africa.

CHAPTER 7

PERSPECTIVES ON LANGUAGE OF LEARNING AND TEACHING MATHEMATICS

INTRODUCTION

What is a Language of Learning and Teaching?

The Language of Learning and Teaching (LoLT) refers to the language or languages used for both learning and teaching across the curriculum and gives equal importance to both learning and teaching. It is also the language or languages used in textbooks, classroom materials and for examination purposes across the curriculum. LoLT also refers to the language adopted by a particular school for pedagogic purposes (Adler, 2001; Essien, 2013). While there is extensive theoretical literature on this topic, evidence is limited by the difficulties inherent in measuring the causal effect of language of instruction, referred to as the language of learning and teaching in this chapter.

APPROACHES IN LANGUAGE OF INSTRUCTION

To ensure that learners acquire strong foundation skills in literacy and numeracy, schools need to teach the curriculum in a language that learners understand. Mother-tongue based bilingual (or multilingual) education approaches, in which a learner's mother tongue is taught alongside the introduction of a second language, can improve performance in the second language as well as in other subjects (UNESCO, 2016). The pedagogical theory promotes the use of mother tongue as the LoLT until a level of academic proficiency has been attained in that language (which may take three to six years) rather than using a second language from the start of school (Hakuta, Butler & Witt, 2000).

Educationists who promote the immersion approach argue that first-language instruction will delay the acquisition of English (Taylor & Coetzee, 2013). The approach dictates that learners start learning English as early and as comprehensibly as possible. The theoretical foundations of this argument are that language acquisition, including that of a second language, comes more naturally to young learners than older learners. Imhoff, (1990) argues that bilingual education denies the reality that practice makes perfect.

In contrast, educationists who promote bilingual transitional models maintain that a learner must first develop cognitively to a sufficient level in the medium of their home language in order to gain the skills necessary for second-language acquisition (World Bank, 2005; UNESCO, 2016). Cummins (1992) argues that there is a great interdependence between literacy skills across languages, specifically referring to the languages that learners learn. He also points out that academic proficiency in a language takes considerable time to master – at least five years; therefore, once academic mastery in the home language has been attained, a learner will possess the necessary literacy skills to acquire a second language with ease.

The notion of a zone of proximal development developed by Vygotsky (1962) is also relevant to second-language acquisition. This notion suggests that all learning builds on existing skills and knowledge, and that consequently there is a zone between what is known and unknown in which new knowledge and skills can be developed using existing foundations. However, substantially more advanced skills or very different learning areas will be unattainable because the links to existing skills have not been established. Consequently, if a certain level of language proficiency and understanding of the principles of literacy such as decoding a text have not been reached in the first language (in which a learner already possesses a substantial written and oral vocabulary), then academic mastery of a second language will be beyond the zone of proximal development. For these reasons, proponents of bilingual transitional models predict that not only will a later transition to English benefit a learner's first-language proficiency, but it will also lead to greater proficiency in English in the long run (Taylor & Coetzee, 2013).

While the pedagogical theory supports late-exit transitional models, there are often practical realities which may influence the relative effectiveness of alternative models (Taylor & Coetzee, 2013). In multilingual countries such as South Africa, with numerous home languages, the late-exit transitional models are deemed impractical because:

- it may be difficult to provide quality instruction using the various home languages as LoLTs;
- sufficient high-quality teaching materials may not be available in all the home languages;
- in schools where multiple home languages are used within the same classroom, it may be preferable to teach using English as the LoLT rather than any particular home language; and
- there may be a shortage of teachers who are proficient in the various home languages.

On the other hand, the success of an English immersion model may be impeded by teachers who are not fluent in English, thus compromising the delivery of an English immersion programme. Generally, the teachers' capability and motivation may also be a critical mediating factor.

The proponents of structured immersion programmes generally recommend sophisticated instructional methods which may, even if pedagogically sound, be flawed at the level of implementation in certain contexts (Rossell & Baker, 1996). Similarly, the transition to English that must occur in bilingual models may be extremely disruptive and educationally damaging if there is not a high quality of support materials nor teacher expertise to manage this phase effectively, a concern that is often expressed in South Africa (Van der Berg, Burger, Burger, Vos, Randt, Gustafsson, Shepherd, Spaull, Taylor, Van Broekhuizen & Von Fintel, 2011; Taylor & Coetzee, 2013).

Inasmuch as late transition seems to be beneficial, the gist of this chapter is that at one point or the other a transition will take place. Whether the transition to the use of English as LoLT is made early or late, the teacher must be sufficiently equipped to manage this transition. Therefore, for the teacher to function as a competent classroom practitioner, the teacher's competency in using English as the LoLT must be investigated. IP teachers who usually manage the transition from mother-tongue to English-medium education must be well prepared to manage the transition at whatever stage of the education system the transition occurs.

There may be numerous other political or ideological motivations behind the adoption of a particular language in education policy, such as using a single language to promote national unity or developing a diverse cultural heritage (World Bank, 2005). However, the question of which approach is most suitable in a particular context is ultimately an empirical one. As the next section will show, there is a dearth of empirical work using credible methods to identify the causal impact of alternative language-of-instruction models on second-language acquisition or on other educational outcomes. There is an even more acute shortage of such research conducted in developing countries, especially those in Africa. Consequently, ideology often trumps evidence in language policy debates (Slavin, Madden, Calderon, Chamberlain & Hennessy, 2011; Taylor & Coetzee, 2013).

The next sections review studies conducted globally, in sub-Saharan Africa, in South Africa and in the Eastern Cape Province of South Africa. Each section commences by reviewing studies that investigated the language of teaching and learning across the curriculum, before focusing on studies that review the language of teaching and learning in relation to mathematics education.

LANGUAGE OF LEARNING AND TEACHING: THE GLOBAL LANDSCAPE

Global language dynamics are such that 40% of the global population does not have access to an education in a language they speak or understand as their home language (Walter & Benson, 2012; UNESCO, 2016) and this limits the learners' ability to develop foundations for learning. About 221 million learners are estimated to speak a different language at home from the language of learning and teaching in school (Dutcher, 2004). This section provides an overview of international studies that address the language of instruction in the following countries: United States of America, New Zealand, England, Spain, Saudi Arabia and Papua New Guinea.

Chan (1982) compared the differences in discourse patterns between bilingual and monolingual Mexican-American tutors when tutoring mathematics to bilingual Mexican-American learners based in the United States of America. The research revealed that the bilingual tutors used more general explanations illustrated by examples and counter-examples than monolingual tutors. The bilingual tutors also used and received more accepting and acknowledging responses from learners, while monolingual tutors used and received more negating responses with questions. The research concluded that differences in discourse patterns confirm that more communication occurs when a bilingual person is taught by another bilingual person rather than a monolingual person (UNESCO, 2016).

Several studies reviewing the literature on alternative language-of-instruction regimes have been conducted, mainly pertaining to the question of language policy for Spanish-speaking learners in the United States of America. Rossell and Baker (1996) argued that the evidence from studies of sufficient methodological quality suggested no significant difference between bilingual education approaches and English immersion approaches. However, subsequent re-analysis of the same studies (Greene, 1997; Cheung & Slavin, 2005, 2012) demonstrated that many of these studies had serious methodological flaws and that the most credible studies favoured bilingual approaches.

Cheung and Slavin (2012) found 13 studies that met their methodological criteria for inclusion; however, the 13 studies do not all provide strong grounds for causal inferences and the studies reviewed contain samples that are really too small for precise measurement. For example, the research conducted by Slavin et al. (2011) in six schools was described as the only multiyear randomised evaluation of transitional bilingual education. The small sample of participants was considered as a weakness.

Ramìrez, Pasta, Yuen, Ramey and Billings (1991) followed three types of schools in several districts in the United States of America (immersion schools, early exit to English schools and late exit to English schools) for four years. The research revealed two weaknesses: a lack of comparability of teacher and school characteristics across school types; and the English proficiency of those in the late-exit programme was not observed in years subsequent to the varying treatment. Furthermore, as Cheung and Slavin (2005) note, the research did not adequately control for pre-test scores. Nevertheless, the research found no significant differences across the three groups of schools at the end of the treatment period. Cummins (1992) argues that this finding supports bilingual approaches, since it demonstrates that they can lead to equal levels of English proficiency despite less time being spent on English instruction.

Thomas and Collier (2002) conducted research in the United States of America investigating the language of teaching and learning. They concluded that learners who had undergone dual-language instruction performed better at the end of high school than learners who had only experienced immersion in English.

LANGUAGE OF LEARNING AND TEACHING MATHEMATICS: THE GLOBAL LANDSCAPE

Many indigenous communities in high-income countries are marginalised – the marginalisation has been identified through learner assessments – but this has received minimal attention in international education debates. According to an analysis of TIMSS data, in Australia, approximately two thirds of indigenous learners achieved the minimum benchmark in mathematics in Grade 8 between 1994/1995 and 2011, as compared with almost 90% of their non-indigenous peers (Thomson et al., 2012; UNESCO, 2016).

Dawe (1983) investigated the teaching of mathematics using English as the LoLT to bilingual Punjabi, Mirpuri, Italian and Jamaican primary school learners living in England and for whom English is a second language. The research concluded that first-language competence was an important factor in the learners' ability to perform mathematical reasoning in English as a second language.

Aidoghan (1985) investigated the predictive validity of selection measures and related variables used on 1,261 Saudi Arabian university students and concluded that skill in English is a good predictor of success only up to a certain level, above which differences in English proficiency had no major influence on success.

Mestre (1986) compared groups of bilingual Hispanic learners and monolingual learners, all undertaking a degree in engineering at university in the United States of America. All learners completed a reading test in English, an algebra test and a mathematics word problem test. On the word problem test the bilingual learners were slower and less accurate than the monolingual learners. The research concluded that the reading and interpretative demands of the mathematics word problem test was a source of difficulty; knowing the vocabulary in a word problem is no guarantee that the mathematical relationship will be appropriately interpreted by non-native speakers. This evokes the need to investigate further elements such as the syntax and semantics of written mathematical text that may pose problems for second-language speakers of a language.

Cuervo (1991) investigated the effects of mathematics instruction in two languages, English and Spanish, on the performance of bilingual Hispanic college students in the United States of America. The research concluded that the bilingual students who received mathematics instruction in English and Spanish obtained higher academic achievement. The research revealed that bilingual instruction was more effective than traditional instruction provided in English only.

Clarkson (1992) conducted similar studies in Papua New Guinea. The first case confirmed that comprehension errors constitute a large number of the errors made by Grade 6 learners when solving mathematical word problems and concluded that the influence of English, the LoLT, as well as the influence of the learners' home language Pidgin, had a significant impact on their mathematical performance. The second case concluded that bilingual learners competent in both the languages performed significantly higher in mathematics than both the low-competence bilingual and monolingual learners. The studies argued that competence in the mother tongue as well as in English play an important role in the comprehension of mathematical texts.

Bearde (1993) investigated the correlation of oral language proficiency and mathematics achievement in 1,498 primary school learners for the Woodcock-Johnson

Psycho-Educational Battery Revised Test. Results of the test revealed oral language proficiency as a strong predictor of both mathematical reasoning and basic skills. The research thus suggested that a surface-level understanding of language is insufficient for mathematics achievement; a deeper level of language, involving an understanding of relationships, is needed.

Han (1998) investigated the relationship between the clarity of specific mathematical terms and learners' mathematics achievement among Chinese learners living in the United States of America. The sample was comprised of newly immigrated, monolingual Chinese-speaking learners, American-born, monolingual English-speaking Chinese learners and bilingual Chinese/English-speaking learners. The research revealed that Chinese language ability was a strong predictor in mathematics achievement and the relative clarity of mathematical terms in Chinese contributed to the better performance among the Chinese-speaking learners.

Lim (1998) investigated the relationship between language and mathematics among Korean learners living in America. The research investigated the associations between various background factors such as reading skills, English proficiency, parent's educational background, Korean-language school attendance and length of residence in the United States. The conclusion was that language is associated with mathematics achievements, especially in tasks that require substantial amounts of language processing such as problem solving. Background factors that are directly or indirectly related to language proficiency are also associated with mathematics problem solving. These findings suggest that bilingual learners' success in problem solving is interwoven with their level of proficiency in English and factors that relate to English proficiency. What was recommended was greater exposure to the LoLT and the language of mathematics for learners whose proficiency in English is limited.

Gorgorió and Planas (2001) investigated mathematics teaching and learning in schools with large numbers of immigrant learners in Spain. It was revealed that the language practices that learners bring to school inevitably affect how and what they learn. It was also concluded that in mathematics lessons, the learners had difficulties with understanding everyday words within a mathematical context, as well as a false sense of understanding and communication, or lack thereof, with the teacher. From a teaching perspective, it was concluded that the linguistic barrier is extended beyond everyday communication and teachers feel that the absence of a common language amongst learners is a barrier to their mathematical learning. The recommendation was that more research is necessary with a focus on how mathematical language can be taught and on the relationship between the language of the mathematics class, mathematical language and the process of constructing mathematical knowledge.

Dakroub (2002) investigated the role of Arabic literacy in the academic achievement of middle school learners in English in Michigan, United States of America. Learners were tested to determine their level of literacy in Arabic in relation to academic achievement in English reading, language and mathematics. It was confirmed that there was a significant positive relationship between the learners' literacy in the Arabic language and achievement in English reading, language and mathematics. Learners with high levels of English literacy performed better than the learners with high levels of Arabic literacy.

Barton and Neville-Barton (2003) investigated the dynamics of learning mathematics using English as the LoLT for 80 university students whose mother tongue is not English in New Zealand. The research concluded that in comparison to students who are native speakers of English, students who are not first-language speakers of English experience between a 10 and 15% disadvantage in overall performance through a lack of understanding of mathematical texts. The research provided an insight into how language affects mathematical understanding. It was reiterated that mathematics is not a language-free phenomenon (Reborn, 1995) and the particular vocabulary, syntax and discourse of mathematics challenges learners when English is used as the LoLT. Yushau and Bokhari (2005) investigated Saudi Arabian preparatory-year mathematics students for whom English was a new language of instruction. Prior to entry into university, the students would have learnt entirely through the medium of Arabic – their first language. The general consensus at the university is that these students perform below expectation, contrary to the level of mathematics ability on selection for entry into the university, and this low performance is attributed to lack of English proficiency. The researchers implemented a mediated teaching approach with the primary focus on providing language support in order to improve the students' understanding in mathematics by providing handouts with translations of the important terminology. The findings of the experiment showed that the use of translation encouraged students to connect previous learning with new learning, and that there was increased participation in class and improved performance in examinations. The researchers found that this approach minimised the language barrier evident in the classroom and, because of the improved performance in examinations, it demonstrated the existence of a positive connection between language proficiency and mathematics learning.

Latu's (2005) research in New Zealand found that Pasifika learners' learning of mathematics through the medium of English was hindered by an under-developed mathematical discourse in both Tongan and Samoan languages and accordingly their ability to deal with complex mathematical sentences, phrases and mathematical terms. This demonstrates the importance of a speaker's first language of learning for the transition to English-medium mathematics education.

Clarkson (2007) investigated high-ability Australian-Vietnamese bilinguals and their use of language(s) when engaged in mathematical problem solving. He found that the learners rely on language switching and thus their competencies in both languages are of importance in how they perform on mathematics problems. When language switching (English to Vietnamese) did occur, it was mainly translating entire problems (as opposed to individual words). This may be a result of all learners having a well-developed mathematical register in Vietnamese, suggesting that it is more than just vocabulary that plays a significant role in the transition to learning mathematics through the medium of English.

According to UNESCO (2016), when language, ethnicity and poverty interact, they can produce an extremely high risk of exclusion and being left behind. Learners from poor households who speak a minority language at home are among the lowest performers. In 2012 poor 15-year-olds in Turkey speaking a non-Turkish language, predominantly Kurdish, were among the lowest performers. Around 50% of poor non-Turkish speakers achieved minimum learning benchmarks in reading, against the national average of 80%. In 2006 of the poor learners in Guatemala speaking an indigenous minority language only 38% learned the basics in mathematics while 77% of the rich learners who speak Spanish reached that level. In Guatemala learners in bilingual schools have higher attendance and promotion rates, and lower repetition and dropout rates. Moreover, they have higher scores on all subjects, including mastery of Spanish. A shift to bilingual education would reduce repetition and result in a significantly more cost-effective education system (Patrinos & Velez, 1996; UNESCO, 2016).

The literature reveals that the challenges regarding language in education are most prevalent in regions where linguistic diversity is greatest, as in Asia, the Pacific and sub-Saharan Africa (UNDP, 2004).

LANGUAGE OF LEARNING AND TEACHING: THE SUB-SAHARAN LANDSCAPE

There is a lot of work that needs to be done to increase studies on language in education in developing countries – especially countries in Africa require attention. More studies focusing on the African continent still need to be conducted in order to increase the number of studies that have a large enough sample for precise estimations, and a multi-year duration so as to shed light on ultimate educational outcomes. The majority of studies on language of learning and teaching in African countries are written by linguists in favour of mother-tongue instruction. This section provides an overview of studies conducted in the sub-Saharan region that address the language of instruction in the following countries: Mali, Uganda, Kenya, Burkina Faso, Nigeria, Ethiopia, Tanzania, Cape Verde, Cameroon, Mozambique, Botswana, Zimbabwe and Lesotho.

According to Bühmann and Trudell (2008), in Mali's Pédagogie Convergente programme learners commencing school with mother-tongue instruction ended up with better mastery of the official language, French. Between 1994 and 2000, learners who began their schooling in their mother tongue scored 32% higher on French proficiency tests at the end of primary school than those in French-only programmes.

Piper and Miksic (2011) investigated the relationship between language of instruction and reading acquisition in Uganda and Kenya using observational data and regressing reading scores on a set of observed characteristics, including language of instruction, which varies across schools. Piper and Miksic (2011), however, concede that the cross-sectional nature of their data precludes causal inferences.

Nikiema (2011) recorded the results of tests in French carried out with learners from bilingual schools in Burkina Faso and found them to be comparable to or higher than those of learners in traditional French-instruction schools. The benefits of bilingual/multilingual programmes extend beyond cognitive skills to enhanced self-confidence and self-esteem. In Burkina Faso mother-tongue instruction facilitated the use of effective teaching practices in the classroom and encouraged learners to be active and become involved with the subject matter.

Earlier research and common practice in the past have supported earlier transitions from home language to official language. However, recent evidence now claims that at least six years of mother-tongue instruction – increasing to eight years in less well-resourced conditions – is needed to sustain improved learning in later grades for minority language speakers and reduce learning gaps (Heugh, Benson, Bogale & Gebre Yohannes, 2007; UNESCO, 2016). However, many countries in sub-Saharan Africa that support bilingual education continue to favour early transition to the official language, usually by Grade 4 (Alidou et al., 2006).

Fafunwa (1989) documented the Six Year Primary Project (SYPP), also called Ife project, conducted in Nigeria which is regarded as the most authoritative case study on the use of the mother tongue in formal education (Bamgbose, 2000, 2004; Alidou, Boly, Brock-Utne, Diallo, Heugh & Wolff, 2006). The project began in 1970 with two experimental schools using Yoruba as the medium of instruction for the six years of primary education, while one control school used only three years of mother-tongue instruction and then switched to English as LoLT. Evaluations of the project found that learners who switched to English after six years of mother-tongue instruction performed better in English and in other subjects compared with those who did so after only three years. The project claimed to clearly demonstrate the positive effects of late-exit transitional models relative to early exit from the mother tongue. Although, several additional schools were added to the project in subsequent years, the small sample size remains a weakness in comparison with recent standards in randomised control trials. Secondly, apart from receiving more years of mother-tongue instruction, experimental schools received other instructional materials as well as a specialist English teacher. Therefore, as Akinnaso (1993) contends, it was not possible to separate the effects of language of instruction from other aspects of instructional quality.

Adetula (1990) investigated the extent to which the language of instruction contributes to performance in mathematics in Nigeria. The findings revealed that learners performed better when the mathematical problems were presented in their home language than when presented in English. It was concluded that the teaching and learning process is compromised when learners are forced to learn mathematics in a foreign language. This is because they have to learn new vocabulary as well as being able to express themselves mathematically in the new language.

Ethiopia has gone further than many countries in seeking to combine mother-tongue instruction with Amharic and English in Grades 1 to 8. Learners' participation in bilingual programmes for eight years improved their learning in subjects across the curriculum and not just in the language of instruction. Primary school learners learning in their mother tongue performed better in Grade 8 in mathematics, biology, chemistry and physics than learners with English-only schooling (Heugh et al., 2007; UNESCO, 2016).

Walter and Chuo (2012) conducted an experiment in Cameroon in which 12 schools received instruction in the mother tongue (Kom) for Grades 1 to 3, switching to English instruction in Grade 4. Twelve matched schools received the traditional instruction in English. After five years of schooling (i.e. two years after the switch to English for experimental schools), those in experimental schools were performing better in English reading than those in control schools. Learners taught in Kom showed a marked advantage in achievement in reading and comprehension compared with learners taught only in English. Learners taught using Kom as the LoLT also scored twice as high on mathematics tests at the end of Grade 3. However, the learning gains were not sustained when the learners switched to English-only instruction in Grade 4. There was no significant difference in mathematics skills between experimental and control schools at the end of the experiment. Thus, the early exit from a mother-tongue environment prevented them from sustaining this performance across the curriculum (Walter & Chuo, 2012). The merits of this research lie in its 5-year duration (2007 – 2012). However, a major weakness is that the experimental schools were recommended by local education officials. Therefore, despite the matching process, these schools may differ in certain unobserved ways from the control schools.

Benson (2000) reports on an experiment in Mozambique comparing a bilingual programme to the traditional Portuguese-only programme. Although sample size was again small (four treatment and four control schools) and there were admittedly several design problems, Benson maintains that the project pointed to increased classroom participation, self-confidence and language proficiency amongst learners in the experimental schools.

Teachers in Botswana view code switching negatively and refer to it as a compensatory strategy. Unlike Lesotho learners, Botswana learners have not been reported to switch codes. In Grade 1 and Grade 2, there is an effort on the part of some Lesotho teachers to make learners listen to English, while others do not do this. The former group speaks English and switches to Sesotho for clarity. The latter group speaks Sesotho and switches to English only to introduce a few new English words. Drills are a common feature in most lessons. The switching of codes in this manner has been observed in Lesotho schools and is thought to facilitate understanding of new and difficult concepts (Akindele & Letsoela, 2001; Arthur, 2001).

Teachers' and learners' communicative competence and experience warrant attention. Most teachers are competent in English for the classes they teach; there are, however, some instances of broken grammar or use of wrong expressions that could lead learners to giving incorrect answers. One teacher, for example, asked the class the question: '***How much do you pay apple?***' Most learners lack communicative competence in English, except in the English-medium schools, where English is the medium of instruction from the first grade upwards. Most of these English-medium schools depend on fees as well as public funds. So the greater linguistic competence of learners may be partly attributed to social selection. Nevertheless, the contrast is informative. Learners in the English-medium schools do use English for communicative purposes. They do not merely mimic what teachers say, but use English to analyse the content of the lesson. They make rich individual comments, based on grounded analyses, and use creative language. To some extent, this indicates what could be achieved in other schools where Sesotho is in practice the main medium of communication among teachers and learners, even in the higher grades and learners hardly ever speak English. This environment is shaped by teachers' attitudes and mechanical approach to their work which effectively denies the learners the chance to improve their communication skills in English and appreciate its structure. The task of raising learners' competence in English, both spoken and written, requires a communicative approach to both the teaching of the subject and the use of the language across the curriculum.

Mufanechiya (2011) assessed the language of instruction in Zimbabwean primary schools with the aim being assessing the challenges faced by both teachers and learners when using English as the LoLT in Masvingo Province. It was concluded that English is perceived to be the economic gatekeeper by both learners and teachers. Although the Language in Education Policy of 1987, which stipulated English as the LoLT in schools in Zimbabwe was revised in 2006 to include other dominant languages in the country, teachers still preferred to use English in their classroom. Learners also strove to use English at all times and do not shy away from using 'broken English' as long as their efforts increase their mastery of the language. A lot of code switching was used by the teachers to help learners access the content taught. Learners also conducted their group discussions in English, but made an effort to report back to the class in English. For a country that has suffered huge amounts of economic turmoil, teachers and learners viewed English as a means to getting a job, a chance to get educated, the ability to participate more fully in the life of their own country, or have the opportunity immigrate to another country. Fluency in English also translates into an expansion of one's literary and cultural horizons and the expression of one's political opinions or religious beliefs. It was noted, however, that changes in the country's language in education policy were not accompanied by sufficient teacher preparation, and as a result teachers used methods that worked best for them in the classroom.

In Lesotho Moloi, Morobe and Urwick (2008) noted that the LoLT in the first three grades is Sesotho, the mother tongue of most learners; from Grade 4 onwards the prescribed LoLT becomes English. However, they noted that the actual situation in most schools is one of either using Sesotho predominantly or of switching between Sesotho and English.

LANGUAGE OF LEARNING AND TEACHING MATHEMATICS: THE SUB-SAHARAN LANDSCAPE

Taole (1981) investigated the effect of studying a selected mathematics topic among 444 Lesotho learners and ten teachers from six secondary schools in Lesotho. In each school the learners were divided into three groups: the first group was taught in English, the second group was taught in Sesotho using translated resources, while the third group has access to both the English and Sesotho materials. The results showed that the learners taught in Sesotho performed slightly better than those taught in English, while learners taught bilingually performed slightly better than those taught in Sesotho. It was also revealed that among the learners taught in English, those who performed better had a higher level of English proficiency than those who failed.

Ferro (1983) investigated the influence of language on the mathematical achievement of 89 tertiary-level bilingual students enrolled for a two-year basic mathematics course in Cape Verde, southeast of Senegal. The students were divided into three groups. The first group was taught entirely in English, the second group was taught in a mixture of Cape Verdean and English, while the third group was taught in a mixture of Portuguese and English. The conclusion indicated that students whose mother tongue is Cape Verde, and were taught mathematics using a mixture of Cape Verde and English, performed better than those taught entirely in English and those taught in a mixture of Portuguese and English.

Maro (1994) investigated the ability of Tanzanian secondary school learners to reason using English, their second language. English and Kiswahili versions of a mathematics reasoning test were used to test the learners' reasoning ability in the two languages. The conclusion revealed that the performance of the learners on the mathematical reasoning tests varies depending on the language of the test. Better performance was recorded on the Kiswahili tests.

According to Arthur (2001: 354), the practice of using mother-tongue instruction in the first three grades of school and then switching to using English as the LoLT from Grade 4 onwards is prevalent in many sub-Saharan countries including Botswana, Tanzania and South Africa.

LANGUAGE OF LEARNING AND TEACHING: THE SOUTH AFRICAN LANDSCAPE

This section provides an overview of studies within the South African context that address the language of instruction. The studies have literacy transition and literacy-related foci.

In South Africa prior to the Nationalist Party rule in 1948, there was a relatively loose policy of mother-tongue instruction which varied from province to province (Hartshorne, 1992; Sepeng, 2015). When the Nationalist government came into power in 1948, Afrikaans was established alongside English as a fully-fledged official LoLT (Adler, 2001) for all learners in minority white, coloured and Indian schools. The Bantu Education Act of 1953 changed the language policy and extended the use of mother tongue and Afrikaans as LoLT and by 1959 all eight years of primary education for black learners was done in the mother tongue and secondary education used English and Afrikaans as LoLT in a ratio of 50:50. To implement the new policy, all teachers in black schools had five years to become competent in Afrikaans through the intensive in-service Afrikaans language courses that were offered by the government (Hartshorne, 1992). This LiEP was specifically and explicitly designed to serve the apartheid regime, and it was vehemently resisted, leading to the 1976 Soweto uprisings.

Mbude-Shale (2013) describes the Bantu Education System (1953 – 1976), which offered African learners mother-tongue instruction for eight years, as an impoverished curriculum whose basis was to educationally disable black learners. The current chapter contends that some products of the Bantu Education System who ended up as teachers, specifically primary school teachers in public schools, could have transferred their impoverished curriculum knowledge to their learners.

Nel and Swanepoel (2010) investigated the language errors of English second-language teachers and how these affect their learners at the University of South Africa (UNISA), a large, open, distance-education university in South Africa. They found that the majority of the practising teachers who enrolled as students for the practical component of the Advanced Certificate in Education (ACE): Inclusive Education (Learning Difficulties) for the year 2008 lacked English proficiency. Because the pre-service teachers' home language is not English, they found it difficult to master the course, as reflected by the low throughput rate of 44 per cent in 2008. The main question investigated was: *Does poor proficiency in English of English second- language (ESL) pre-service teachers influence English second-language learners' progress during learner support lessons taught by pre-service teachers as part of their teaching practice for the ACE: Inclusive Education (Learning Difficulties)?* In an attempt to answer the question, they reported on typical errors made by the learners and the pre-service teachers, and on the similarities between teacher errors and learner errors, against the background of a literature overview that included the relationship between input and output and prominent theories of second-language acquisition. A qualitative analysis of the pre-service teachers' portfolios was undertaken. Documentary analysis was conducted by means of error analysis of the pre-service teachers' portfolios, which included ELL support lessons and ELL evidence submitted by the pre-service teachers.

The fact that language errors occurred in the portfolios of qualified practising English second-language teachers teaching ELLs was disconcerting. Research findings revealed that practising teachers who were enrolled for the ACE were themselves in need of corrective measures and interventions that would enable them to acquire English proficiency before they could be expected to help their ELLs. The findings concurred with Cloud (2005) who notes that:

> The practical knowledge base of teachers should be grounded in theory and principles and should be informed by effective language and content teaching that is appropriate for the different stages of a programme and teachers' development.
>
> Cloud (2005: 279 – 280)

Thus, various concepts and descriptive accounts should be considered to enable teachers to understand their classroom experiences and offer pedagogic choices to their learners (Mitchell & Myles, 2004: 261). In addition, teachers should be trained to teach through the medium of English as they have been seen to lack the knowledge and skills to teach the four language skills and ignore the importance of methodological skills (Uys et al., 2007: 77). The reading ability of students at tertiary-level institutions should be tested and their reading skills developed (Pretorius, 2002: 101). Given the fact that vast amounts of research focus on primary school learners, the research highlighted the fact that more research into the language proficiency of practising English second-language teachers is needed, including those who use English as LoLT, as well as a more comprehensive investigation into the effect that a low level of language proficiency of English second-language teachers has on ELLs.

The results of Nel and Swanepoel's investigation are relevant, timely, significant and in line with the present chapter because they alert academia to the extent and impact of the qualified practising English second-language teachers' low level of English proficiency and its possible effect on the ELLs' ability to develop English proficiency. Nel and Swanepoel further recommended that research is necessary in order to identify the critical contributing factors and to develop a multifaceted solution to the problem, including a re-evaluation of teaching methodologies. A two-tiered approach is needed: firstly, the pre-service teachers' level of English proficiency should be raised; and secondly, guidelines should be developed to improve learner support lessons presented in pre-service teachers' practical teaching portfolios to ensure that ELLs are taught effectively (Nel & Swanepol, 2010). Inasmuch as many higher academic institutions support first-year English second-language students by assisting them in upgrading their cognitive academic language skills, the research indicated that even students with three-year qualifications may require assistance in upgrading their CALP (Nel & Swanepoel, 2010). Thus, Nel and Swanepoel's investigation serves as a springboard to explore the language proficiency of IP mathematics teachers.

South African language dynamics are such that 90% of South African learners are not native English speakers (Spaull, 2016); therefore, English becomes their first additional language in school. The teaching of English is key to the overall success of learners as their ability to speak English inevitably impacts on all areas of their learning – not only listening, speaking, reading and writing, but also in mathematics (*as emphasised by this book*), sciences and all other content areas where an understanding of concepts and effective communication skills are required. Figure 7.1 shows the frequency of the use of the 11 official languages in South Africa as languages of learning and teaching as well as assessment at primary school level. Different home languages are used in the first three grades while from Grade 4 onwards, English dominates as a language of learning and teaching.

Figure 7.1: Language of Learning and Teaching in South African primary schools

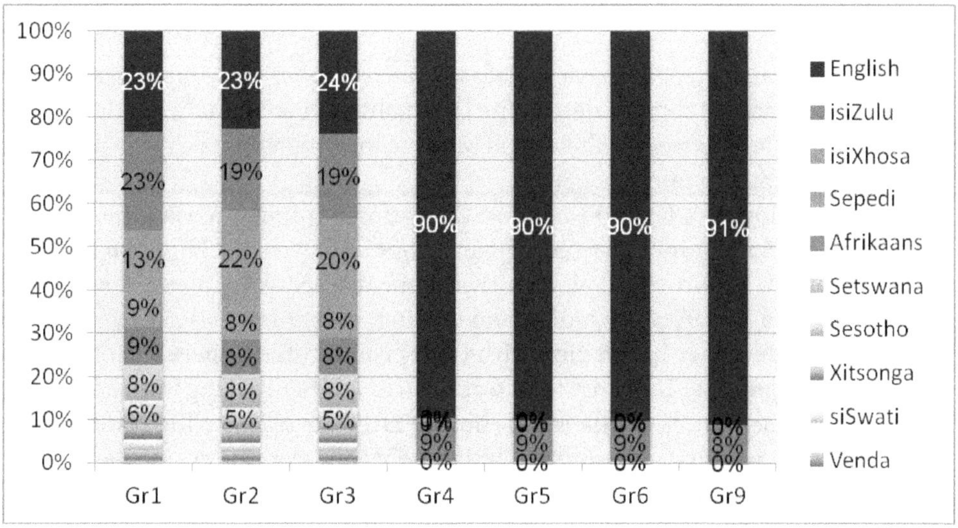

Spaull (2016)

Reports have shown that learners' language competency across all phases is disturbingly lower than the expected standards and consequently learners progress to higher grades and phases without the prerequisite knowledge and language skills (Mohlabi-Tlaka, 2016: 7; Krugel & Fourie, 2014: 220; Van Dyk, 2005: 38). The reality in many public schools is that learners transition from the Foundation to the Intermediate phase while still battling to read, let alone write in the first additional language. Learners' poor and inadequate language skills create a perpetual and recurring problematic situation which contributes to the lack of intellectual growth necessary for national socio-economic development.

MacDonald (1990) and Hoadley (2010) describe research conducted in pre-independent South Africa's Bophuthatswana homeland which examined learner challenges in coping with 'crossing the threshold' to learning all subject areas using English as a LoLT. It was revealed that learners had a vocabulary repertoire of 700 words when the curriculum demands could only be met by a minimum of 7,000 words. The most proficient among the learners had only about 10% of the requisite vocabulary and this compromised learners' reading with comprehension and effective learning in general. According to Hoadley (2010: 8), "the sudden transition resulted in most learners resorting to rote learning content which they did not understand". There was significant loss of meaning on the part of the learners. Learners' pedagogical experiences were connected to the high dropout rate at Grade 4 at the time.

> Chorusing and rhythmic chanting in classroom and absence of individual, evaluated performances was a strategy to mask both teachers' and learners' poor command of English and their lack of understanding of academic content. In a sense it represented a form of learning that enabled them to hide the absence of substance.
>
> Hoadley (2010: 8 – 9)

The quality of teacher language use and classroom dynamics is an important feature of this book. Teacher talk may determine the quality of instruction, particularly in an IP mathematics classroom. Taylor and Moyane interrogated classroom dynamics regarding language competency.

Taylor and Moyane's (2004) Khanyisa Education Support programme baseline assessment of 24 randomly selected primary schools in two rural Limpopo Province districts confirmed chorusing, low cognitive demand, weak assessment, slow pacing, poor quality and quantity of reading and writing as characteristic at the Grade 3 level. Individual learner interaction with books constituted a mere 3% of time in the literacy classrooms. The most manifest form of learner reading was the reading of three or four sentences on the board chorally after the teacher. The scant writing that was evident was that of decontextualised words and not sentences. The existent classroom practices were exposed as very limiting in terms of developing learners' proficiency in the target language. Taylor and Moyane's investigation is instructive because it indicates the extent to which attempts were made to enhance Grade 3 isiXhosa-speaking learners' proficiency in the English language as the key to establishing the extent to which they were prepared for the challenges of reading in Grade 4.

Since the transitional challenges are related to reading, a review of research in reading was necessary. Hoadley (2010) reports on an investigation of the teaching of reading conducted by the Human Sciences Research Council (HSRC) in twenty schools in Limpopo Province in Grades 1 to 4 classes. The large-scale investigation involved two-hour observations in 77 classes. Results indicated very little evidence of reading: in 12% of the classrooms virtually no reading was evident. In the few instances where it was evident, teacher modelling was absent and the reading was confined to words in isolation and not extended texts. Only picture discussions were evident. In 69% of the time learner responses were not elaborated on. The conclusion indicated that "the scale of exposure to vocabulary and text falls way below what should be expected at each grade level observed" (Hoadley, 2010).

Taylor and Coetzee (2013) interrogated language of instruction practices in South African primary schools. A positive correlation between English instruction in the first three grades and subsequent performance in Grades 4, 5 and 6 was revealed.

They also found that mother-tongue instruction in the early grades significantly improves English acquisition, as measured in Grades 4, 5 and 6.

Vorster, Mayet and Taylor (2013) conducted a national assessment to estimate the disadvantage of writing a test in English compared to writing in mother tongue. The research was composed of two sets of test scores for the same learners in the same year on the same test administered in English on one occasion and in the mother tongue on another occasion. However, the research does not provide specific recommendations on when the language of learning and teaching should switch from mother tongue to English.

Taylor and Coetzee (2013) used longitudinal administrative data from the Annual National Survey of schools as well as test scores in English and maths from standardised tests written as part as the Annual National Assessments (ANA). The baseline estimates indicated that instruction in English is beneficial for the performance in both the English and the mathematics tests. However, after controlling for school fixed effects, they also found that there was a significant disadvantage to receiving instruction in English rather than the learner's home language. They also interrogated the robustness of the results against a list of factors in order to make sure that the results were not driven by changes in school quality correlated with the change in language of instruction over this period. Their findings were in line with the pedagogical theory which promotes the acquisition of a first language before moving on to a second language. The importance of their assessment is in being the first robust research which attempted to disentangle the impact of language on the performance of learners in South Africa.

Taylor and Coetzee's (2013) results indicate the advantage of additional years of home-language education to learners in the under-resourced schools in South Africa. However, although the results are robust, they are also very heterogeneous and do not hold for all schools in South Africa. They therefore concluded that the results could be used as suggestive evidence that the current language in education policy, where schools have the autonomy to make their own decisions regarding the language of instruction adopted (DoE, 1997), may be the most appropriate. However, as long as this is the case, there is little incentive to switch to an African language as the LoLT in schools (Taylor & Coetzee, 2013). With a view to promoting multilingualism in the post-apartheid period, parents have been given the right to choose the language of instruction at their children's schools. In practice though, the majority have chosen English as the language of teaching and learning, because they consider it to be the language that will afford their children the greatest success and status.

> '...in selecting a strategy to have their children learn English, they demonstrably take the worst route, namely to choose English as the language of instruction from as early a grade as possible.'
>
> Weideman and Van Rensburg (2000: 157)

There is substantial evidence in the language and education literature to suggest that learners need to have reached an adequate competency in reading and writing in their mother tongue first before they are able to acquire competency in a second language. In support of this view, it emerged from Taylor and Coetzee's findings that adults in their sample who were proficient in their home language were also significantly more likely to be proficient in English. The analysis revealed that home-language proficiency is one of the most important determinants of English-language proficiency among Africans. It was recommended that it would be difficult to implement mother-tongue instruction in a country where African languages have low economic value and status and are rarely used in public life. It was further noted that the consequences of not tackling the mismatch between parents' perceived intentions and language outcomes in the South African LiEP are apparent in the alarmingly low throughput rates of learners within the ordinary school system as well as the poor language skills among those who have just completed school. Thus, parents and teachers need to be made aware of the benefits of mother-tongue instruction so that they can make informed decisions regarding the best way for learners to achieve both English-language proficiency and a suitable level of cognitive development (Taylor & Coetzee, 2013).

LANGUAGE OF LEARNING AND TEACHING MATHEMATICS: THE SOUTH AFRICAN LANDSCAPE

In South Africa, learners transition from learning mathematics (*and other subjects*) through mother-tongue instruction to learning mathematics through the medium of English as shown in Figure 7.2.

Figure 7.2: Language of instruction in the South African ordinary school education system

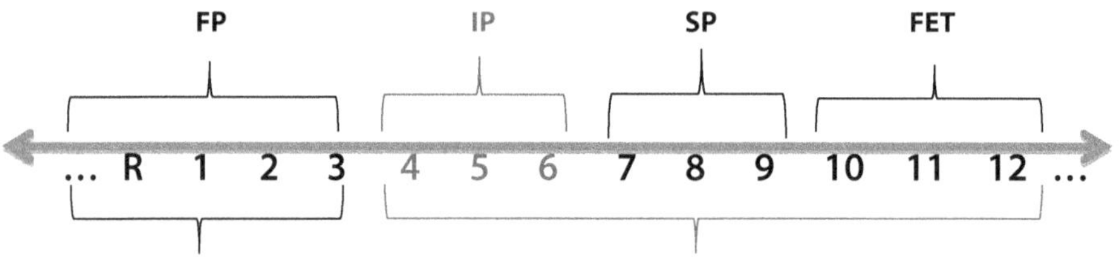

Adapted from Tshuma (2016, 2017)

Rakgokong (1994) argued that using English only as the LoLT in multilingual primary mathematics classrooms in South Africa has a negative effect on the learners' meaning-making and problem-solving ability. The research highlighted that in classrooms where English was the only language used for teaching and learning, learners were unable to conceptualise the relevant discourses. The conclusion was that when English was the sole language used in the learning environment, learners were constrained in engaging in the discourse of the classroom, thus affecting the development of their mathematical knowledge and understanding.

Varughese and Gencross (1996) found that South African first-year mathematics university students using English as LoLT had difficulty in understanding terms such as *integer*, *perimeter* and *multiple*. These findings supported the contention that language influences mathematics learning and understanding. This chapter contends that these students' transition from mother-tongue instruction at primary level might have been mishandled; consequently, misconceptions of these terms which are introduced at primary level are carried forward into higher levels of education.

Naudé (2004) compared Afrikaans learners who attended English lectures and Afrikaans learners who attended Afrikaans lectures, and found that there was no significant difference in performance between the two groups, even though the Afrikaans learners attending the English lectures were academically stronger. This finding supports the view that the students were disadvantaged as they were not proficient in mathematical English. Similarly, when examining the influence of language on the mathematical performance of learner, Dawe and Mulligan (1997) concluded that teachers need to encourage learners to recognise and distinguish between 'mathematical' English and 'natural' English, as these are sources of confusion and lead to errors in performance.

LANGUAGE OF LEARNING AND TEACHING: THE EASTERN CAPE PROVINCE

> The Eastern Cape was the hub of education for black people for a long time, a province with a wonderful history of production of educated heroes and heroines of struggle; individual teachers, learners, parents and all community based organisations should play a leading and pivotal role in rebuilding the Eastern Cape Education Department; no individual or an organisation should be allowed to act in a manner that is against the spirit of rebuilding the Eastern Cape Education Department.
>
> Ms Makgate in Ncanywa (2014)

The Eastern Cape Department of Education (ECDoE) is divided into 24 districts namely, **Bisho, Butterworth, Cofimvaba, Cradock, Dutywa, East London, Fort Beaufort, Graaff-Reinet, Grahamstown, King William's Town, Lady Frere, Libode, Lusikisiki, Maluti, Mbizana, Mt Fletcher, Mt Frere, Mthatha, Ngcobo, Port Elizabeth, Queenstown, Qumbu, Sterkspruit** and **Uitenhage**. The education districts overlay with district municipalities and metropoles as shown in Figure 7.3.

Figure 7.3: ECDoE District Boundaries

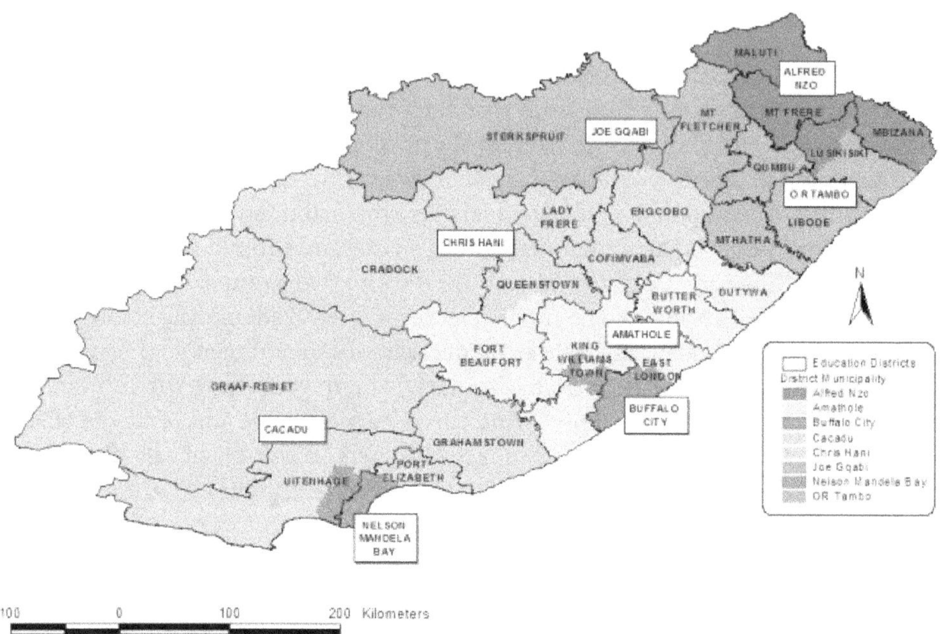

ECDoE (2014)

With regard to the LoLT, which is usually the mother tongue at the FP level and English and/or Afrikaans in the former *Model C schools*,[1] Afrikaans learners who are taught and examined in their mother tongue perform better than other learners who have no choice in being instructed in English (ECDoE, 2010). Even former Model C schools are populated by non-whites as a large population of black middle-class households enrol their children in these schools. The province has about 5% of teachers in REQV 15 and 16; and a high percentage (52%) of under-qualified teachers in REQV 10, 11 and 12 (ECDoE, 2010).

Ncanywa (2014) analysed the state of the ECDoE schools during the second decade of democracy, and revealed among other factors that, as an important human resource, teachers need to be: properly trained and well paid, and attracted to and retained within the teaching profession. In addition, it was revealed that the curriculum should address the language used for instruction as well as cater for all types of learners, so that there are no learners left behind, and that learner performance is largely influenced by teacher quality, school and community characteristics (Ncanywa, 2014). Figure 7.4 shows matriculation pass rates for the 10 worst districts in the country in 2012. Eight of these districts are found in the Eastern Cape Province, one in Limpopo Province and another one in Northern Cape Province.

1 Model C schools are well-resourced, exclusive schools which were formerly reserved for white learners under the apartheid government.

Figure 7.4: Matriculation pass rates for the 10 lowest districts in 2012

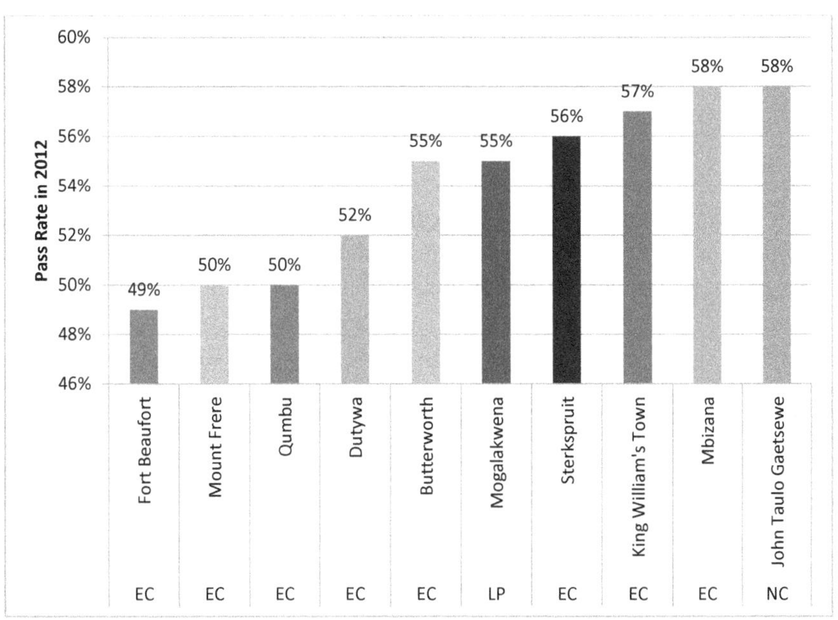

DBE: Atlas of Education Districts in South Africa (2013)

It is worth noting that the significantly low learner performance that surfaces at matriculation level could be emanating from the lower levels of the ordinary school education system, such as the IP level as illustrated in Figure 7.1; hence improving IP mathematics teachers' qualifications through upskilling their competency in the language of instruction could contribute to positive matriculation results in the ECDoE and contribute to rebuilding the province into the hub of education it once was.

Studies on the impact of language of instruction in the ECDoE are even scarcer. Most existing studies are small-scale qualitative studies. Brock-Utne (2007) used observations from two classes to illustrate that IsiXhosa-speaking learners learn better when instructed in their home language. The few existing English second-language studies conducted amongst isiXhosa speakers in the ECDoE (Sepeng, 2014a, 2014b) reveal that learners are not interested in learning to read or write in their home language. These educationists attributed this notion to the fact that isiXhosa has long been marginalised and devalued. Hence the current evidence on the impact of language of instruction on learning outcomes is insufficient.

PEDAGOGICAL IMPLICATIONS: LANGUAGE OF LEARNING AND TEACHING MATHEMATICS

Because of the difficulties experienced by teachers and learners when a language that is not their home language is used as LoLT, and after considering the benefits of learning in one's home language, organisations such as the United Nations

Educational, Scientific and Cultural Organisation (UNESCO) and German Agency for Technical Cooperation (GTZ) have been at the forefront of promoting the use of home languages in African classrooms (GTZ, 2005; Chitera, 2016).

The research and other attempts that have been made in the ECDoE to promote mother-tongue instruction beyond the stipulated FP level is hereby acknowledged. Earlier sections of this book noted that fluency in the mother tongue is essential for better acquisition of additional languages. Case studies referred to in this book also highlighted that teachers who over-rely on mother-tongue instruction usually do so as a substitute for incompetence in the prescribed LoLT. To combat such incompetent teacher practices, teacher education programmes should include language competency modules so that teachers can effectively use code switching and translanguaging to make informed decisions regarding the use of learner multilingualism as a resource in their mathematics classrooms. Otherwise, an unnecessarily prolonged use of mother-tongue instruction may seem a disservice to the learners who will at one point or the other have to change into using English as LoLT. Continuing mother-tongue instruction unnecessarily is only delaying a transition which is inevitable and therefore emphasises that the teachers should be competent in using English as LoLT, so that whenever the transition eventually occurs, the teachers will be better able to manage it. In a nutshell, teachers and the entire ordinary school system need to confront the challenges posed by using English as LoLT instead of evading them and producing learners who would find it difficult to cope with science, technology, engineering and mathematics (STEM) careers at university level. Such unintended eventualities of seemingly good practices in education are bound to keep the Eastern Cape as the least satisfactorily performing province in education, warranting government administration more often than other provinces in the country. With the world becoming a global village, it is necessary to afford all learners equal opportunities that can make them compete fairly beyond the borders of where their home languages are spoken.

As illustrated through the above-mentioned studies conducted globally, in the sub-Saharan countries, in South Africa and in the ECDoE, the factors that affect achievement are multidimensional (Setati, 2003). As a result, it is a complex task to determine the most appropriate LoLT for use in classrooms. However, complex as the task may be, teacher competency in the prescribed language of instruction is a crucial component which deserves more attention than it is currently being awarded. In addition, poor performance of multilingual learners cannot be solely attributed to the learners' language competency (which is affected by the teacher's language competency in the prescribed LoLT) in isolation from the wider spectrum of social, cultural and political factors that pervade the teaching and learning spectrum (Setati, 2003). The studies reviewed in this section of the chapter emphasise that language is one of the many factors that are related to learner performance; and teacher language competency in the prescribed LoLT needs to be prioritised in order to positively influence achievement across the curriculum, including achievement in mathematics.

There are many debates among researchers regarding the appropriate language to be used as LoLT, whether to use the languages with European roots or the Indigenous African Language (IAL)(s) of the learners. There are some who are in favour of the languages with European roots or second languages, while others favour IAL. For those in favour of the former, the argument is that the use of languages with European roots, such as English, Portuguese and French, as LoLT has more benefits for the learners, because these languages are often spoken widely elsewhere in the world. In addition, these languages are seen as a symbol of power, status and prestige (Baldauf & Kaplan, 2005; Gutierrez, 2002; Hameso, 1997; Pennycook, 1998; Setati, 2005; Tollefson, 1991; Trewby & Fitchat, 2001). They are mostly used to gain access to social status, tertiary education and business opportunities.

Chitera (2016) reveals that teaching and learning of mathematics in home languages is as complex as teaching and learning in foreign languages. It is a well-known fact that mathematical language is not well developed in most African languages and Chitera argues that mathematics teaching using a language that is not well developed undermines the confidence of teachers. As a result, teacher-centred approaches, chorus answers and use of one-word answers still dominate the mathematics classrooms even though both teachers and students can communicate freely.

Studies have also shown that there are tensions between formal and informal mathematical language in a classroom where LoLT is the home language. It emerges that textbooks use words which they referred to as 'formal mathematical language', yet the words are uncommon to both teachers and students. As a result, teachers are faced with a dilemma of choosing what words to use between the common ones referred to as informal versus the formal ones. It was seen that teachers preferred the informal terms rather than the formal ones, since learners were unfamiliar with them. However, even though they preferred the informal terms, they were not sure which ones to use as they still switched between the two terms being used (Chitera, 2016).

Thus, even though the LoLT is the home language, teachers tend to dominate classroom discussions. One of the reasons for introducing the local languages was that teachers were using teacher-centred approaches because the LoLT was English; however, even in the classrooms where the LoLT is a local language, this approach still dominates. The implication here is that introducing the home languages as LoLT when the mathematical language is not developed does not make the teaching and learning of mathematics easier. There is more than just the LoLT. Furthermore, because of a lack of technical terms and proper mathematical discourse in the local languages, both teachers and books tend to use many ambiguous words in the lessons. The context does not enrich the development of the teachers' nor the learners' mathematical discourse. In this way, learners' attention risks being drawn away from the concept by the context. Inasmuch as the literature notes that the use of a second language as LoLT limits learners' participation, lack of explanations of mathematical concepts in home languages also limits both the teachers' and learners' interaction (Chitera, 2016).

Chitera (2016) also argues that mathematical concepts and discourses produced when teaching and learning in home languages are distorted when the concepts and discourses are not well developed in the home languages. Being fluent in the LoLT does not mean that one will be able to explain the concepts without distorting them. Chitera noted that textbooks and policy makers assume that when one is fluent in the home language, then the teaching of mathematics would be straightforward. However, the literature has shown that this is not the case. This indicates that being fluent in the LoLT and being able to teach mathematics correctly are different things that need to be intertwined skilfully in order not to distort the mathematics. Chitera (2016) thus argues that teaching mathematics in home languages is not as easy as it is assumed, more especially if the mathematical discourse is not well developed. However, being fluent in the LoLT, the same language that is used in textbooks to explain mathematical concepts helps the teacher to better understand the concepts and to explain these concepts to the learners, a view endorsed by the current chapter.

CONCLUSION

This chapter reviewed global, sub-Saharan and local literature on perspectives of English as a medium of mathematics instruction. The reviewed studies were conducted in contexts similar to the South African context – specifically the under-resourced multilingual settings in the Eastern Cape Province. The LoLT used in primary school is one of the most important inputs into the education system. In many African countries, the predominant indigenous home language spoken by the majority of learners is not well developed for academic purposes, leading to the adoption of English as the language of instruction from a very early age. This is also the case in South Africa, where some primary schools use African indigenous languages as LoLT for the first three years and then switch to English at the beginning of the fourth year. Thus, most primary schools use English as the LoLT, even though the majority of the learners in the school are ELLs.

Having reviewed language development theories including global, sub-Saharan and local literature on perspectives of English as a medium of mathematics instruction in contexts similar to the South African context, (Chapters 6 and 7); **Case Study C** sheds more light on the linguistic practices in under resourced multilingual mathematics classrooms. This case study aims to investigate the extent to which Eastern Cape Department of Education (ECDoE) Intermediate Phase (IP) teachers' classroom linguistic practices promote meaningful mathematics instruction, guided by a framework for interpreting language factors in mathematics learning. Non-participant classroom observations were conducted at ECDoE schools where 5 IP mathematics teachers are based, followed by individual teacher interviews. The qualitative data from classroom observations and teacher interviews were qualitatively analysed. The data revealed text-deficient classrooms, teachers' inconsistent use of English, the prescribed LoLT as well as inconsistent, unstructured use of code switching and translanguaging strategies meant to improve learners' understanding of mathematical concepts, all of which contribute to poor linguistic practices in IP mathematics classrooms. It is crucial to conduct similar studies in other under resourced multilingual mathematics classrooms to determine whether the findings apply to other areas.

CASE STUDY: C

LINGUISTIC PRACTICES IN INTERMEDIATE PHASE MATHEMATICS CLASSROOMS: THE EASTERN CAPE PROVINCE

INTRODUCTION

The teaching and learning of mathematics, like any other subject, requires that both the teacher and learners communicate effectively. The primary function of language in mathematics instruction is to enable both the teacher and the learners to communicate mathematical knowledge with precision.

Picture the scene: When I taught elementary school mathematics and science in Fort Lee, New Jersey, I had access to a computer laboratory, printers, photocopiers, and a whole library of textbooks and easy reading to give to my learners. Then I moved to Bathlaros village outside Kuruman and I was just thankful to have electricity twenty-four hours a day. I knew that teaching primary school mathematics and science in Africa would mean I wouldn't be able to have my learners do interactive lessons on the internet or give them each their individual textbook to work from, but by no means would having limited classroom resources make things impossible. In fact, it was a fun challenge that forced me to get creative with my lessons. (Adapted from: https://www.gooverseas.com/blog/tips-for-teaching-with-limited-classroom-resources.)

The extract above is an exposition of an expatriate teacher's experiences comparing the availability of teaching and learning resources between a well-resourced environment and an under-resourced one. Although the exposé highlights general teaching and learning resources, the case below focuses on teaching and learning resources that specifically influence the quality of linguistic practices in some under-resourced mathematics classrooms in the Eastern Cape Department of Education (ECDoE). Another stark difference is that in the exposé, the teacher's competence in the prescribed language of learning and teaching (LoLT) is quite high compared to teachers in the ECDoE. Furthermore, the teacher in the exposé does not share a home language with the learners in Bathlaros while the teachers in the current case share a home language (isiXhosa) with their learners. A lot can be inferred from comparing and contrasting the surreal expatriate teacher's experiences and the realistic experiences of participant teachers investigated in this case. The bottom line, however, is that teaching somewhere with limited classroom resources is tough; more so if there is an underlying language barrier compounding the situation.

To investigate how Intermediate Phase (IP) mathematics teachers improvise on the use of limited classroom resources at their disposal, navigate the development of the prescribed LoLT, the development of mathematical language and the teaching of subject content, it is crucial to investigate the classroom linguistic practices in the ECDoE. This case study briefly explains the ideal classroom linguistic practices against the backdrop of the prevailing practices in sub-Saharan Africa, and presents the research questions, conceptual framing, research design and methodology, findings and discussion.

CLASSROOM LINGUISTIC PRACTICES

As is widely reported in the literature, in multilingual classrooms, teachers and learners frequently draw on more than one language for a range of functions; and these practices may be part of a planned bilingual curriculum or may arise spontaneously in response to particular needs. The following classroom linguistic practices apply to any teaching and learning environment. In this case study, they are used to benchmark the linguistic practices taking place in under-resourced mathematics classrooms.

Classroom Organisation – Learners construct solutions for mathematics through discourse. Therefore, it is important to organise learners in groups in order to encourage learner participation through talk. Dialogic teaching encourages learners' active participation (Mercer, 2006: 507; Alexander, 2000). **Classroom Resources** – Learners understand content better if they use objects derived from their own environment in their learning. The use of resources encourages active learning by learners. This makes the learners active respondents in their own learning (Carruthers & Worthington, 2006: 74; Willis, 1998; Crawford & Witte, 1999). **Teacher Activity** – If a teacher and the learners understand concepts and procedures, the teacher will have better control of the classroom activities. In addition, learners should learn by understanding and not by rote. Therefore, the learners' understanding of concepts helps to facilitate learning (de Corte 2004: 280; Even & Tirosh, 1995: 164). **Teacher-Learner Interaction** – Teacher-learner interaction in which learners are actively engaged in the process of constructing numerical knowledge and understanding is effective in the learning process. The learners learn by doing and thus help to find the solutions to problems themselves. Learners construct mathematical knowledge as advised by teachers. Teachers therefore do not give learners solutions to problems, but guide the learners to find these solutions. Success gained this way would lead to more interest in participating in learning (Van de Walle, 2007: 3; Schunk, 2005: 412). **Learner-Learner Interaction** – Learner interaction enhances social interaction as learners can share how they arrived at the answers. Dialogue between learners and teachers is good for the learning process. This interaction helps the introverted learners to open up and participate in the learning process too (Van de Walle, 2006; Vithal, 1992).

Classrooms in sub-Saharan Africa are often overcrowded, under resourced, with teacher-centred patterns of interaction that provide little scope for learners to discuss ideas through discussions with each other or with the teacher. Specifically, with reference to mathematics classrooms in public schools in Tanzania, Kajoro maintains that teaching practice in the classroom is characterised by teaching of algorithms, procedures and facts that are rote-memorised and reproduced by the learner; classroom teaching is mainly teacher-led and informed by theoretical perspectives that position the teacher in the role of a transmitter and the learner in the role of a passive recipient (Halai, 2016; Kajoro, 2016). Is this the best path (linguistic practice) to follow in the production of literate and numerate 21st century learners and global citizens well equipped with the necessary skills to solve problems?

OBJECTIVES

Analysing IP mathematics teachers' classroom linguistic practices in ECDoE schools seeks to:

i. Determine the frequency English, the prescribed LoLT is used by the teacher in mathematics classroom communication.
ii. Determine the strategies used by the teacher to teach/explain specialised mathematics vocabulary for the concepts covered during lessons.
iii. Determine the language(s) used by the teacher for chalkboard summaries during mathematics lessons.
iv. Establish the availability of DBE workbooks, textbooks or other audiovisual learning and teaching resources written in English for use in mathematics lessons.
v. Establish the availability of charts/word walls around the classroom which indicate the teaching of mathematics in English.

CONCEPTUAL FRAMING

The conceptual framework grounding this case is Ellerton's (1989) framework for interpreting language factors in mathematics learning.

A Framework for Interpreting Language Factors in Mathematics Learning: Ellerton (1989)

Ellerton's framework for interpreting language factors in mathematics learning shows the need to link the various aspects of language factors in mathematics learning (Ellerton, 1989). The framework can be viewed from a 3-dimensional perspective as illustrated in Figure C.1.

Viewing the framework from a 3-dimensional perspective implies viewing mathematics instruction holistically. When assessing the framework holistically and from an all-encompassing perspective, it can be seen that culture occupies the entire classroom, and that communication within this culture is of key importance. Language, **thinking,** and **mathematics learning and understanding** cannot be discussed solely from a cognitive perspective. Cultural and pedagogical associations also need to be explored, given the central role of language in both of these phenomena (Ní Ríordáin, 2011).

Figure C.1: A framework for interpreting language factors in mathematics learning

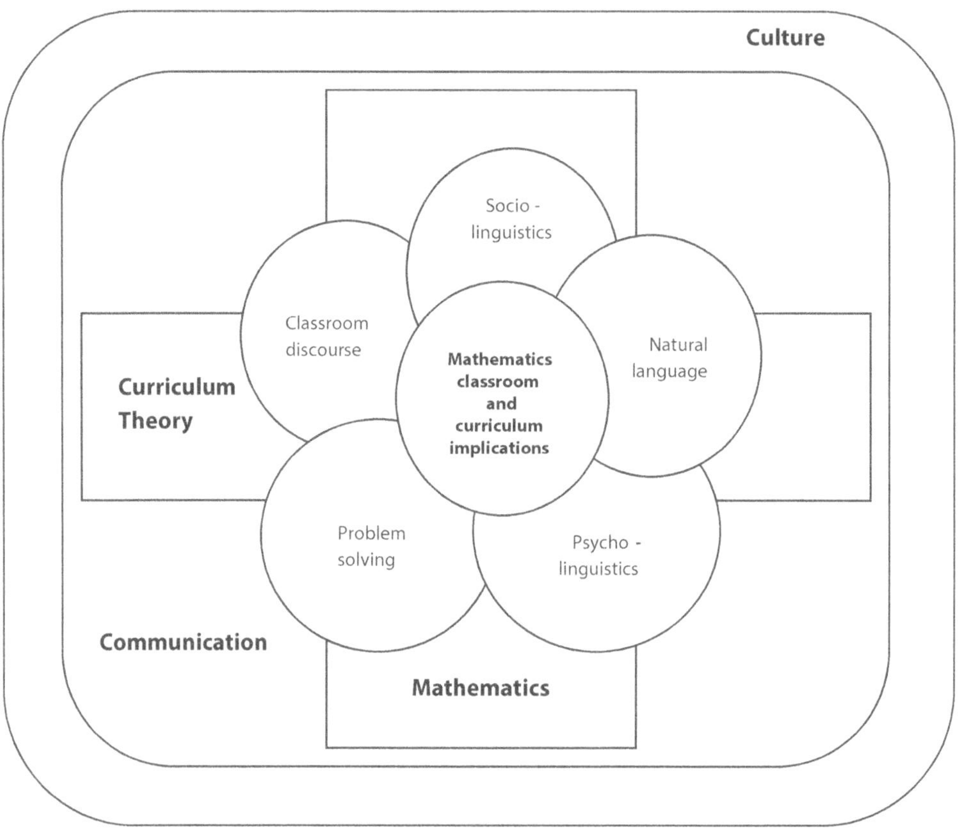

(Ellerton, 1989)

Sociolinguistics is an aspect of language that stresses the interrelationship between language and social life, rather than focusing narrowly on language structures (*Sociolinguistics*, 2004); psycholinguistics refers to the psychological states and mental activity associated with the use of language (*Psycholinguistics*, 2013). Communication and language become central factors in matters such as sociolinguistics, natural language, psycholinguistics, problem solving and classroom discourse, all of which intersect with each other, as well as with most parts of the framework. This model indicates the centrality of the teacher, the mathematics classroom and the curriculum in acknowledging the importance of language issues in mathematics education. These were key factors identified by the case in the transition from FP to IP level in the ordinary school system.

METHODOLOGY

The case used a descriptive research design to enable the researcher to make conclusions and obtain information that describes the existing phenomena. The case employed qualitative methods using non-participant classroom observations and interviews in collecting data to allow the researcher more probing of the existing phenomena (Golafshani, 2003). To identify the linguistic practices prevalent in under-resourced IP mathematics classrooms in the ECDoE and to facilitate a broader analysis of the availability of teaching and learning resources written in English for use in IP mathematics lessons, a screening tool checklist was prepared. The screening tool collected specific data on the following variables: language of instruction in the mathematics lesson, learner participation, the teaching and learning of specialised mathematics vocabulary, chalkboard summary, learning and teaching resources used in the classroom, and the classroom environment.

Classroom observations were conducted for five days at all the schools where the sampled IP mathematics teachers are based. During this phase the researcher observed lessons with the aim of getting a clear picture of what approaches teachers used in the teaching of mathematics, how they interacted with the learners and also how the learners participated during mathematics teaching. The major aspect of the observation entailed active listening. The researcher sat at the back of the classroom taking field notes describing the classroom environment and the teachers' and learners' linguistic practices during the mathematics lesson using an observation guide. The researcher observed three mathematics lessons at each school: one class at the entry level (Grade 4), one class in the middle (Grade 5 or 6) and one class at the exit level (Grade 7) of IP.

After the classroom observations, the researcher held interviews with the teachers observed. The interviews were semi-structured and allowed the teachers to further explain any decisions and actions they had taken during the lessons. These interviews were conducted at the earliest convenient time after the lesson, and were audio taped to capture as much detail as possible. The qualitative data collected from classroom observations and teacher interviews were analysed qualitatively.

SAMPLING

The unit of analysis was classroom linguistic practices in of 5 IP mathematics teachers from the ECDoE. For the purposes of this research, the teachers were studied specifically as to whether they were using and promoting appropriate linguistic practices in mathematics instruction. The section below details the IP mathematics teacher participants' demographic information.

Participant Demographic Data

The section below presents the profiles of the observed teachers, followed by the linguistic profiles of the teachers and their learners and a brief contextual description of the schools and lessons observed before the presentation of the different linguistic practices observed in the classrooms. Table C.1 provides the profiles of the observed teacher participants.

Table C.1: Observed teacher participant profiles

S	P	G	AGE	RACE	ETM (YEARS)	Q
S1	T1	F	4	Black	15	▪ Primary Teachers' Diploma ▪ Advanced Certificate in Education
S2	T2	F	4	Black	7	▪ Primary Teachers' Diploma ▪ Bachelor's Degree in in Education ▪ Advanced Certificate in Education
S3	T3	F	4	Black	9	▪ Primary Teachers' Diploma
S4	T4	F	4	Black	6	▪ Primary Teachers' Diploma ▪ Advanced Certificate in Education
S5	T5	M	2	Black	1½	▪ National Diploma in Adult Basic Education and Training ▪ Post Graduate Certificate in Education

Key:
S – School
P – Participant
G – Gender
AGE – Age Group: {1: 25 and below; 2: 26 – 29; 3: 30 – 34; 4: 35 and above}
ETM – Experience Teaching IP Mathematics
Q – Qualifications

As illustrated in Table C.1, the majority of the teachers were middle-aged females. Most of the teachers had been teaching other subjects such as Technology and Social Sciences at primary school level before they began teaching IP mathematics. The teachers' experience of teaching IP mathematics varied from several years to only a few years. In addition, all the teachers were qualified, having obtained pre-service diplomas formerly offered at Teachers' Colleges as their initial teacher education qualifications; some teachers had upgraded these initial teacher education qualifications to obtain further qualifications through in-service programmes offered at different universities in the country. Table C.2 provides the linguistic profiles of the observed IP mathematics classes.

Table C.2: Observed classes' linguistic profiles

	Grade	Teacher's 1st language	LEARNERS' 1ST LANGUAGES				Total No. of learners in class	Topic of the lesson
			English	Afrikaans	isiXhosa	Other (Specify)		
T1	4A	IsiXhosa	✗	✗	✓	–	67	Quadrilaterals
T2	6A	IsiXhosa	✗	✗	✓	–	50	Numeric Patterns
T3	5A	IsiXhosa	✗	✗	✓	–	99	Transformations
T4	6	IsiXhosa	✗	✗	✓	–	42	Input – output Machines
T5	7	IsiXhosa	✗	✗	✓	–	22	Multiplication

As indicated in Table C.2, the lessons observed were all at the IP level (ranging from Grade 4 to Grade 7). One lesson (Grade 4) was at the entry level of the phase, one lesson (Grade 5) was at the second level of the phase, two lessons (Grade 6) were at the third level and one lesson (Grade 7) was at the exit level of the phase.

In all the lessons observed the teachers' home language is isiXhosa, which further confirms the point made by Tshabalala (2012) that the majority of South African teachers are not first-language speakers of English. None of the learners in the observed lessons spoke English, Afrikaans or any other language as their home language. In all the lessons observed the learners' home language was isiXhosa, which meant that the teachers and the learners shared the same home language. The total number of learners in the classes varied, with the smallest class having 22 learners, while the largest class had 99 learners. Thus, the class size in these schools is relatively large. **T1** presented a lesson on quadrilaterals, **T2** presented a lesson on numeric patterns, **T3** presented a lesson on transformations, **T4** presented a lesson on input-output machines, and **T5** presented a lesson on multiplication.

To get a clearer insight into the lessons taught, a brief description of the context of the observed classrooms is here presented. Although the observations where done in under-resourced schools in the Eastern Cape Province, the setting and environment of each school is unique. The contextual details of the observed classrooms thus provide information on the enrolment of each school as well as the staff complement and the conditions which may have a direct or indirect impact on what actually happens inside the classrooms regarding how each teacher presents his or her lesson. Table C.3 provides an overview of the lessons that were presented in each of the IP mathematics classrooms observed against the backdrop of the contextual setting of each school.

Table C.3: Brief description of schools and the observed lessons

CONTEXT	P	LESSON DESCRIPTION
S1 **Enrolment:** 1,022 Learners (Grades R – 9) No Grade 8 **Staff Complement:** 30 Teachers	**T1**	**Topic: Quadrilaterals** Description of different polygons summarized on the board, the shapes described are drawn. Discussion of concepts to be learnt takes place before group work, practical activities and presentations of the words done in groups. Written work is done individually towards the end of the lesson.
S2 **Enrolment:** 903 Learners (Grades R – 9) No Grade 8 **Staff Complement:** 28 Teachers	**T2**	**Topic: Numeric Patterns** Very impressive rhymes of multiples done to introduce the lesson. Number patterns written are on the board and the teacher explains the difference between constant difference sequence and constant ratio sequence.
S3 **Enrolment:** 802 Learners (Grades R – 9) **Staff Complement:** 25 Teachers (5 Mathematics Teachers)	**T3**	**Topic: Transformations** The lesson begins with an oral mental exercise. In addition, the teacher poses several questions to the learners as part of the introduction. The teacher reviews the learners' assumed knowledge by asking clarity-seeking questions.

CONTEXT	P	LESSON DESCRIPTION
S4 **Enrolment:** 454 Learners **Staff Complement:** 10 Teachers (Grade R – 9)	**T4**	**Topic: Input – Output Machines** The teacher indicates right at the beginning of the lesson that the purpose of the lesson is to revise content covered earlier in preparation for examinations. After the announcement, the teacher continues to explain input and output machines.
S5 **Enrolment:** 103 (1 multi-grade class) **Staff Complement:** 8 Teachers	**T5**	**Topic: Multiplication** The teacher does not engage much with the learners at the beginning of the lesson, and goes on to explain the multiplication strategy that is going to be covered during the lesson.

Key:
P – Participant
S – School
T – Teacher

As indicated earlier in this section, the schools in which the teachers are based are under-resourced and are characterised by enormous class sizes due to a shortage of classes. In schools where there is *'No grade 8'* as indicated in Table C:3 the ECDoE is currently separating primary and secondary schools, by moving Grade 8 and 9 learners previously based in primary schools to nearby secondary schools, a process which also impacts on teacher deployment.

Figure C.2: Break time at S3

At **S3** the school's administration block had been converted into classrooms and a smaller room was built by parents adjacent to the block to function as the principal's office and secure storage for the school's photocopying machine, screen and projector. The school is challenged by having a serious shortage of furniture and the large class sizes; there was no space to accommodate a teacher's table.

At **S4** an educational development partner was in the process of constructing a road connecting the school to the main road. The road construction included a bridge which would ensure that learners can safely commute to school during the rainy season when rivers are flooded. Although the schools had

Figure C.3: School furniture situation at S4

well-maintained buildings, there is a serious shortage of furniture and in some classrooms learners sit on broken chairs.

S5 is a relatively small school compared to other schools in the vicinity, with one block of classrooms for all grades and another separate block housing the principal's office, administration office and a staff room. With only 103 learners and one multi-grade class (Grades 2 and 3 taught by one teacher in one classroom), parents in the community seem to prefer to spend money on transport to send their children to neighbouring schools with a reputation for better performance. The contextual settings of the schools where classroom observations were conducted provide more information on the under-resourced schools' environment against the backdrop of the South African education system.

The mathematics teachers identified in Table C.1 are generalist teachers who are assigned to teach mathematics, not specialised to teach the subject. The teachers presented lessons on different topics. Although all the classroom observations were conducted in one week, teachers were all teaching different topics, including teachers teaching the same grades, such as the two Grade 6 teachers observed. These variations could be attributed to the fact that the pace of the learners in the different schools was different. The brief lesson descriptions also provide a brief synopsis of how each teacher presented their lesson.

In general, the majority of the lessons were teacher-centred (**T3**, **T4** and **T5**) and this could be attributed to the relatively large class sizes. Where learners were actively involved (**T1** and **T2**) there was a lot of chanting and rhyming, which indicated that rote learning dominated the classrooms. However, some teachers did bring practical activities into their lessons and their classrooms were organised into groups to promote discussion and interaction amongst learners.

The aim of conducting the classroom observations was to gather more information on the classroom environment in which the participant teachers practised as well as to assess the linguistic practices in each classroom observed. As a non-participant observer, the researcher broadly assessed the following: classroom organisation, teaching and learning resources, teacher activities, and interactions between the teacher and the learners as well as interaction amongst learners themselves.

FINDINGS

The research question was: **To what extent are ECDoE IP teachers' classroom linguistic practices promoting meaningful mathematics instruction?** and the key findings are that sampled teachers exhibited poor linguistic practices while presenting mathematics lessons, their classrooms are relatively text-deficient and there is little evidence of teachers developing learner linguistic practices. The section below presents, analyses and discusses data collected from the classrooms observed, followed by data from the teacher interviews.

NON-PARTICIPANT CLASSROOM OBSERVATIONS

Language of instruction in mathematics lesson

The first variable focused on the language of instruction used in the IP mathematics classroom, with particular focus on the frequency of using English as the LoLT. Observations are presented in Table C.4.

Table C.4: Language of instruction in IP mathematics lessons observed

	Variable 1: Language of instruction in mathematics lesson *{Language(s) used by the teacher to give instructions and ask questions}; comment on the frequency English is used as the LoLT.*	
	Language(s)	**Comment**
T1	English	▪ The teacher uses English to give instructions to learners.
T2	English	▪ The teacher uses English to give instructions to learners. ▪ The teacher prompts learners to explain the difference between two ratios in English.
T3	English and isiXhosa	▪ The teacher uses English to present the lesson. ▪ The teacher responds to some of the learners' questions in isiXhosa. ▪ The teacher reads the questions together with the learners in English.
T4	English	▪ The teacher uses English as the medium of instruction; however, the teacher's utterances have some grammatical errors.
T5	English and isiXhosa	▪ The frequency of the teacher's use of both English and isiXhosa throughout the lesson is equally balanced.

Pedagogical implications: Language of instruction in mathematics lesson

As illustrated in Table C.4, three of the teachers (T1, T2 and T4) explicitly used English during the lesson presentation, while T3 and T5 used both English and isiXhosa as languages of instruction. While T1 used English to give instructions to the learners, the teacher's utterances were limited (did not speak much), and the researcher wondered whether the exclusive use of English was an everyday occurrence or whether it was influenced by the presence of an observer in the classroom. The researcher also wondered if T1's exclusive use of English was the reason why the teacher did not speak much. T2's use of English as language of instruction was more extensive than T1's and the teacher prompted learners to give explanations in English. The teacher seemed to be at ease in conversing with the learners in English as well as in promoting interaction with the learners in English.

T4 also used English as the language of instruction; however, the teacher's utterances had some grammatical errors. Without making generalisations, T4's linguistic competencies confirm Tshabalala's (2012) observation that the majority of South African teachers who are not first-language speakers of English are not fluent in the language, a point also confirmed in the English teacher language competency assessment presented in Case A. Although T3 presented the lesson in English, isiXhosa was used occasionally to respond to the learners' questions or to provide further clarity on certain aspects of the lesson. T5 used English and isiXhosa equally throughout the lesson.

Inasmuch as the majority of the teachers are commended for conducting their lessons in English, it was evident that the teachers need more support in keeping up with the stipulations of the LiEP. Firstly, the teachers' proficiency in the language in which they are supposed to teach needs to be improved, so that the teachers can teach comfortably in English themselves. Considering that the learners may not have much access to the English language outside the classroom, the teacher should model best practice by speaking grammatically correct English. For the teachers who used both English and isiXhosa in their instruction, it is clear that they were practising code switching and using the shared language as a resource in teaching IP mathematics. However, the practice needs to be standardised, so that the teachers know when and how to code switch and translanguage systematically, as suggested by Chitera (2016), for the purposes of meaningful mathematics teaching and learning.

Learner Participation

The second variable focused on learners' involvement in the learning activities, particularly focusing on the language(s) learners used when talking amongst each other and when communicating with the teacher, or when participating in class activities. How the learners used English in classroom activities was noted. Observed practices are presented in Table C.5.

Table C.5: Learner participation in IP mathematics lessons observed

	Variable 2: Learner participation *Learners' involvement in the learning activities (focusing on the language(s) they use when talking amongst each other/when communicating with the teacher); or when participating in class activities. To what extent are the learners using English in classroom communication?*	
	Language(s)	**Comment**
T1	English	▪ Learners respond to the teacher's instructions in English. ▪ Learners use well-constructed and complete English sentences to respond to the teacher's questions. ▪ Excellent learner participation – during presentations of work done in groups, learners present their work in English.
T2	English	▪ Learners respond to the teacher's instructions in English. ▪ Learners use ***chorus answers*** to respond to the teacher. ▪ During revision, learners attempt to explain their answers in English.
T3	English and isiXhosa	▪ Learner participation is above average. ▪ As learners respond to the teacher's questions, the teacher discourages the use of chorus answers and asks learners to respond to questions individually.
T4	English	▪ Few learners who respond to the teacher's attempt to use English. ▪ When the whole class responds to the teacher's questions, chorus answers in English are used.
T5	English and isiXhosa	▪ There is limited interaction among the learners. ▪ The teacher demonstrates work on the board, and when questions are asked, learners respond in chorus answers. ▪ Learners respond to the teacher's questions in isiXhosa and the teacher translates the responses into English.

Pedagogical implications: Learner Participation

The linguistic patterns observed in learner participation as illustrated in Table C.5 above, loosely correlate with their respective teachers' frequency in using English as a language of instruction as noted earlier. In classrooms where teachers made an effort to use English as the language of instruction, learner participation was higher on average than in those where teachers used a mixture of English and isiXhosa. In addition, in classrooms where the teacher used English more frequently, the learners also strove to respond to the teacher's questions in English. Therefore, if the learners tend to emulate the teachers' linguistic practices, teachers always have to model best practices. Thus, in order for better English proficiency to filter down to the learners, teachers' proficiency must be of higher quality.

Learners in T1's class made an attempt to respond to the teacher's questions in English, using well-constructed sentences. Although there is a possibility that the learners could have been drilled in preparation for the observation, the mere fact that they attempted to answer in English is commended. Learners in T2's class produced a lot of chorus answers, another indication that they could have been drilled beforehand. The chorus answers in T2's class went unchecked, while in T3 the teacher discouraged them and prompted learners to answer individually. The use of chorus answers could be a strategy used to manage extremely large classes (e.g. a class of 99 learners). Lee (2006: 36) describes chanting as an old-fashioned method that affords learners the opportunity to:

> ... use the mathematical language themselves, in order to get used to the way that expressions are used and to begin to use mathematical terms in order to express the web of concepts and ideas that are encompassed by those terms.

Thus, chanting allows learners to vocalise words that are often not easy for them (Lee, 2006) and also allows learners to think about the words and phrases that are used to express these mathematical ideas. For chanting to produce these positive results, the teacher (who is usually proficient in the language of instruction) would write on the board these unfamiliar phrases and initiate the chanting by voicing the words in unison with the learners. However, in the case of T4 as observed in Variable 1 (Table C.4) where:

> '... the teacher's utterances have some grammatical errors'

there is a danger in having learners chant grammatically wrong answers, as they would always remember these chants and choruses even in their higher grades and this creates further misconceptions.

In T5's class there was limited interaction among the learners and in most cases the learners responded to the teacher's questions in isiXhosa, even in cases where the question was posed in English. This practice could be an indication that the amount of communication that takes place in isiXhosa is usually higher. Considering the fact that T5 was teaching Grade 7 learners, who are at the exit level of IP, more utterances in English could have been expected. Although the teacher translated the learner responses that were provided in isiXhosa, the practice needs to be regulated and standardised.

When comparing and contrasting the ***limited interaction*** among learners to the ***chorus answers*** provided by the learners, it is worth highlighting that although whole-class aural work is important, it is also important for learners to practise working on their own, particularly reading and writing on their own, as these are the skills that are expected of them and assessed in examinations. It is also important for learners not only to give the correct answer, but also to be able to explain their thinking. One way of encouraging this would have been to allow learners to talk to the learner next to them and explain their thinking. In addition, there is a need to support mathematical thinking. This could include strategies such as asking learners to notice patterns (what stays the same and what is different) and relationships and linkages across taught concepts, which can be used effectively to encourage learners to think mathematically. Thus, the teacher needs to promote learner participation in such a way that a web of mathematical discourse is created and discourse through the prescribed medium of instruction needs to be cultivated and promoted.

Specialised Mathematics Vocabulary

The third variable focused on the strategies used by the teacher to teach and explain specialised mathematics vocabulary for the concepts covered during the lesson. Observations are presented in Table C.6.

Table C.6: Strategies used to teach specialised mathematics vocabulary in IP mathematics lessons observed

	Variable 3: Specialised mathematics vocabulary *What strategies does the teacher use to teach/explain specialised mathematics vocabulary for the concepts covered during the lesson?*
T1	▪ The teacher uses practical examples to revise the naming of different polygons.  **Figure C.4:** Models of 2-dimensional shapes used to reinforce content and mathematical vocabulary in T1's class
T2	▪ The following new terms (***constant difference sequence, constant ratio sequence***) are introduced to the learners, but not much is done to explain the meanings of these new terms.
T3	▪ The teacher uses a chart to explain the following terms: ***rotation***, ***reflection*** and ***translation***.
T4	▪ There is no new mathematical vocabulary introduced to the learners during the lesson.
T5	▪ There is no new mathematical vocabulary introduced to the learners during the lesson. ▪ The teacher explains the mathematical procedures involved in multiplication in isiXhosa.

As indicated in Table C.6, in the majority of the lessons observed there was no specific focus on specialised mathematics vocabulary. T1's lesson involved some practical activities (Figures C.4, C.5) that could be manipulated for mastery of mathematics register.

Figure C.5: Models of polygons constructed by learners in groups in T1's class

Pedagogical implications: Specialised Mathematics Vocabulary

The lesson was well presented and learner-centred and the teacher is commended for all the efforts in planning and preparing for the lesson. However, a lot more could have been done to assist learners to master the terms: ***quadrilateral, rhombus*** and ***parallelogram***. In addition to merely writing these words on the board, the teacher could have further provided learners with the meanings of these words and asked them to write them at the back of their exercise books or written these words on a separate section of the chalkboard or even a word wall. As detailed in Chapter 2 of this book, such words are not part of everyday English and the chances of learners encountering them outside the mathematics classroom are very slim – and even slimmer for a mathematics classroom in an under-resourced community.

Similarly, T2's lesson involved terms like ***constant difference sequence*** and ***constant ratio sequence*** (Figure C.6) but there was no attempt to provide explicit meanings for these compound terms that do not occur in everyday English.

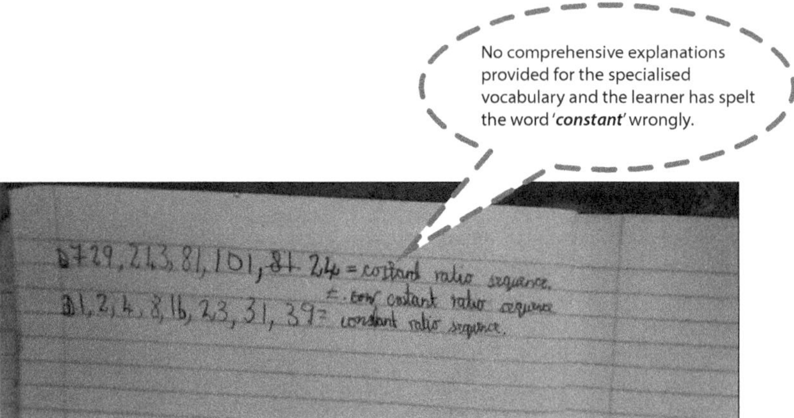

Figure C.6: A sample of a learner's written work on '***constant ratio sequence***'

Barwell (2012) states that a mathematics teacher teaching ELLs must fulfil three main functions: to teach mathematical concepts and procedures, to teach the mathematics register, and to teach the language of instruction. In the observed classrooms, it may appear that the teacher focused only on the first function of teaching mathematical concepts and procedure without paying much attention to the register or to the language of instruction. It may be inferred that the teachers were not exposed to the other two linguistic functions during their own initial teacher education, and as a result do not afford them the warranted attention. In the observed classes, the least the teachers could have done was to make learners aware of the specialised mathematics vocabulary. In better-resourced settings, teachers could have made use of mathematics dictionaries to help learners understand the meanings of the words, and in such under-resourced classrooms the teachers could have encouraged learners to add these words in their ***learner-made dictionaries*** as illustrated in Chapter 2 of this text.

T3 used a chart to explain new mathematical vocabulary to the learners; however, the use of the chart was not exhaustive. T3's reference to a chart to explain the terms ***rotation, reflection*** and ***translation*** is commended; however, more could have been done to ensure that the learners understand the vocabulary. Thus, the chart could have been used comprehensively to ensure that the learners fully understand the meanings of the new words. T4 could have reiterated the implications of the words ***input*** and ***output*** to assist the learners in understanding the mathematical concepts entailed; the same applies to T5. Ideally, during lesson preparation, the teacher should identify specialised mathematical vocabulary that may lead to misconceptions among ELLs and ensure that some attention is given to these words during lesson presentation.

Chalkboard Summary

The fourth variable focused on the language(s) the teacher used when writing on the chalkboard during the lesson. Observations made are presented in Table C.7.

Pedagogical implications: Chalkboard Summary

Variable 5 focused on the language(s) used to write on the chalkboard; all the teachers are commended for using English, the stipulated language of instruction, to write on the chalk board. As illustrated in Table C.7, the chalkboard was mainly used for diagrams; there was not much text written on the board in all the five classrooms. However, there is always room for improvement.

It was noted that in all classes none of the teachers wrote the date and topic of the specific lessons taught at the beginning of their respective lessons. As trivial as this may sound, writing the date and topic of each lesson taught is one of the simplest ways of creating a text-rich classroom (and appreciation of it), and it also helps struggling learners to master spelling. In addition, learners have a tendency to copy whatever the teacher has written on the board; therefore, if the teacher

Table C.7: Languages used when writing on the chalkboard in IP mathematics lessons observed

	Variable 4: Chalkboard summary: *Comments on the language(s) the teacher uses when writing on the chalk board during the lesson.*	
	Language(s)	**Comment**
T1	English	▪ The chalkboard is used extensively to illustrate drawings of different 2-dimensional shapes.  **Figure C.7:** A virtual geoboard drawn on the chalk board in T1's classroom
T2	English	▪ Examples of the two types of ratios are written on the board. ▪ The teacher uses the chalkboard to revise group work given to the learners. ▪ Learners respond to the teacher's questions using chorus answers.
T3	English	▪ Only the new words are written on the chalkboard.
T4	English	▪ There are no new words written on the chalkboard except for calculations. 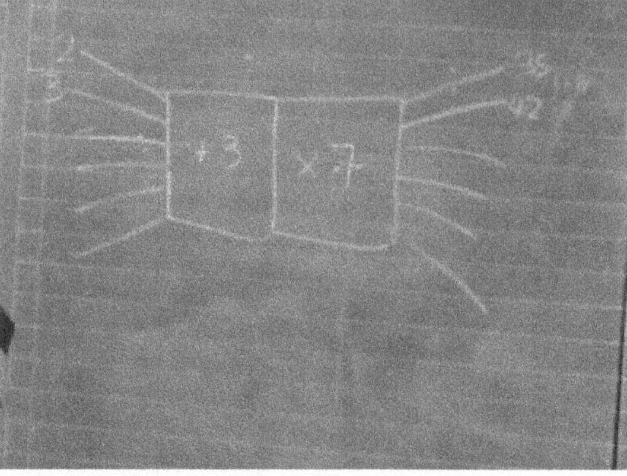 **Figure C.8:** A function machine drawn on the chalk board in T4's classroom
T5	English	▪ The teacher is the only one who writes on the chalkboard. ▪ Generally, the chalkboard is only used for calculations and nothing else throughout the lesson.

writes the topic on the board, the learners will copy the correct spelling of the topic into their exercise books.

T3 used the chalkboard to write new words and there is a possibility that the learners copied what the teacher had written verbatim. In the observed lessons the chalk board could have been used extensively to note key aspects of the lesson, which the learners would then copy into their books for future reference or for revision purposes. Educational research reveals that learners have different learning styles. Learners who appreciate visual images benefited from T1's and T4's lessons, where the chalkboard was used for diagrammatic representations of concepts taught.

Although T1 is commended for simulating a geoboard on the board, had the names of the different quadrilaterals drawn on the virtual geoboard been written alongside the corresponding shapes, learners would have benefited from the labels. In addition to using straight lines to produce neat and accurate diagrams, the chalkboard could have been used more extensively to illustrate the key terms that are used to describe quadrilaterals such as ***diagonal, parallel, perpendicular, bisect, opposite***, as illustrated in Figure C.9 so that the defining properties of different quadrilaterals are understood with greater clarity.

In T4's classroom, if the words ***input, rule, output*** had been used to label the function machine, learners could have made associations between the labels and the different parts of the function machine respectively; this would have made the diagram more meaningful than simple computations of numbers on the machine. Similarly, learners may have benefited from chalkboard summaries highlighting key points from the lesson.

Another reason for the chalkboard summary not being as text-rich as expected is inadequate pedagogical content knowledge. Thus, there is a possibility that the chalkboard summary was not as text-rich as the researcher expected because of the teachers' inadequate mathematical content knowledge rather than poor linguistic skills. Also, the teachers could have covered these mathematical terms in lessons preceding the observation, or were intending to cover the mathematical terms in the lessons after the observation. Either way, this information was not part of the review of assumed knowledge or the review of concepts learnt in the preceding lessons. As critical as this may be, a consideration of mathematical content knowledge falls beyond the scope of this case (and the book).

Learning and Teaching Resources

The fifth variable focused on the presence and use of DBE workbooks, textbooks and other audio-visual learning and teaching resources written in English used during the lesson. Observations made are presented in Table C.8.

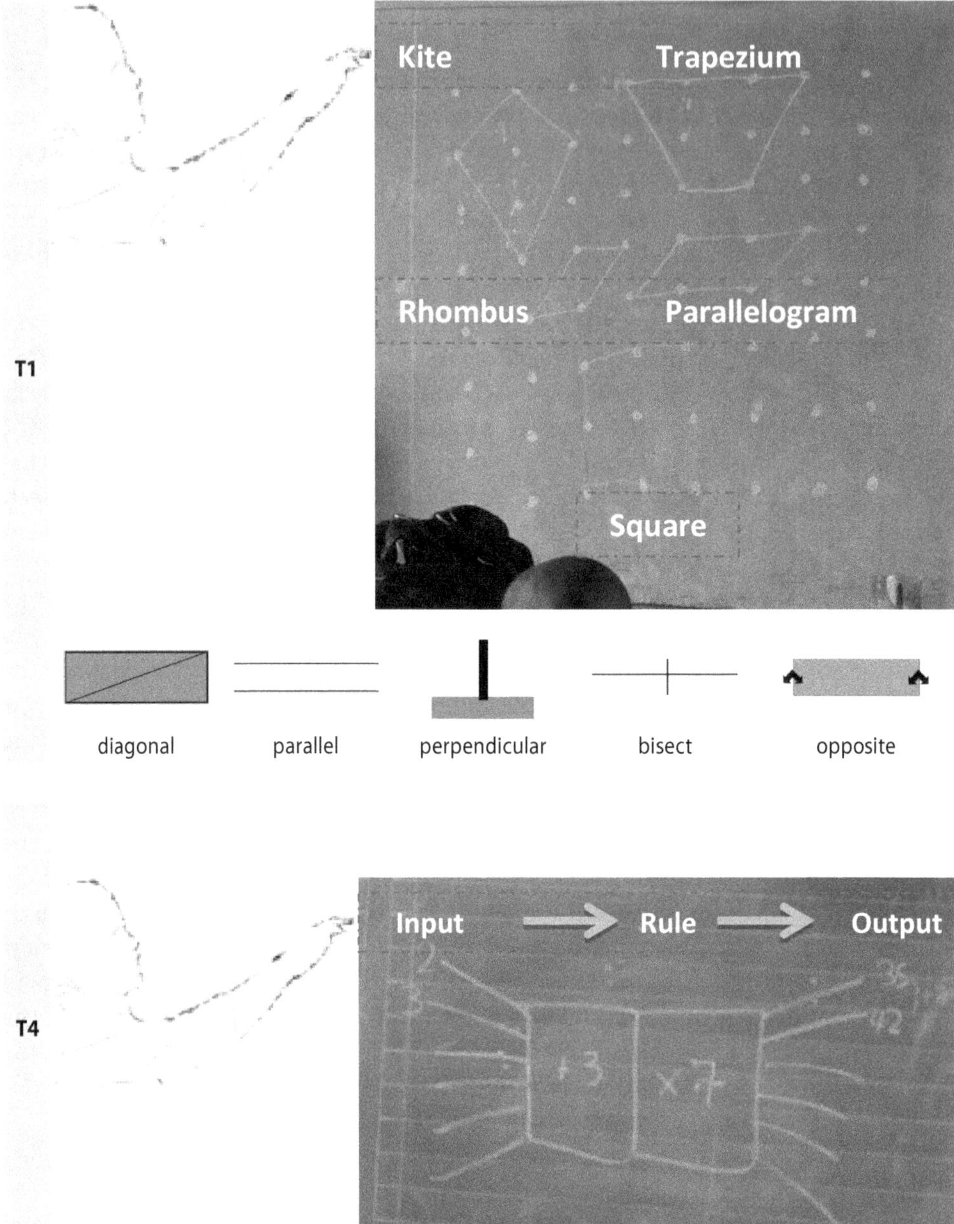

Figure C.9: Examples of text that could have been written on the chalkboard to improve understanding of mathematical concepts in T1's and T4's classrooms

Table C.8: Use of learning and teaching resources in IP mathematics lessons observed

Variable 5: Learning and teaching resources
Comments on any DBE workbooks, textbooks or other audio-visual learning and teaching resources written in English used during the lesson.

	(a)	*(b)*	*(c)*	**Comment**
T1	✗	✗	✓	▪ In groups, learners use straws and wool to construct different models of 2-dimensional shapes.
T2	✗	✓	✗	▪ The group work done by the learners is from a textbook. ▪ The teacher reads the questions together with the learners before learners attempt answering the questions in their groups.
T3	✗	✗	✓	▪ The teacher uses a chart to explain the new concepts. ▪ The chart used during the lesson seems to have been made specifically for teaching the day's lesson. ▪ There is no evidence that the chart has been on the classroom walls before. ▪ Towards the end of the lesson, learners attempt questions on a worksheet provided. 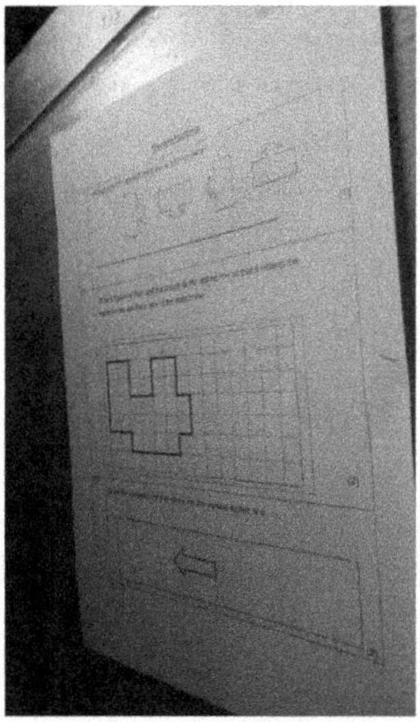 **Figure C.10:** Learner worksheet on Transformations in T3
T4	✓	✗	✗	▪ The work done by the learners is from the DBE workbooks.
T5	✗	✗	✗	▪ There is no textbook or any form of written text for use throughout the lesson. ▪ Learners follow what is written by the teacher on the chalkboard.

Key:
a = DBE Workbooks
b = Textbooks
c = Audio-visual learning and teaching material

Pedagogical implications: Learning and Teaching Resources

As indicated in Table C.8, T1 used audio-visual teaching and learning resources to engage learners in the practical activity of constructing models of two-dimensional shapes using straws and wool. The teacher is highly commended for drawing on this kind of creativity. After constructing models in groups, learners presented their work to the class by describing the models they had made; thus, the activity prompted learners to use mathematical language in their class presentations. Since the lesson was practical, there was no reference to text-based teaching and learning resources, except for the names of the shapes that were written by the teacher on the chalkboard. Perhaps in the following lesson, consolidating content learnt practically, the teacher used text-based teaching and learning resources.

T2's lesson revolved around a textbook written in English. All the activities done by the learners were from the particular textbook. The teacher read through the textbook instructions together with the learners. Although this promotes chorus answers, the teacher was at least modelling the correct pronunciation of the words written in the book. The teacher could also have asked different learners to take turns to read different sentences and correct their pronunciation if needs be. Although time consuming, in this way the teacher would have been able to identify learners who struggle to read and provide them with the relevant support. Such learners would be difficult to identify when the whole class reads the same text at once.

T3 used a chart to demonstrate different types of transformations. Although the teacher is commended for the efforts in preparing the chart, there was no evidence of any other charts displayed on the classroom walls. Text on the chart was written in English and the chart could possibly have been prepared for the observation session. In addition to the chart, a worksheet on transformations was provided to learners towards the end of the lesson.

T4's lesson on input and output machines was extracted from a DBE workbook written in English. The DBE workbooks are provided to schools to supplement textbooks and other teaching and learning resources available in schools. Different authorities have different views regarding the DBE workbooks; however, in under-resourced schools without textbooks, the DBE workbooks provide the absolute minimum form of text that can be used for teaching and learning purposes. What remained unestablished was whether T4 had access to other textbooks to supplement the workbooks.

T5's lesson on multiplication did not make reference to any form of written text whatsoever. The lesson was a typical ***'chalk-and-talk'*** lesson and the learners had to passively listen to the teacher throughout. The teacher did not even have any textbook at hand and at the end of the lesson learners were not given any homework to practise what had been taught during the lesson. There were no charts in the classroom and it was not clear whether the school had received any DBE workbooks, even though these are provided to all schools free of charge by the Department of Basic Education. Or was it sheer lack of preparation by the teacher?

Classroom Environment

The sixth and last variable was linked to Variable 5 above and focused on the presence or absence of charts/word walls around the classroom which indicated that mathematics is taught in English. Observations are presented in Table C.9.

Table C.9: Classroom environment in IP mathematics lessons observed

	Variable 6: Classroom environment *Are there any charts/word walls around the classroom which indicate that mathematics is taught in English?*	
	Yes / No	**Comment**
T1	Yes	▪ There are several charts on the classroom walls.  **Figure C.11:** Some of the charts displayed on the walls in T1's classroom
T2	Yes	▪ There is only one isolated chart on multiplication tables in the entire classroom.  **Figure C.12:** A chart on multiplication tables in T2's classroom
T3	Yes	▪ There are very few charts in the classroom on different subjects, and among these few charts, two of them are on mathematics. 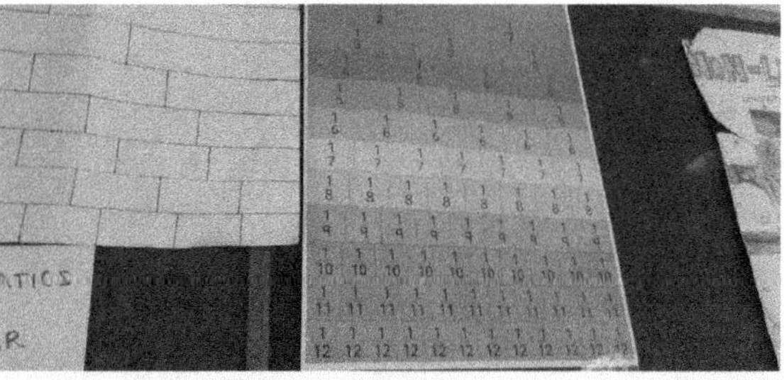 **Figure C.13:** A fraction wall chart in T3's classroom
T4	No	▪ Classroom walls are bare; there are no charts.
T5	No	▪ Classroom walls are bare; there are no charts.

Pedagogical implications: Classroom Environment

As illustrated in Table C.9, the classroom environment in the different classes observed varied, ranging from many charts to no charts at all. T1's classroom had several charts on the walls, which created a text-rich environment for learners. The charts were for different subjects, including mathematics, and they were all written in English. T2's classroom had only one chart on mathematical tables and there is a possibility that this single chart had been displayed in class in preparation for the observation session. T3's classroom had a few charts on the walls. The charts were on different subjects, and among these, two were on mathematical concepts. Both mathematics charts represented the fraction wall; the only difference was that one was 'handmade', while the other was printed on glossy paper. Thus, the two charts were presenting exactly the same information. T4's and T5's classrooms were bare; in the absence of learners and the chalkboard, the classrooms would have resembled any other ordinary rooms.

It is a norm that teachers use charts as (basic) teaching and learning resources, and that teachers usually receive charts as corporate gifts and memorabilia from textbook sellers or during CPTD workshops and conferences. Therefore, the researcher was expecting to see some charts in the observed classrooms. The absence of charts in classrooms indicates that learners have very limited access to written texts within their learning environment. The teacher does not have to be the one who makes charts to be displayed in the classroom; learners' work which they produce in different lessons can be celebrated by being displayed on the classroom walls. In this way learners are motivated, the teachers' work load is reduced and the classroom environment becomes text-rich. The practice of displaying learners' work in class does not have to be limited to IP mathematics only, but cuts across the curriculum.

It would be valuable to establish how well teachers are trained in the production of authentic, text-rich charts for use as learning and teaching resources in their classrooms – another crosscutting practice that is not exclusive to IP mathematics education.

Presentation of an IP mathematics lesson

The section on general comments was used to capture any noteworthy observations made in each of the classrooms observed. This section accommodated the differences in the classrooms, teachers and learners observed regarding the use of English as a LoLT. Observations are presented in Table C.10.

Pedagogical implications: Presentation of an IP mathematics lesson

As illustrated in Table C.10, generally some teachers use English as a medium of instruction and encourage their learners to do the same, while other teachers practise code switching and encourage learners to ask questions in their home language, such as T3 who prompted learners to ask questions ***ngeSintu***, which means ***home language***. The general comments noted in the different classrooms indicate the prevailing challenges in under-resourced schools. These challenges include the prevalence of

Table C.10: General comments in IP mathematics lessons observed

	General comments on linguistic practices in observed IP mathematics classrooms
T1	▪ The teacher moves around the classroom to distribute resources for constructing models of 2-dimensional shapes. ▪ When learners speak to each other, they use English.
T2	▪ The conclusion of the lesson was not done. ▪ There is very limited interaction amongst learners – learners do not discuss much amongst themselves, most of the work is done individually. ▪ The teacher does not walk around the classroom to check learners' work. ▪ Learners are seated in groups and the teacher explained that the seating arrangement is for facilitating learner interactions, though there isn't much discussion that takes place amongst learners.
T3	▪ Learner participation is above average. ▪ Learners are encouraged to move to the front of the classroom to showcase their answers. ▪ The lesson is relatively practical – learners have a chance to go to the chalkboard and illustrate their answers. ▪ The teacher solicits questions from the learners and prompts learners to ask questions ***ngeSintu*** – to ask questions in isiXhosa. ▪ The teacher promotes discovery learning throughout the lesson. ▪ Learners use examples from the classroom to explain the term 'transformation'.
T4	▪ Both the teacher and learners seem to be demotivated throughout the lesson.
T5	▪ Without textbooks or any reference materials; learners seem to be struggling to follow and keep up with what the teacher is explaining on the board. ▪ The teacher cannot get to the back of the classroom due to the over-crowdedness and does not seem to be concerned about how the learners are faring. ▪ The lesson is rather teacher-centred: dominated by the teacher talk, and the teacher does not pause to check if the learners are following. ▪ A quick check through the exercise books of learners closest to the researcher reveals some misspelt words in the learners' work which have not been corrected.

chanting and chorus answers, poor classroom organisation and large class sizes hindering learner-learner interaction as well as teacher-learner interaction, lack of adequate teaching and learning resources, text-poor classrooms and poor linguistic practices modelled by the teachers. The linguistic practices displayed by the observed teachers confirm data collected from other chapters in this text which indicate that teachers' proficiency in the language of instruction needs immediate attention. The section below presents, analyses and discusses the data obtained from the teacher interviews after the classroom observation.

Teacher Interviews

The teacher interview schedule was used to gather specific data on the following aspects: comparisons between the teacher's and respective learners' home languages; the language(s) spoken by the learners on arrival at school; teacher perceptions of whether English is a barrier in learners' understanding of IP mathematics; and the teacher's frequency of using code switching.

Responses to the teacher interviews are presented sequentially question by question in the section below. Table C.11 presents the responses to the first question on the teacher interview schedule.

Table C.11: Comparison between teacher and learner home languages

	Question 1: *(a) Is your mother tongue and that of your learners the same?* *(b) Is this an advantage or a disadvantage in your teaching IP mathematics English as LoLT?* *(c) Give a reason for your answer in (b) above.*		
	Yes / No	**Advantage / Disadvantage**	**Reason**
T1	Yes	Disadvantage	▪ In FP learners use isiXhosa and in IP they switch to English, so most of the time I have to explain, for slow learners I have to teach them English first.
T2	Yes	Disadvantage	▪ Because they cannot switch to English when they reach IP.
T3	Yes	Disadvantage	▪ Because learners read numbers in their home language.
T4	Yes	Disadvantage	▪ Learners are unable to understand English and mathematics terminology.
T5	Yes	Disadvantage	▪ Sometimes English is difficult to understand for learners.

As illustrated in Table C.11, in response to part (a) of the question, all the interviewees indicated that their mother tongue is the same as that of their learners. In response to part (b) of the question, all the interviewees also indicated that sharing the same mother tongue with their learners is a disadvantage. On reflection of these answers which state that it is a disadvantage to share the same language with the learners, the researcher acknowledges interplay of the Hawthorne Effect (Mouton, 2001) on the participants, based on the following facts. Firstly, the interviewees could have imagined this as the kind of response the researcher was anticipating; secondly, the interviewees were oblivious of the fact that the home language that they shared with their learners could be a resource that could be used effectively in IP mathematics instruction; and thirdly the interviewees did not receive comprehensive instruction on using English to teach IP mathematics. In response to part (c) of the question, the interviewees provided reasons that confirmed the above comments and that they were encountering numerous challenges in handling ELLs while teaching mathematics. Also, as noted in earlier sections on the quantitative data collection, the interviewees found problems with the learners' English-language competency and none whatsoever with their own.

Having established whether the teachers' and learners' home languages were the same or not, it was important to document the different home languages spoken by the learners on arrival at school. Table C.12 presents the responses to the second question on the teacher interview schedule.

Table C.12: Learner linguistic profiles on arrival at school

	Question 2: On arrival at school, what home language(s) do the learners already speak? (You may mention more than one).			
	English	**Afrikaans**	**isiXhosa**	**Other (specify)**
T1	✗	✗	✓	–
T2	✗	✗	✓	–
T3	✗	✗	✓	–

	Question 2: On arrival at school, what home language(s) do the learners already speak? (You may mention more than one).			
	English	**Afrikaans**	**isiXhosa**	**Other (specify)**
T4	✗	✗	✓	–
T5	✗	✗	✓	–

As indicated in Table C.12, all five respondents indicated that when the learners arrived at school, the only language they could speak proficiently was isiXhosa. The information presented was expected, as the teachers are based in the Eastern Cape Province's under-resourced schools where learners do not have much exposure to languages other than their home language, isiXhosa. The information presented also confirmed the data gathered through the quantitative data collection.

Having established a general overview of the learner linguistic profiles, interviewees were requested to state their opinions on whether English is a barrier to their learners' understanding of the mathematical concepts taught in class. Table C.13 presents the responses to the third question on the teacher interview schedule.

Table C.13: Teacher perceptions of English as a barrier in learners' understanding of IP mathematics

	Question 3: Would you say English is a barrier to IP learners' understanding of mathematics?	
	Yes / No	**Reason**
T1	No	▪ Not to all the learners, to some learners it is a barrier because they are not familiar with the language. At the back of the learners' exercise books, I always ask learners to provide space for new vocabulary and dictionary words.
T2	Yes	▪ Because they start school in Grade R – 3 using their home language when learning.
T3	Yes	▪ Because leaners cannot read and write numbers in English; e.g. *'zimbini'* in isiXhosa instead of 'two'.
T4	Yes	▪ It is a barrier since they fail to understand the symbol and notation of mathematics.
T5	Yes	▪ They are not good in English.

As indicated in Table C.13, the majority of the interviewees indicated that they perceive English to be a barrier to IP learners' understanding of mathematics. Only one interviewee indicated that English is not a barrier to some of the learners. This particular interviewee, who possibly because of extensive experience in teaching IP mathematics, alluded to a strategy used to ensure that English is not a barrier during mathematics instruction.

> *'At the back of the learners' exercise books I always ask learners to provide space for new vocabulary and dictionary words'.*

Although the listing of new vocabulary and dictionary words only contributes to the compilation of passive vocabulary, as noted by Hinkel (2007), the teacher respondent is commended for at least trying to help learners better understand mathematics by ensuring that English is not a barrier in the teaching and learning process, compared to other respondents who merely acknowledge that English is a barrier and do nothing much to assist the learners.

For the interviewees who indicated that English is a barrier to IP mathematics instruction, challenges related to the switch from mother-tongue instruction at FP to English instruction at IP were cited. As stated in earlier sections, although the home language for learners based in rural schools is isiXhosa, the language of instruction is English. Many of these learners are not frequently exposed to English, which becomes a challenge for them because their language proficiency and level of reading comprehension play an important role when solving mathematics word problems. Reynders, (2014) discovered that language barriers for learners who are not taught in their mother tongue lead to misunderstanding regarding mathematical word problems.

Frequency in using code switching when teaching mathematics lessons

Having established teacher perceptions on whether English is a barrier in learners' understanding of IP mathematical concepts and procedures, interviewees were requested to rate their frequency in using code switching during instruction. Table C.14 presents the responses to the fourth question on the teacher interview schedule.

Table C.14: Teacher frequency on using code switching in IP mathematics lessons

Question 4:
(a) Would you rate your frequency in using code switching when teaching mathematics lessons as never, sometimes, always?
(b) Give a reason to your answer in (a) above.

	Never	Sometimes	Always	Reason
T1		✓		▪ Even though I explain in English, learners would not understand. I give a new word, I give its meaning in isiXhosa without explaining further because I want the learners to familiarise themselves with the maths concepts in English.
T2		✓		▪ Because they do not understand English.
T3		✓		▪ Because learners will be unable to follow instructions during the lesson.
T4			✓	▪ To make them understand mathematical language.
T5		✓		▪ To help the learners.

Pedagogical Implications: frequency in using code switching when teaching mathematics lessons

When posing the question, the researcher briefly explained what the process of code switching entails to provide interviewees with clarity on the question. The explanation was given in such a way that it would not influence the interviewees' responses. As indicated in Table C.14, the majority of the respondents indicated that they 'sometimes' use code switching during IP mathematics instruction, while one interviewee indicated that he 'always' uses code switching. The reasons cited in support of the responses presented indicate that the teachers code switch in order to help learners better understand the mathematics concepts, since the learners are not proficient in English. Noteworthy is **T1**'s response highlighted below:

> *'I give a new word, I give its meaning in isiXhosa without explaining further because I want the learners to familiarise themselves with the maths concepts in English'.*

In this response the interviewee indicated that, although they use code switching to help the learners understand the mathematical concepts and procedures, the code switching is done bearing in mind that the LiEP requires instruction to be done in English. As such, the teacher only provides the meaning of the new word in isiXhosa, but does not go further to explain the meaning in isiXhosa as way of encouraging the learners to master the English version of the new word. Although the interviewee does not categorically state that this is a systematic way of using code switching, the way the whole process is done indicates a systematic use of the practice. Again, considering the fact that the interviewee is an experienced IP mathematics teacher, it can be inferred that the practice emerged out of a number of trial-and-error methods, hence the need to introduce systematic strategies of code switching and translanguaging during and after initial teacher education curricula.

DISCUSSION OF FINDINGS

The section above presented, analysed and discussed data gathered through the classroom observations conducted in the sampled under-resourced schools the Eastern Cape Province of South Africa. Data from the classroom observations indicated relatively poor linguistic practices, ranging from teachers speaking in grammatically wrong English sentences; teachers code switching haphazardly; teachers encouraging learners to speak or ask questions in their mother tongue, isiXhosa, in order to promote teacher-learner interaction; lack of text-based teaching and learning resources such as textbooks and charts; as well as uncorrected wrong spellings in learner exercise books possibly because of the extremely large class sizes. Although teachers may not have control over the availability of resources such as textbooks and dictionaries, the relative absence of charts and word walls which can be made by the teachers with the help of their learners was a particular cause for concern (Tshuma, 2017). The data reflected text-deficient classrooms, teachers' inconsistence use of English, the prescribed LoLT as well as inconsistency in the use of code switching and translanguaging strategies meant to improve learners' understanding of mathematical concepts, all of which contribute to poor linguistic practices in IP mathematics classrooms.

RECOMMENDATIONS

To improve the quality of linguistic practices in under-resourced mathematics classrooms, the following are recommended:

- In addition to providing computers and computer software, mathematics kits, radio cassettes and tape recorders, DBE, in partnership with corporates and civil organisations, should take responsibility in providing ***text-rich*** teaching and learning resources such as charts, word walls and bilingual mathematics dictionaries which are vital teaching and learning resources in primary schools effective for mathematics instruction;

- Coordinated and effective use of semiotic resources including physical manipulatives (as discussed in Chapter 3) would add value to the teaching and learning of mathematics;
- Findings reveal that the majority of teachers in under-resourced schools are not confident enough to implement effective linguistic practices in mathematics instruction. Therefore, in addition to prioritising literacy courses for pre-service teachers, university teacher education departments should also upskill pre-service teachers on how to develop learner linguistic practices for meaningful mathematics instruction; and
- Similar studies should be conducted in other under-resourced regions of the country as well as in well-resourced classrooms to establish whether the findings apply to other areas.

CONCLUSION

This case analysed IP mathematics teachers' classroom linguistic practices in the multilingual under-resourced settings of the Eastern Cape Province. South Africa advocates a single, democratic education system with equal education opportunities for all learners without recognising under-resourced schools (such as those in the ECDoE) as a separate category with unique needs. However, in under-resourced schools the learners' contexts and needs differ from those in urban and peri-urban schools. Du Plessis (2014: 1109) notes that "many rural areas are characterized by various factors that negatively influence the delivery of quality education". As such, the socio-economic realities of under-developed schools disadvantage learners and the classroom observations conducted in this case confirmed that the quality of teaching, particularly focusing on the linguistic practices, is compromised. The few semiotic resources used in some classrooms were not fully exploited to enhance meaning making (see Chapter 3). Isolated conditions in under-resourced communities do not attract highly qualified teachers; in this case, classes are taught by teachers not proficient in English, the prescribed language of instruction at IP level.

CHAPTER 8

MULTILINGUAL BASED BILINGUAL EDUCATION FOR MATHEMATICS IN THE EASTERN CAPE PROVINCE

INTRODUCTION

Does a learner's mother tongue play any significant role in the learner's primary mathematics education?

Learners' home language plays a major role in primary schooling, particularly in the foundation and intermediate phases. Most learners come to school with their home language as the only instrument in which they can express themselves. The learners' sociocultural background incorporating indigenous knowledge, or local knowledge, is directly linked to their home language. As a result of the latter, the use of the learners' home language can also shorten the amount of time which learners need in order to cognitively move from what they currently know to what they need or are ready to know. This cognitive distance is referred to as the Zone of Proximal Development (ZPD) of the child (Vygotsky, 1978).

Many recent studies have shown that most learners who speak African languages start failing mathematics when they reach the higher phase levels; and there are several speculations about the causes of this failure. The higher phase levels of education tend to point a finger at the lower phase levels and vice versa. Afrikaans schools usually perform better because learners are taught in their mother tongue from Grade R (Foundation Phase) to Matric (Further Education and Training). This chapter seeks to understand how mathematics is taught in isiXhosa in the intermediate phase, and to show how the learners' home language, isiXhosa, can be used alongside English to enhance conceptual understanding in science and mathematics. It investigates how dual-medium education models could be established using the concept of mother-tongue-based bilingual education (MTBBE).

In South Africa, learners, particularly those who speak African languages, struggle to deal with word problems and with fractions and sums in which geometry has to be used to calculate area. In general, learners experience substantial problems in communicating their answers in the language of assessment (usually English). Research has revealed that the vast majority of Grade 6 learners in the Western Cape Province have not even mastered the literacy and numeracy levels expected of Grade 4 learners.

Most textbooks and other teaching resources, particularly in mathematics and science, are still written mainly in English and most teachers are expected to teach in English. However, most teachers in Xhosa-dominant schools in the Western Cape Province use isiXhosa to teach or engage in code mixing (the mixing of two or more languages or language varieties in speech) and code switching in order to be understood by their learners. Speakers use code mixing when they are fluent in both languages.

The Western Cape Department of Education's (WCED) Literacy and Numeracy Strategy (LitNum, 2006), and the Language Transformation Plan (LTP, 2007) sought to address these challenges by promoting the use of learners' mother tongue in the classroom, wherever practicable, at least until Grade 6.

WHAT IS MOTHER-TONGUE EDUCATION?

There has been an on-going debate globally about the best terms to describe education which is based on the language most familiar to the learner and which aims to ensure strong competency in other languages as well. The result was the coinage of an umbrella term 'mother-tongue-based' programmes encompassing varieties of instruction primarily based on the mother tongue. The terms listed below are used throughout this chapter and they are explained against the backdrop of the future of mother-tongue education in South Africa.

A person's *mother tongue* is the main language used constantly from birth to interact and communicate with a child by their family. If more than one language is used in this way throughout childhood, a child can be considered bilingual or multilingual (Pinnock, 2009). In the South African context one language is usually used more frequently than the other(s). The mother tongue is also referred to as the ***first language*** or home language, and these three terms are used interchangeably in this chapter. *Mother-tongue-based education* is based on, and begins teaching in, the language used by the learner at home since birth (Pinnock, 2009). *Mother-tongue-based multilingual education* is learner-centred, active basic education which starts in the mother tongue and gradually introduces one or more languages in a structured way, linked to the children's **understanding in their first language or mother tongue**. If only one other language is gradually introduced, it is called *mother-tongue-based bilingual education* (Pinnock, 2009). *Mother-tongue-based programmes* involve teaching and learning conducted in the learner's first language, usually with a gradual transition to a second language or foreign language; learners have the opportunity to learn core concepts primarily in a familiar language and later learn the vocabulary for those concepts in a new language. Mother-tongue education is especially beneficial in Early Childhood Development (ECD) programmes, pre-school and the early grades (South & Lall, 2016). In the South African context the reception year of the ECD programmes and the first three grades in the ordinary school education system are called the Foundation Phase (FP).

SOUTH AFRICA'S MOTHER-TONGUE POLICY

South Africa is one of a number of countries in the world with high levels of linguistic diversity. The 11 official languages which are stipulated and recognised in the country's language policy include nine indigenous languages (isiZulu, isiXhosa, isiNdebele, Sepedi, Setswana, Sesotho, SiSwati, Xitsonga, Tshivenda) and two languages with European roots (English and Afrikaans). Afrikaans and English are the only languages with a developed academic literature and in which it is possible to write school leaving examinations. The country's Constitution stipulates that everyone has the right to receive education in the official languages of their choice in public educational institutions where that education is reasonably practicable. The current Language in Education Policy (LiEP) encourages the use of mother-tongue instruction during the FP, followed by a transition to English or Afrikaans as from the IP onwards. However, the LiEP allows School Governing Bodies (SGB) to make the final decisions.

> '... the governing body must stipulate how the school will promote multilingualism through using more than one language of learning and teaching, and/or by offering additional languages as fully-fledged subjects, and/or applying special immersion or language maintenance programmes...'
>
> DoE, (1997)

The country's current mother-tongue education policy needs to be viewed in the light of the above-mentioned policies.

WHAT IS MTBBE?

Mother-tongue-based bilingual education (MTBBE) is, in the South African context, a more persuasive and more easily comprehensible rendering of the meaning of 'additive multilingualism'. MTBBE entails the following:

- using the mother-tongue (home language) of the learner as a *formative language of learning and teaching (LoLT)* from Day 1 in Grade R or Grade 1 up to and including the last day of the school year in Grade 6;
- introducing the first additional language (FAL) as a subject as soon as possible in the foundation phase, including Grade R;
- assuming that a dual-medium approach is preferred by the parents or guardians, gradually using the FAL as a *supportive LoLT* as and when the children have adequate competence; and
- ideally, using mother tongue/home language + FAL as *complementary LoLTs* at a 50:50 level by the end of Grade 6. Normally, however, other permutations of this dual-medium model can be expected to prevail because of teachers' limited language proficiency and subject knowledge as well as other constraints of a material or managerial nature.

MTBBE is a more complex term than any of its predecessors, combining elements of a pedagogic or learner-oriented dimension (mother-tongue based) with a term that carries a heavy load (bilingual education). The extended definition is quite specific about the favoured language distribution model.

THE IMPACT OF MOTHER-TONGUE EDUCATION: MTBBE IN THE EASTERN CAPE PROVINCE

Although quite a number of the provinces in the country have below-average achievements in school-leaving assessments, the result of several factors including language of instruction, this chapter makes particular reference to the Eastern Cape since the province has been lagging behind academically for several years: in 2011 the province was placed under administration and in 2014 Mt Frere district was the lowest-performing district nationally (DBE, 2015). Furthermore, in 2012 the Eastern Cape pioneered the implementation of Mother Tongue Based Bilingual Education (MTBBE) in one district and by 2018, 10 least-performing schools in each district in the Eastern Cape were implementing the programme (Isaac, 2018)[1]. By September 2020, the MTBBE initiative saw 61,465 matrics from quintile 1-3 schools (the under resourced schools) in the Eastern Cape Province; writing their preliminary mathematics examinations in isiXhosa. Below are some perspectives regarding the implementation of MTBBE in the Eastern Cape Province.

According to Govender[2] (2015), a unique project involving learners in 81 schools in rural Cofimvaba District in the ECDoE is paying huge dividends. Learners from Grades 4 to 7 in these schools are studying mathematics, natural sciences and technology in isiXhosa, their mother tongue. The practice of teaching learners in their mother tongue is largely confined to learners who are in Grades 1 to 3 (DoE, 1997). However, the ECDoE supports the initiative, which is known as mother-tongue-based bilingual education. In 2014 Grade 6 learners at Luzuko Junior Secondary School scored 100% in mathematics in the Annual National Assessment (ANA), compared to a dismal 40% obtained in 2012. The school does not have a library or science laboratory and learners still use pit latrines. The school principal, Nkosinathi Mvumbi, said learners' results had improved dramatically in the two subjects since being taught in their mother tongue.

> Previously, their results were really poor in those subjects because they were taught in English. Now we are getting marvellous results.
>
> Mvumbi (2015)[3]

Learners are still being taught their other subjects using English as the LoLT.

> English is not their mother tongue and that is why they do so badly. They understand mathematical concepts better when they are taught in Xhosa.
>
> Mvumbi (2015)

Mvumbi was hoping that learners could be taught all their subjects in their mother tongue. Mayizole Skama, Comfimvaba District Director, expressed the view that

1 Isaac, J. in *Mail and Guardian*, 2018 – 09 – 28.
2 Govender, P. in *Sunday Times*, 2015 – 05 – 17.
3 Mvumbi, N. in *Sunday Times*, 2015 – 05 – 17.

mother-tongue instruction was having a positive impact on learners' academic performance in the district.

> There is no doubt that it has helped improve our Annual National Assessment results.
>
> Skama (2015)[4]

Contrary to these assertions, Sepeng (2015) suggests that isiXhosa does have a place alongside English, playing a dual role of language of teaching and learning mathematics in multilingual classrooms (Sepeng, 2013, 2014a). While investigating the impact of isiXhosa as LoLT in multilingual mathematics classrooms in four Port Elizabeth Township schools in the ECDoE, Sepeng (2015) made two key findings: firstly, both English and isiXhosa are used together and secondly both learners and teachers lacked competence in their home language. As such, although the teachers acknowledged the dilemma of using English as LoLT, because of the difficulties encountered when persuading learners to speak English during mathematics lessons, the teachers preferred and continued to use English as the language of instruction (Barkhuizen, 2002; Brock-Utne, 2002).

These key findings shed significant light on the questions posed: if both learners and teachers lack competence in their home language, it may be inferred that their competency in the LoLT, which is a second language to them, is even more dire. In addition, if the teachers prefer and continue to use English as LoLT, they should be supported to improve their own English-language competency and to be upskilled in effectively using different languages as resources in mathematics classrooms for their own as well as the learners' benefit.

HEGEMONY OF ENGLISH IN SOCIETY VS MTBBE

Although there are 11 official languages in South Africa,[5] Afrikaans and English are the only languages with a developed academic literature and in which it is possible to write the external assessments. There are a number of challenges associated with this hegemony of English, the most important being that not all South African learners have equal access to English in their schools, leading to a situation where those accepted to enter university with a limited English proficiency exhibit lower confidence and participation levels, and ultimately lower performance and success rates in their chosen degree course (Hurst & Mona, 2017). According to StatsSA (2012), only about 23% of South Africans speak Afrikaans or English as their first language. In order to achieve educational and hence labour market success, the majority of South African learners therefore need to become fluent in either English or Afrikaans as their second language.

In reality, the vast majority choose to learn English rather than Afrikaans as the second language, given its status as a global language. English language

4 Skama, M. in *Sunday Times*, 2015 – 05 – 17.

5 The 11 official languages include nine indigenous languages (isiZulu, isiXhosa, isiNdebele, Sepedi, Setswana, Sesotho, SiSwati, Xitsonga, Tshivenda) and two languages with European roots (English and Afrikaans).

proficiency therefore influences life chances through its influence on educational success. However, Casale and Posel (2011) demonstrate that English proficiency also improves labour market returns directly. Using a traditional earnings function methodology controlling for an individual's level of education, they find a significant wage premium for black South Africans associated with being able to read and write English fluently.

This situation presents a difficult policy question to countries like South Africa: when and how should the teaching of English be introduced in schools, and when and how should a transition to English as the primary language of instruction in non-language subjects occur? South African legislation and education policy do not prescribe which of the 11 official languages should be used, but leaves the choice of language of instruction to School Governing Bodies (SGB)s, which are comprised of a parent majority as well as the school principal, several staff members and, in the case of secondary schools, learners (DoE, 1997).

Currently, most schools in which the majority of learners are not English- or Afrikaans-speaking opt to use First Language in Grades 1, 2 and 3, and then transition to English as the language of instruction in Grade 4. This approach, though not compulsory in policy, has been encouraged by the national and provincial departments of education. Some schools, however, have chosen to go "straight-for-English" as the language of instruction from the first grade. The Curriculum and Assessment Policy Statements (CAPS) also prescribe that English should be introduced and taught from Grade 1 in all schools (DBE, 2011). Consequently, all schools should have some English being taught from the first grade, but for some schools English is also the language of instruction from Grade 1, whereas in most schools this is only the case from the fourth grade. Despite some reasons why mother-tongue education is an educationally sound policy, the majority of South Africans prefer English and not their home language as LoLT. Reasons perpetuating this belief and practice are explored below.

There is a lack of suitable textbooks and material for the specialised language needs of ELLs (Lemmer, 1995: 92; Chick, 1992: 284; Reagan, 1985: 76; de Wet, 2002). This empirically demonstrated notion is supported by Indigenous African Language (IAL) teachers. In a press interview (Jones, 2001: 1) IAL teachers accused the South African government of not making African language textbooks available. Initiatives to produce texts for learners in IALs have been cautious and have had mixed results. Gough (1994: 10 – 11) however, is of the opinion that the development of learning material in IALs is not an insurmountable problem. In his view, there are piles of IAL learning materials gathering dust somewhere in Department of Education storerooms. He suggests that publishers and teachers should use these textbooks (which probably display the influence of the apartheid era) as a point of departure for preparing new publications.

The lack of IAL learning material does not mean that teachers do not use IAL in their classrooms. African-language-speaking teachers often use their bilingual or multilingual competencies as they grapple to interpret a syllabus resourced in English to African-language-speaking children (Bloch & Edwards, 1998).

In the light of the insistence of learners on English as LoLT, cognisance must be taken of De Wet's (2002: 119) and Lemmer's (1995: 91) observations that teachers in traditional black schools often lack the English proficiency that is necessary for effective teaching. Teachers do not have the knowledge and skills to support English-language learning and to teach literacy skills across the entire curriculum. A large number of African teachers are educated in "an English dialect" (De Wet, 2002: 119) and this may have negative consequences for the learners – learners often imitate their role models' (teachers') (wrong) pronunciation, grammar and vocabulary.

If the most important obstacle in the use of IALs as LoLT, namely a lack of educational material, is compared with the problems facing learners who use English, and not their home language as LoLT, the former problem should not be insurmountable (Plüddemann, 2010).

PEDAGOGICAL IMPLICATIONS: MTBBE

Despite the on-going debates and substantial progress achieved over the past two decades, South Africa still faces major challenges regarding mother-tongue education. While these debates for and/or against mother-tongue education rage on, there are learners in South African classrooms today who need to be catered for. What can be done to cater for these learners and for future generations to ensure that they are not linguistically disadvantaged?

- **Aligning the language of instruction in teacher education institutions with the language of instruction in schools**

Aligning the languages of instruction in teacher education institutions with the languages of instruction in the ordinary school education system is a critical element that is supposed to underpin MTBBE in South Africa. MTBBE assumes that the language of instruction in the ordinary school education system be the learner's first language (Alexander, 2006). At present most learners whose first language is an African language take up mother-tongue instruction from Grade R up to 3 before the unstructured, ***on-paper*** transition to using English as language of instruction in Grade 4. It is well documented that the (oral) use of the home language continues ***de facto***, as the transition to English as LoLT in the fourth year of schooling is premature in most under-resourced multilingual settings. Both teachers and learners often engage in a range of compensatory behaviours, including code switching, code mixing, rote learning, chanting, chorusing and safe talk to mask the absence of meaningful learning (Arthur 2001; Brock-Utne 2004; Alidou et al. 2006). It should therefore be acknowledged that to use a late-exit transition model to structurally and incrementally effect introducing an additional/second language as a language of instruction while systematically decreasing the use of the first language of instruction will help to ensure meaningful teaching and learning of mathematics (DoE, 2005). This acknowledgement should also inform how teachers' University Education departments are instructed. Specifically, teacher educators should demonstrate how the teacher should implement MTBBE in schools and handle the late-exit transition (Tshuma, 2017).

- **Material development process**

If MTBBE is fundamentally grounded on using the learners' first language, it goes without saying that the materials used in MTBBE classrooms should also be based on the learners' first languages. There is therefore a great need for developing relevant and/or adapting the existing teaching and learning resource materials to meet the requirements of MTBBE. Some of the material development processes that have been involved in MTBBE include:

- development of a few teaching and learning resources inadequately covering the subject content;
- translating existing materials into isiXhosa in order to form bilingual teaching materials;
- translating learning area work schedules and key technical terms for wall posters;
- conducting workshops to show the teachers how to plan bi- or unilingual lessons and develop appropriate materials; and
- guiding teachers on how to choose textbooks that have, among other things, a good balance of content and teaching methodology.

Although these processes are a positive start towards developing MTBBE-compliant teaching and learning resources, a lot more still needs to be done. The development of new materials and adaption of existing materials should go beyond merely translating key terminology to encompass the development of reading material that learners may use outside the MTBBE classrooms. Producing glossaries or resource packs made up of translated key mathematical terms undermines the whole teaching and learning process and reduces it to memorisation of key terms without understanding the background knowledge. Since mathematics has its own register, concepts and procedures that are usually found only within the mathematical context, there is a lot of investment that should be channeled towards the MTBBE-compliant teaching and learning resources (Plüddemann, 2010).

In general, there a need to standardise the development of MTBBE teaching and learning resources by publishers. The standardisation will avoid producing direct translations, which are not meaningful, and translated out of context – a product of content translating by linguists who are not mathematics subject specialists. The writers of most of the published textbooks are not speakers of an African language. It is essential that textbooks be originated in the mother tongue. Publishers should, therefore, engage dedicated subject specialists in their translation units working with experienced linguists.

▪ Policy: the umbrella body

First and foremost there is a need for a clear and concise policy on mother-tongue education. The current policy promotes mother-tongue education in the primary years and then mandates the SGBs to make the final decision in the later years. This democracy allowance would be plausible if SGB members were prepared adequately to make informed decisions that are not influenced by the aftermath of the Bantu Education System, which promoted mother-tongue education based on a narrow political agenda. Therefore, government should educate the masses on the benefits of delivering foundation skills to primary school children in their mother tongue (Le Cordeur, 2010). Government initiatives in educating the masses must begin with clear policy stipulations followed by active implementation of the stipulated policies.

Mismatch between policy and implementation

There is surely a mismatch between the declarations of the LiEP and what is being implemented in schools. The LiEP, which is informed by research, declares that learners need to receive their formative education in their mother tongue, yet parents continue to enrol their children in English immersion schools.

> '...in selecting a strategy to have their children learn English, they demonstrably take the worst route, namely to choose English as the language of instruction from as early a grade as possible.'
>
> Weideman and Van Rensburg (2000: 157)

Why is there a mismatch? Why are parents not heeding the declarations of the country's LiEP? To reach a consensus, the following are ***non-negotiables*** as it would be difficult to implement mother-tongue instruction in a country where African languages have low economic value and status, and are rarely used in public life:

- Using African languages in school examination boards and higher institutions of education;
- Using African languages in support of the Eurocentric pedagogy in emulation of powerful economies with powerful educational systems primarily conceived in their national languages (Marnewick, 2015);
- Increasing awareness among parents and teachers of the benefits of mother-tongue instruction, so that they can make informed decisions regarding the best way for learners to achieve both English-language proficiency and a suitable level of cognitive development (Taylor & Coetzee, 2013; Le Cordeur, 2010);
- Having adequate teaching and learning resources in all the official languages in schools (including under-resourced schools) (Marnewick, 2015); and
- Using planned and purposeful multilingual teaching and translanguaging (Tshuma, 2017; Chikiwa, 2016; Sepeng, 2015), which allow African indigenous languages to support the main language of instruction.

Prioritising these ***non-negotiables*** within the South African education system would go a long way in promoting mother-tongue education.

Mother-tongue education and teacher education: the missing link

Chitera (2016) states that most of the language practices in multilingual classrooms that teachers produce cannot be traced back to their teacher education institutions. In the South African context these practices are manifested in two ways: FP pre-service teachers are taught in English, yet after the completion of their studies they are expected to teach in mother tongue; there is thus a considerable mismatch between teacher education and reality which requires urgent attention by teacher education curricula planners and institutions.

For teachers to be well-equipped foot soldiers of the education system they need to be well trained and the onus is on teacher educator institutions (Tshuma, 2017) to ensure that teachers are well trained to promote multilingual instruction in the classrooms. Fundamentally, teachers should cultivate a culture of using learner languages as a resource in the teaching and learning process (Van de Walt, 2016). If teachers are not well trained and supported during and after initial teacher education, they will not be able to deal with the diversity in their multilingual classrooms.

To counter the perceived inadequacies of some indigenous African languages which are not yet fully developed to support meaningful instruction, teachers can use audio-visual teaching and learning resources, gestures and realia to decipher African indigenous languages that:

- have no written systems;
- have very small populations of users; or
- are not widely understood.

However, the ability to do this does not come naturally to teachers, and hence should be part of the teacher education curricula, and best practices in this regard should be modelled by teacher educators during both the pre- and in-service teacher education programmes. Pinnock (2009) and Kooy and Hulshof (2010) suggest including the following benchmarks for teacher education that promote mother-tongue education:

- support and requirements for teachers to understand the language and cultural background of children;
- an understanding of language development (including the importance of mother-tongue education and how children learn a language and learn to read); and
- an understanding of the interdependence of mother-tongue education and second-language development and appropriate first- and second-language teaching.

Evidently, teachers cannot achieve these goals without the support and guidance of the teacher educators, teacher education curriculum developers and policy makers.

The explicit advantage of mother-tongue education is that it provides a firm foundation for learning other languages as well as mastering content knowledge taught in school and in higher education. Looking ahead to 2030, it is apparent that systematic changes in the current teacher education curricula will undoubtedly demonstrate the positive results of mother-tongue education. It is not enough to look only at school education or higher education without focusing on teacher education: teacher education is actually the link between the two.

CONCLUSION

This chapter has outlined the country's existing mother-tongue policy, the impact of MTBBE in the Eastern Cape Province, hegemony of English in Society as well as what the implications of mother-tongue education at present would have for learners, teachers, teacher educators, curriculum developers and policy makers. What is explicitly certain is that mother-tongue education produces cognitive rewards; however, South Africa's readiness to tap into these rewards in the near future remains uncertain. Inasmuch as the small-scale MTBBE project is greatly applauded for yielding results in the Eastern Cape Province, the balancing act lies in how well MTBBE manages to produce a curriculum that is not impoverished, resulting in the majority of the MTBBE learners being illiterate or innumerate.

It would appear that English immersion and early transition models will continue to be popular among South African parents and learners, appearing to be better options in comparison to mother-tongue education; and the status quo will remain and determine future trends until ***all the foot work*** has been done to ensure that the benefits of mother-tongue education are realised in practical terms. To better serve learners in the South African education system today, there is a need for genuine investment in, and commitment to good-quality mother-tongue-based multilingual education, delivered by linguistically empowered teachers who are well prepared during (and supported after) initial teacher education. South Africa needs to move beyond the dialogue on mother-tongue education and start producing teachers who understand and are able to implement mother-tongue education in schools.

> ***"... the critical nature and centrality of mother-tongue education requires finding defensible and shared ways to prepare teachers that reflect the changing nature, knowledge and practices of mother-tongue education..."***
>
> Hulshof & Kooy (2010: 742)

CHAPTER 9

MATHEMATICS TEACHER EDUCATION FOR MULTILINGUAL SETTINGS

INTRODUCTION

Are teacher education institutions preparing mathematics teachers well enough to serve multilingual learners?

As alluded to in earlier chapters, an education system is only as good as its teachers. This chapter reviews different studies with a focus on multilingual teacher education, since multilingualism is a reality in many classrooms. There is a vast array of research on multilingual teacher education; however, not much of it focuses on mathematics teacher education. This chapter provides an overview of practices in teacher education institutions in some European as well as some African countries that have policies targeting teachers who are intended to teach in multilingual classrooms. The chapter also reviews some programmes put in place by various countries in an attempt to prepare teachers for multilingual classrooms.

TEACHER EDUCATION: THE EUROPEAN LANDSCAPE

University teacher education departments' key responsibility is supporting and challenging the development of teachers for the future. However, current global practice reveals that teacher education is caught in the realms of change of school curricula, examination systems, and educational policies instead. Rather than taking a pre-emptive leadership role, teacher education institutions often appear to be reacting to the dynamic changes in education and challenging for their place in the change process. University departments of teacher education in some countries are concerned about how to sustain teacher education programmes when other stakeholders in education such as schools are taking responsibility for initial and continuing professional teacher education (Livingston, 2016).

The complexity and speed of changes in society signal the need for teacher educators and teacher education institutions to be more proactive and prominent in their contribution to the change process. New ways of understanding knowledge and how we learn, new and powerful technologies and new patterns of integration need more dynamic forms of teacher education. Teachers need a strong foundation of initial teacher education, but they also need to understand themselves as learners, ready to learn and adapt their practice throughout their careers, supported and challenged by a different range of teacher education opportunities. The section below presents ideals for teacher education based on research conducted in the following five European countries: England, Ireland, Spain, Norway and the Netherlands:

- England: Regarding inclusivity, it is important that teachers continue to learn and develop practice throughout their careers as learners' learning needs change over time.
- Ireland: There is a need to develop a shared understanding of what constitutes good practice between schools and university departments of teacher education.
- Ireland: There is a need to review and challenge traditional views on teaching methods.
- Spain: There is a need to draw on and reinforce mentoring in the professional development of in-service teachers for the purposes of better understanding the dynamic teaching profession. The only bait here is that the mentors should be champions who are willing to share best practices with recently graduated teachers.
- Norway: There is a need to integrate digital competency in education. More than 70% of teacher educators believe that pre-service teachers acquire the relevant digital competence through their teacher education programme while less than 50% of pre-service teachers believe that their education has taught them how to use technology in the classroom.
- Norway: Regarding in-service teachers, there is a need for teachers to continue their learning and develop their practice in a changing international context specifically in alignment with international assessment on education such as the Programme for International Student Assessment (PISA) and Trends in International Mathematics and Science Study (TIMSS). International policies have a positive impact on teaching, and they contribute to changes in the meaning of good teaching and the content and skills that are provided through education. In Sweden there is a convergence between transnational and national policy arenas regarding what is important about school mathematics. It is the teachers' role to compensate for different individual learner backgrounds and therefore a need to deepen teachers' knowledge of pedagogy and subject matter.
- The Netherlands: teacher collaboration is widely regarded as a powerful tool for teacher professional development. However, more needs to be done about the support structures needed in schools to enable teachers to work collaboratively in a way that leads to improvement in learners' learning.
- The Netherlands: teachers need more support in developing the ability and dispositions to work collaboratively. There is particular need for a continuum of teacher education from pre-service to in-service.
- The Netherlands: Inasmuch as teachers are individual learners, learning in collaboration with peers impacts on teachers' learning in different ways. Some teachers consider their concerns as challenges stimulating their learning, while others experienced their concerns as a reason to revert

> to previous teaching methods. Therefore, change involves both teachers' emotions, and their knowledge and skills as well as leadership and facilitation which are important in enabling teacher education.
>
> Adapted from Livingston (2016)

The perspectives above present ideals for teacher education in European countries which can be used for benchmarking current practice in teacher education in other parts of the world such as Africa. However, these perspectives do not focus much on multilingual settings, possibly because European countries are not as linguistically diverse as African countries. The section below presents perspectives on multilingual teacher education specifically focusing on the African continent. The countries reviewed include Burkina Faso, Ethiopia, Niger, Ghana and Malawi.

MULTILINGUAL TEACHER EDUCATION: THE AFRICAN LANDSCAPE

In Burkina Faso teachers who receive regular pedagogical support from linguists at the University of Ouagadougou and are familiar with the country's official languages used as LoLT in schools, teach in *Ecoles Bilingues* (Brock-Utne & Alidou, 2005; Chitera, 2009a). *Ecoles Bilingues* cater for learners aged 9 years and above, who have not had a chance of enrolling in formal primary schools. The learners are more mature and usually have fully developed home-languages skills before enrolling in schools classified as Ecoles Bilingues which provide instruction in two languages.

In Ethiopia the prescribed LoLT for primary school teachers being prepared to teach in the first four years of schooling is the same as the LoLT for the first four years of schooling in primary schools; however, this LoLT stipulation is not followed in actual practice (Mekonnen, 2005). There is a mismatch between the LoLT of primary education of the second cycle (Grades 5 – 8) and the LoLT of primary teacher education for the second cycle: the LoLT for Grades 5 – 8 is the mother tongue, but the LoLT being used in the teacher education institutions is English. Without the teacher educators implementing the policy on the ground, pre-service teachers graduate with little or no knowledge of how to teach in their local languages. Thus, mathematics teacher educators need to upskill pre-service teachers with methods to teach in home languages, since teaching in the learners' first language differs considerably from teaching in a second or third language.

In Niger there are significant numbers of untrained teachers and new graduates from secondary schools awaiting other employment opportunities (Traore, 2001; Benson, 2002; GTZ, 2005; Chitera, 2009a). Large numbers of practising teachers awaiting training are not found in Niger only, but also in other African countries and they usually have inadequate school-based support. Though enthusiastic, they need to be trained to teach effectively in the mother tongue and the respective official languages:

> Teacher's enthusiasm cannot substitute for qualification required for teaching in mother tongues and official languages. Many bilingual

> teachers face serious professional challenges. They may be able to speak the LoLT, but they have not mastered reading and writing in that language.
>
> GTZ (2005: 120)

In Ghana Addabor (1996) found that there was no teacher education in the mother tongue or in bilingual teaching methodology. However, Alidou and Brock-Utne (2005) reported the emergence of GTZ in the teacher education scenario working to strengthen teaching in local languages in many teacher education colleges in the country. Through the GTZ, teachers receive ongoing training for the classes they are going to teach later, and at the same time new teachers are educated in using the local languages for instruction.

In 1996 Chilora (2000) reported that the Malawi government invested significantly in teacher training programmes to assist teachers to cope with the implementation of the then LoLT, Chichewa. Teachers were trained in teaching in Chichewa as the LoLT. Textbooks were also produced in Chichewa with the exception of teachers' guides produced in English to accommodate teachers not fluent in Chichewa (Chilora, 2000). Teacher training colleges did not do much to help teachers cope with the implementation; therefore, mathematics teachers struggled to cope with the demands of LoLT when teaching mathematics in bi/multilingual classrooms. Teacher education programmes do not have standardised LoLT competency programmes.

Teaching behaviour is frequently moulded by prior educational experiences. Pre-service teachers bring into the programme their prior knowledge, beliefs and experiences, which affect their assimilation and construction of new knowledge; teacher educators are themselves products of their own prior experiences in traditional settings. Therefore, it is crucial to monitor and regulate what happens in primary teacher education institutions (Shiundu & Mohammed, 1996; Gay & Ryan, 1999).

MATHEMATICS TEACHER EDUCATION: THE SOUTH AFRICAN LANDSCAPE

In South Africa specifically; it is generally accepted that there is both an absolute shortage of teachers and a relative shortage of teachers qualified and competent enough to teach specific subjects or learning areas (primarily mathematics, the sciences, technology and languages, but also arts and culture, and economic and management sciences), in specific phases (especially but not only the FP), in specific languages (African languages in particular, and also sign language and Braille), in special needs schools, in early childhood development (ECD), and in rural and remote schools (DoE, 2011).

Figure 9.1 shows mathematics teachers' competencies compared to 13 other Eastern and Southern African countries and reveals that the competency of South Africa's Grade 6 (Intermediate Phase level) mathematics teachers is at the lower end of the spectrum. When looking at inequalities within South African schools, there are some inconsistencies. Whilst South African mathematics teachers in quintile five (the richest areas) schools compete with the average Kenyan mathematics teachers, the bottom three quintiles (approximately 60% of the population representing the generally under-resourced areas) perform slightly worse than Lesotho and Zambia and this is a cause of serious concern.

Figure 9.1: South African Mathematics Teachers' Competencies in Relation to Eastern and Southern African Counterparts (Spaull, 2013)

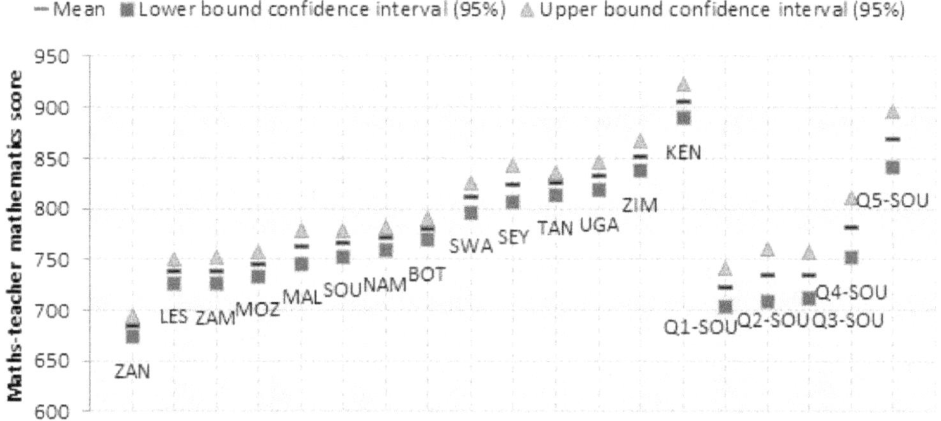

South Africa is an economic powerhouse compared to Lesotho and Zambia which are lower-income economies. Most of the countries with similar results to South Africa have a much lower per capita spending on education. Countries performing much better than South Africa such as Tanzania, Uganda and Zimbabwe have per capita GDPs which are a small fraction of South Africa's and also have high numbers of learners enrolled in low-fee private schools.

In South Africa, teacher education is located in the higher education sector and there is strong emphasis on the completion of degree studies. However, while the country already possesses a similar system of institutional providers, programmes and teacher development centres at various levels; the system is not adequately coordinated or integrated, and there is a need to improve the capacity, cooperation and reach of existing institutional providers and development centres to enhance the quality and relevance of both initial and continuing, as well as formal and informal, teacher education and development programmes (Parker, 2013).

It is important to note the historical context of teacher education in South Africa as this has directly impacted on the current capacity, quality and quantity of mathematics teachers in schools today.

TEACHER EDUCATION COLLEGES

In 1997, there were 109 public colleges of education in South Africa: 95 state and 5 private colleges providing contact teacher education, 8 state colleges and 3 private colleges providing distance education. The 109 teacher education colleges were well distributed nationally and catered for 150 380 pre-service teachers spread across all disciplines studying towards a four-year Diploma in Teacher Education, equivalent to a bachelor's degree (Hall, 1996). Towards the end of 1997, there was widespread uncertainty about the future of the 109 teacher education colleges resulting in the ventilation of different options: closure; rationalisation; conversion into polytechnics; conversion into community colleges; integration into university departments of education; and amalgamation or clustering. The uncertainty was mitigated by the report of the Technical Committee on the Incorporation of Colleges of Education into Higher Education published in December 1997.

Ten years later all of these colleges had ceased to exist as independent entities. Between 1994 and 2000 many colleges of education were closed, merged or incorporated into larger entities, as part of provincial rationalisation processes aimed at overcoming the educational inequalities of apartheid and reducing an identified over-supply of primary teachers. In December 2000, 25 colleges of education, including two distance-education colleges, were declared by the Minister of Education as subdivisions of various universities and technikons, with effect from 31 January 2001 (DoE, 2000). These 25 colleges were intended to function under the name and auspices of the particular university or technikon to which they were transferred along with their assets, plant and property. A 2008 survey (Parker, 2013), however, found that only 9 of the 25 were still listed on the books of, and being used by, a higher education institution. The remaining approximately 76 colleges were retained by the provinces, destined for the most part to continue to carry out education-related functions as campuses of FET colleges, teacher development institutes, education resource centres, schools and/or provincial education offices, with a small remainder being utilised by other government departments.

In 2013 the Department of Higher Education and Training (DHET) initiated the process of reopening[1] teacher education colleges as an attempt to work towards South Africa producing between 20,000 – 30,000 better qualified teachers every year. The first scheduled to be re-opened is the former Ndebele College Campus in Mpumalanga for Foundation Phase teacher education in 2013, and from 2014/2015 there are plans to open one former teacher education college each in KwaZulu-Natal and the Eastern Cape provinces. The process is still underway. Current updates on these colleges indicate that they are open to students for registration and courses in some of the academic modules have commenced, but no qualified teachers have yet been produced as the time period is still insufficient for that measurement. These colleges focus on pre-service teachers with none of them currently providing

1 http://www.sanews.gov.za "SA to re-open teacher training colleges" Accessed 25 January 2020.

any professional development or resource support to in-service teachers – this is only planned for the future.

On the other hand, private education institutions such as Embury Institute for Higher Learning and Two Oceans Graduate Institution have opened their doors to the public, offering 4-year degrees in education and some small-scale education institutions are offering Grade R teaching diplomas. As of 2018, university teacher education departments have progressively been offering 4-year degrees specialising in Foundation Phase Teaching (Grade R) for the purpose of improving the quality of Foundation Phase teaching and learning.

DEPARTMENTS OF MATHEMATICS EDUCATION IN SOUTH AFRICAN UNIVERSITIES

All universities in South Africa that have a Department of Education provide a 4-year Bachelor of Education degree, which prepares pre-service teachers. Figure 9.2 shows the distribution of the 25 South African universities.

Figure 9.2: South African universities offering degrees in teacher education

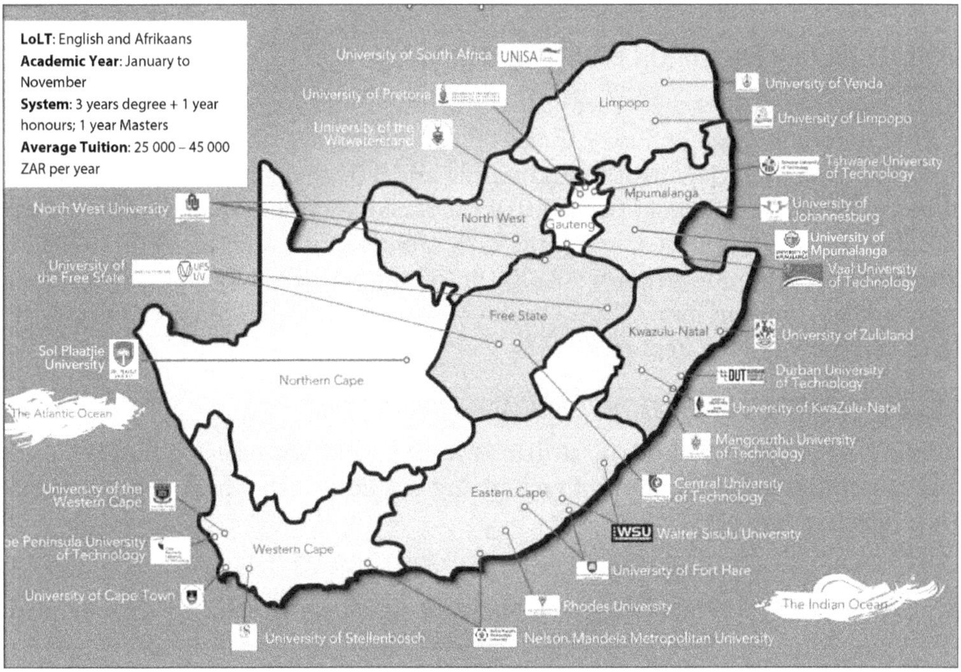

British Council (2013)

The University of Mpumalanga opened its doors in 2013 and Sol Plaatje University, which opened in 2014, are the latest additions to the South African universities. Some of the university departments extend their services on an ad-hoc basis to the professional development of in-service teachers; however, the interventions are not government-subsidised and have to be fully paid by the participants. 40% of all South African teachers are produced by UNISA. The Department of Mathematics

Education at UNISA offers courses across the undergraduate and postgraduate levels for both pre- and in-service teachers across the phases. The department also engages in community education and outreach projects: there is currently one project underway entitled "Maths Teacher Professional Development". It is a research project and does not involve any direct teacher education, but it observes the mentoring of other institutions to capture best practices. Witwatersrand University operates the Marang Centre for Mathematics and Science Education, a division of the School of Education which focuses on pre-service and in-service mathematics and science teacher development. They accommodate teachers from many African countries, not just South Africa. Their programmes are all linked to degrees, both undergraduate and postgraduate (BSC, MSc and PhD in mathematics and science education). They do offer some certificate programmes, but this is not their core focus. Similarly, the University of Stellenbosch operates the Stellenbosch University Centre for Pedagogy (SUNCEP), formerly called the Institute of Mathematics and Science Teaching (IMSTUS), while the North-West University operates the Unit for Open and Distance Learning (UODL).

PRIVATE HIGHER EDUCATION INSTITUTIONS

The number of new teachers currently being produced by private higher education institutions is negligible. Fewer than 100 new teachers graduated from the three private higher education institutions offering initial teacher education qualifications. The teacher attrition rate in the country has remained stable at around 5 – 6% per annum since the late 1990s. Assuming a 5% attrition rate, and taking into account that there are currently just over 400,000 practising teachers in the schooling system as a whole, a minimum of 20,000 teachers need to be replaced every year (DBE, Annual School Survey Report, 2014).

IN-SERVICE MATHEMATICS TEACHER EDUCATION INSTITUTIONS

Table 9.1 documents some institutions in South Africa offering mathematics teacher education programmes to practising teachers in the form of professional development courses.

Table 9.1: In-Service Mathematics Teacher Education Institutions

INSTITUTION	DESCRIPTION
Sci-Bono Discovery Centre	As part of their commitment and work to improve the results of school learners in maths and science, they offer a well-structured teacher development programme. As part of this programme the teachers also have access to free online support and teaching tools. These courses, however, are not accredited or linked to any university. All their work is currently within the Gauteng Province only.
Umlambo Foundation	This foundation's goal is to achieve and sustain improvement and high-standard learner outcomes in public schools, especially in rural and underprivileged areas. Their initiatives consist of targeted interventions in disadvantaged schools with a focus on improving school leadership to drive overall school effectiveness in curriculum delivery. Their focus is on rural schools in South Africa and in developing teachers as part of their whole school approach.
Dinaledi Schools Project	In 2001 the Department of Education established the Dinaledi Schools Project to increase the number of matriculants with university-entrance mathematics and science passes. The strategy involves selecting certain secondary schools for Dinaledi status that have demonstrated a potential for increasing learner participation and performance in mathematics and science, and providing them with the resources and support to improve the teaching and learning of these subjects. Part of the support given to these schools is the professional development of their mathematics HODs and teachers, as well as equipment to use in class and access to ICT platforms for support.
LEAP Future Leaders Programme	LEAP is a school in Cape Town that educates high school learners in mathematics, science and English. LEAP runs a "Future Leaders Programme" that is specifically a teachers' training programme. It focuses on disadvantaged communities where there is a significant lack of adequately trained teachers in South Africa. It mainly focuses on pre-service teachers who are further supported after the initial training programme.
African Teacher Education Network (ATEN)	This network is about encouraging understanding, use and sharing of Open Education Resources (OER) to support teacher education and development in Africa. The network is a loosely connected group of teacher educators, with participants currently from Botswana, Ghana, Kenya, Malawi, Mauritius, Mozambique, Nigeria, South Africa, Tanzania, Togo, Uganda, Zambia, UK and USA. Specific tools in this network relate to the ACE Maths programme.
TESSA	TESSA is an international research and development initiative that brings together teachers and teacher educators from across sub-Saharan Africa. It offers a range of materials (OER) in 4 languages to support school-based teacher education and training. These materials are authored primarily by academics from across Africa. TESSA is currently part of a case-study project with Fort Hare University's teacher training programme.
SAMF – South African Mathematics Foundation	SAMF is registered as a non-profit organisation aiming to advance the mathematics development and education of South African learners and young people through improved quality teaching and learning of mathematics as well as through public awareness activities. Founded in 2004 by the Association for Mathematics Education of South Africa (AMESA) and the South African Mathematical Society (SAMS), SAMF serves as a national office for mathematics to promote the effective co-ordination, administration and advancement of mathematics in South Africa. SAMF has four main pillars: learner development, teacher development, advocacy and research.

INSTITUTION	DESCRIPTION
CASME – Centre for the Advancement of Science and Mathematics Education	CASME is a non-profit education development agency that was established in 1985. The centre possesses vast experience and expertise in teacher professional development and implements a range of interventions in partnership with the Department of Education, parastatals, government agencies, Corporate Social Investment programmes, universities and national development initiatives amongst others. The Centre's focus is on the advancement of mathematics and science teachers' subject competence and teaching skills in South African schools through hosting vacation schools and resource portals.
Thandulwazi-Rokunda Teacher Development Programme	Thandulwazi is derived from the isiZulu, meaning "the love of learning". The Thandulwazi Maths and Science Academy is an education programme run by the St Stithians Foundation. It was established in 2006 out of the need for effective action in addressing the critical issues pertaining to the teaching of Maths and Science in schools in and around Gauteng Province. The Thandulwazi-Rokunda Teacher Development Programme focuses on up-skilling the teaching methodologies of teachers currently working in schools in previously disadvantaged areas and providing professional development for these teachers. In addition to teacher development, Thandulwazi also provides extra mathematics lessons for learners, learner sponsorship programmes as well as an internship teacher training programme.
Numeric	Numeric was founded in 2011 with a goal to create exciting and high impact learning environments and in the process help learners excel in mathematics. Initially, the target group was Grade 9 and 10 learners preparing to study pure mathematics at Grade 10 level and the program was offered online. In 2013, focus shifted exclusively to Grade 8; and eventually to Grade 7 learners from low-income areas in order to help the learners in Grade 7 establish strong foundations in mathematics. The online program was also changed to face-to-face interactive sessions incorporating peer learning and games. Through their program of teacher interns, Numeric develops well-equipped and passionate mathematics teachers prepared to teach in public schools.
Bala Wande & Funda Wande	Bala Wande: Calculating with Confidence was established in 2020 as a sister programme for Funda Wande funded by Allan Gray Orbis Foundation. Its aim is to develop fully bilingual learner activity booklets and video-based teacher guides for Grades 1-3 in all South Africa's official languages and equip teachers in no-fee schools with the resources and training they need to teach numeracy and reading for meaning by age 10. It is a contemporary initiative that seeks to decolonise early literacy and numeracy by availing world-class teaching resources in under-resourced settings.
AIMSSEC	The African Institute for Mathematical Sciences (AIMS) is a pan-African network of centres of excellence for postgraduate training, research and outreach in mathematical sciences. Its mission is to enable Africa's brightest students to flourish as independent thinkers, problem solvers and innovators capable of propelling Africa's future scientific, educational and economic self-sufficiency. AIMSSEC is the teacher education wing of AIMS South Africa which delivers mathematics professional teacher development courses. The courses offered at AIMSSEC are structurally similar to those offered by other teacher education institutions, except that AIMSSEC provides bursaries for the teachers' tuition. These courses are designed for the majority of mathematics teachers who are mathematically underqualified, preparing them to be effective teachers.

The scope of some of the institutions detailed in Table 9.1 (*not exhaustive*) is within the general arena of teacher development. The South African Council for Educators (SACE) is responsible for monitoring and managing the Continuing Professional Teacher Development (CPTD) System. CPTD is a system for recognising all teacher development activities. SACE does not deliver the professional development interventions itself, but it endorses the teacher education courses offered by the various in-service teacher education institutions. The National Education Collaboration Trust (NECT) is an organisation dedicated to strengthening partnerships among business, civil society, government and labour in order to achieve the education goals of the National Development Plan. It strives to both support and influence the agenda for reform of basic education. NECT does not deliver on teacher training itself, but it is a key stakeholder in the larger network and influences policy and implementation frameworks for teacher training. Workers' Unions like SADTU have recently established teacher-training programmes. The Department of Basic Education's Teacher Education and Development Units like the Cape Teaching and Learning Institute (CTLI) in Cape Town, offer professional development courses for practising teachers. It is worth highlighting that none of the mathematics courses offered by the highlighted institutions includes any structured language proficiency modules for (mathematics) teachers; hence the existence of a gap in mathematics education.

LANGUAGE PROFICIENCY TRAINING FOR IN-SERVICE MATHEMATICS TEACHERS IN SOUTH AFRICA

In 2012 the DBE and the British Council signed a co-operation agreement which led to the training of over 300 subject specialists and advisors in the FP, IP and FET bands. These three teacher training programmes – CiPELT, CiSELT and LEAP programmes – are aimed at strengthening and supporting curriculum delivery in English First Additional Language (EFAL). These teacher education programmes have been rolled out nationally and the Eastern Cape Province is one of the targeted provinces because of its poor performance (British Council, 2013).

In 2015 the DBE identified 183 high schools that do not offer mathematics as a subject choice at Grade 10 level. This shortage of high schools offering mathematics as a subject choice compounded several problems, namely a continued shortage of skilled and qualified teachers able to teach MST and a shortage of skilled professionals in the science and mathematics-based disciplines such as medicine, engineering, and other technology-related fields. The DBE decided to reintroduce mathematics to these 183 schools and embarked on a multi-pronged strategy to enable this and to help achieve the targets set in South Africa's National Development Plan (2030), which aims to increase the number of learners eligible to enrol for maths and science based degrees to 450,000 by 2030.

The DBE collaborated with the British Council to pilot an in-service teacher training programme using a "work together and work away" model supported by classroom observation technologies and remote feedback. This model will establish a system of on-going, contextualised support for mathematics teachers at schools that are reintroducing mathematics as a subject in Grade 10. The teacher education programme will focus on "problematic areas" of the South African Mathematics Grade 10 curriculum and will be aligned to the South African Curriculum and Assessment Policy Statements (CAPS) (British Council, 2013).

Pearson is a multinational company that provides textbooks as well as professional development services through its Teacher Education and Leadership Academy. One of the courses provided by Pearson's teacher education academy is Teaching Mathematics to Additional Language Learners, a six-day course intended for Intermediate Phase teachers of Mathematics. The course outcomes include provision of deeper knowledge of teaching techniques and classroom strategies: active, hands-on learning and practice; peer teaching, observation and feedback; improved content knowledge, especially in traditionally "difficult" topics as well as fun, interactive approaches with homework DVDs. By the end of the six-day course, teachers are expected to understand how to teach particular mathematical concepts to increase learner understanding and engagement; how to teach academic vocabulary that learners need to be successful; and appreciate the practical experience of teaching the concepts through demo lessons. Pearson's teachers' training courses are guided by the Sheltered Instruction Observation Programme (SIOP), a method which was initiated, tested and implemented in the United Stated of America.

PEDAGOGICAL IMPLICATIONS: MATHEMATICS TEACHER EDUCATION FOR MULTILINGUAL SETTINGS

This context reveals that:

- Prior to 2000 there was a vast array of institutions producing qualified teachers in South Africa, and that since 2000 these numbers have significantly decreased;
- All teachers who qualified through teacher training colleges up to the early 2000s obtained a four-year teacher training Diploma in Education;
- Since the colleges were closed down or incorporated into the universities and technikons, the Diploma was changed to a bachelor's degree in Education;
- There is a greater need for newly qualified teachers at this point in the SA context than for the professional development of existing teachers – although this does not imply that the professional development of teachers is not needed;
- Mathematics teachers are part of the critical skills set that is currently needed in the South African context; and
- Teachers in the rural and poorer schools require more support than those in urban areas.

Inasmuch as the environmental scan of the South African teacher education landscape has established that the teaching profession is characterised by drastic shortages of qualified teachers (across all learning areas and mostly mathematics) (DBE, 2014) in public and under-resourced schools, it is worth highlighting that the available teachers need support regarding language proficiency skills. This is a significant factor which seeks to alert educationists and inform curriculum design about the crucial but largely ignored fact that qualified practising mathematics teachers are not necessarily proficient in English and that this may have an effect on the quality of instruction through using English as LoLT. There may be several other contributing factors and further in-depth research is required. Nevertheless, a re-evaluation of teaching methodologies and the upgrading of both pre-service and in-service teachers' levels of language skills are suggested as a matter of urgency.

Chitera (2009b, 2012) states that most of the language practices in multilingual mathematics classrooms that teachers produce cannot be traced back to their teacher education institutions. In the South African context, this practice is manifested twofold: FP pre-service teachers are taught in English, yet after the completion of their studies, they will be expected to use mother-tongue instruction. IP pre-service teachers who are also taught in English do not master the language well enough to be competent to handle the transition from mother tongue to English-medium instruction. The linguistic dilemma faced by FP teachers is beyond the scope of this chapter; however, there seems to be a linguistic disjuncture between teacher education institutions and classrooms. One of the ways of addressing this gap between the LoLT in primary schools and the LoLT in teacher education institutions would be to acknowledge and enforce linguistic strategies that can improve mathematics instruction in multilingual classrooms during teacher training.

CONCLUSION

Having good Mathematics teachers in the school classroom should be the central focus of university teacher education departments. Teaching and learning is a two-way process, involving both the teachers and the learners; therefore, practices involved in the school classroom and teacher-training institutions need to be two-way processes (Chitera, 2016). What happens in schools informs the teacher training practices, while teacher-training practices inform the school practices so that in-service teachers need to graduate from the university teacher education departments better prepared to function effectively when tasked to teach in any environment (including multilingual under-resourced settings). In the same manner, whichever policy the schools choose to adopt (mother-tongue instruction or using English as LoLT), it is important to simultaneously equip the teachers for the adopted changes in order to provide a constructive balance between language competency and the LoLT. Wolff (in Alidou et al., 2006: 49) sums this up best by stating that "Language is not everything in education, but without language everything is nothing in education" and this chapter extrapolates Wolff's assertion:

> ***Language is not everything in mathematics education, but without language, everything is nothing in mathematics education.***

While Chapter 8 focuses on the use of mother tongue for mathematics instruction, Chapter 9 specifically focuses on mathematics teacher education for under resourced multilingual settings. Furthermore, Cases A and B focus on IP mathematics teachers, while Case C focuses on classroom linguistic practices; therefore, **Case Study D** focuses on Intermediate Phase (IP) mathematics lecturers by interrogating the extent to which South African university teacher education departments prepare IP mathematics teachers in using English as LoLT. This case study is informed by the Pedagogical Translanguaging theory. Questionnaires and interviews were administered on 10 IP mathematics teacher educators purposefully selected from 26 universities in South Africa. Data were quantitatively and qualitatively analysed. Findings reveal that although the majority of university teacher educator departments use English as the official language of instruction, there is not sufficient consideration given to teachers who are non-native-language speakers of English in terms of equipping them with adequate skills to use English as language of instruction in their practice. Furthermore, there is a dire need to provide guidance on the use of translanguaging in mathematics instruction.

CASE STUDY: D

UNIVERSITY EDUCATION DEPARTMENTS' GUIDANCE ON USING ENGLISH AS LOLT: SOUTH AFRICA

INTRODUCTION

The use of different languages together to enhance communication (translanguaging) can be a powerful tool for teaching and learning, but it can also work against the grain for *language* teachers who are used to supporting learners to master the intricacies of a single language. This brings us to the question how translanguaging can be used to support English Language Learners (ELLs), specifically in under-resourced multilingual settings of South Africa.

Picture the scene: Two learners are sitting together, working intently on a worksheet. They have different mother tongues but some shared knowledge of the words and phrases of each other's languages, so they are moving in and out of English to get their message across. Another two students are sitting together nearby. Both of them are isiXhosa speakers, but have a strong command of English and often use it as their main language. At other times, as now, they blend isiXhosa and English together.

Are either of these examples of translanguaging?

In both cases, yes. The learners are using resources from different languages together, with very little regard for the '*boundaries*' of named languages such as 'isiXhosa' or 'English'. They are using elements of each language together to communicate more effectively. This is translanguaging: it is about using all the language resources to communicate. Translanguaging is about *communication*, not about the *language* itself. There are times when educators need to be language teachers, focusing on accuracy in English so that their learners can pass exams and be taken as proficient speakers of English in wider society. In most cases, though, teachers work with learners to explore concepts, add to their knowledge, make connections between ideas and to enable the learners to voice their ideas and opinions to others. This is about *communicating*, and this is where using *all* our language resources can be very valuable. (Adapted from EAL (2016)).

English and Afrikaans are the prescribed languages of instruction from IP level up to tertiary studies, however contemporary debates on language in education highlight the benefits of translanguaging pedagogies. The aim of this case study is to investigate how university departments of teacher education prepare pre-service teachers in using English as language of instruction in multilingual under-resourced mathematics classrooms. Of great importance is whether pre-service teachers are prepared well enough in using the prescribed language of instruction so that they may readily assimilate the contemporary strategies such as translanguaging to facilitate multilingual teaching and learning. Guided by the pedagogical translanguaging theory, the case study briefly exposes translanguaging as a socially just pedagogy, and presents the research questions, conceptual framing, research design and methodology, findings and discussion.

TRANSLANGUAGING: A SOCIALLY JUST PEDAGOGY?

Hurst and Mona, (2017) describe translanguaging as a socially just pedagogy, but is it specifically referring to the South African context? A socially just approach includes investigating barriers to learners' outcomes and using teaching and learning assessment strategies fairly. A socially just approach also involves critical questioning and resistance, analysis of systems of oppression and positioning and encouraging social action and practising democracy (Hurst & Mona, 2017; Cumming-Potvin, 2009).

Hurst (2016) denotes that high school learners who use an African language at school are regarded as disadvantaged: learners indicated that navigating language and education in high school was not an easy task. Other learners reported experiencing shyness when speaking in class in their home language. English has the power to silence learners, through the fear that they would say something wrong and be ridiculed by fellow learners. In general, there was sadness associated with learners' struggles to speak English and the emphasis on English made the learners doubt their own intelligence as their home languages were seen as inferior. The question here is: Is there any social justice if using one's language brings about feelings sadness and inferiority?

With the prevailing wave of decolonising the higher education system, one of the ways suggested for bringing this about is opening up the education space, using language as a resource and promoting the use of indigenous languages in academic communication– translanguaging. If translanguaging has so many positive spin-offs, how much of this pedagogy is currently being implemented in South African institutions of higher education, specifically university teacher education departments? How much of the pedagogy filters down from university teacher education departments to the classrooms in which teachers are deployed? How do university teacher education departments undo the sadness and inferiority experienced by the pre-service teachers during their own high school years associated with having the same students being champions of translanguaging when they become teachers?

This case denotes that whether translanguaging is a socially just pedagogy in South African university teacher education departments is determined by how well the departments prepare their pre-service teachers in implementing translanguaging after deployment. The case study draws on the pedagogical translanguaging theory specifically because developing pre-service mathematics teachers to proficiently use English as LoLT is only a basic step. To better serve multilingual learners in under-resourced communities, there is a need to go further than the basics by developing pre-service teachers who can effectively use different languages together to enhance communication: translanguaging.

OBJECTIVES

Interrogating the extent to which South African university teacher education departments prepare IP mathematics teachers in using English as LoLT seeks to:

i. Determine whether university teacher education departments include modules focusing on using English as a medium of instruction.
ii. Identify strategies used by teacher educators to present specialised mathematical vocabulary.
iii. Identify resources recommended by teacher educators for addressing language-based mathematical problems.
iv. Identify challenges experienced when lecturing and observing IP mathematics pre-service teachers using English as LoLT.
v. Determine strategies for improving teaching and learning IP mathematics using English as LoLT.

CONCEPTUAL FRAMING

The conceptual framework grounding in this case is Baker's (2011) Pedagogical Translanguaging.

PEDAGOGICAL TRANSLANGUAGING: BAKER (2011)

Translanguaging is the act performed by bilinguals and multilinguals for accessing different linguistic features or various modes of what are described as autonomous languages, in order to maximise communicative potential (García, 2009: 140). Translanguaging fits on the spectrum of work on multilingualism. Included in this range of language activity are code switching, translation and translanguaging. Code switching is usually a relatively short move from the LoLT to the home language of learners and then a switch back to the LoLT. Translation usually entails a repetition of an oral or written text in the more accessible home language of learners (Probyn, 2015). Both of these practices are responsive rather than planned strategies, usually regarded as temporary excursions from the monolingual ideal. In some cases, use of the learners' home language instead of the LoLT is viewed as illicit or transgressive (Probyn, 2001, 2009, 2015). Therefore, translanguaging goes beyond code switching and translation, but includes both. It can be used as a pedagogically sensitive tool to systematically promote learning (Heugh, 2015; Lewis, Jones & Baker, 2012; Probyn, 2015). Both languages are used in an organised way to mediate understanding and learning (Baker, 2011; Garcia & Wei, 2015; Heugh, 2015). Thus, the teacher incorporates into their planning which, why, when and how they are going to use each of the two languages for the purposes of assisting the learners to better understand the concepts taught.

METHODOLOGY

Teacher educator questionnaires were administered at a National Symposium on Primary Literacy and Primary Mathematics Teacher Education hosted by the Department of Higher Education and Training (DHET)'s Teaching and Learning Development Capacity Improvement Programme (TLDCIP), in which the researcher was a participant. Collection of completed questionnaires and interview administration on the teacher educator respondents spanned the entire 12-month period determined by the availability of the sampled teacher educators. The contextual details of the 10 participating institutions are intentionally not provided because the focus is not related to institutional comparison, but instead to understanding and supporting the preparation of IP mathematics teachers in using English as LoLT in the national landscape. Absence of contextual information also ensures anonymity of participating institutions.

SAMPLING

This sample comprised 10 IP mathematics teacher educators purposely selected from the 26 universities in South Africa. These teacher educators were available to the researcher through co-presentation of professional development courses and at national and international mathematics education conferences. The teacher educators ranged from mathematics chairs, professors, senior lecturers, lecturers and didacticians from different Education Departments at the country's universities. The rationale for selecting teacher educators across the different universities is that teachers who end up being deployed to the ECDoE could have trained at any university in the country, including universities lying outside the borders of the Eastern Cape Province. The basis for sampling teacher educators as respondents was to get information from the key agents involved in teacher education, namely the teacher educators. Pedagogic theory denotes instruction as a two-way process which involves the one who is imparting knowledge, and the one to whom knowledge is being imparted. In teacher education, the focus of this case, the teacher educators impart knowledge and skills.

Participant Demographic Data

The only biographical information from the teacher educators that was deemed necessary was gender, position, the official LoLT(s) at their institutions and their experience in years in teaching IP mathematics at university level. The demographic data of the 10 participant teacher educators is indicated in Table D.1.

Table D.1: Demographic information: Teacher educators

	Gender	Race Group	Position	QUESTION What is the official LoLT at your institution?	For how many years have you offered in-service IP mathematics education?
TE1	M	Indian	Senior Lecturer	English and Afrikaans	0 – 3 years
TE2	M	Black	Lecturer	English	More than 10 years
TE3	F	White	Lecturer	English	4 – 6 years
TE4	M	Black	Lecturer	English	0 – 3 years
TE5	F	White	Senior Research Fellow	English	More than 10 years
TE6	M	White	Director Maths Project	English	More than 10 years
TE7	F	Indian	Professor/Numeracy Chair	English	More than 10 years
TE8	F	White	Lecturer	English and Afrikaans	7 – 10 years
TE9	F	White	Professor/Numeracy Chair	English	More than 10 Years
TE10	M	Black	Lecturer	English	0 – 3 years
Totals % where applicable	M (5) 50% F (5) 50%	Black (3) 30% Coloured (0) 0% White (5) 50% Indian (2) 20 %	Lecturer (5) 50% Senior Lecturer (1) 10% Senior Research Fellow (2) 20% Professor / Numeracy Chair (2) 20%	English (8) 80% English and Afrikaans (2) 20%	0 – 3 (3) 30% 4 – 6 (1) 10% 7 – 10 (1) 10% More than 10 years (5) 50%

Key:
TE = Teacher Educator
M = Male
F = Female

The teacher educator participants were distributed equally between male and female. The participants held various positions in their universities ranging from lecturers (50%), Senior Lecturers (10%), Senior Research Fellows in Mathematics Education (20%) and Professors (20%), who are currently holding South African Numeracy Chair positions. This variety of positions enabled the participants to provide well-balanced information, based on their experiences and expertise regarding the language of instruction in primary mathematics education. 50% of the participants had held teacher educator positions for more than 10 years, while the rest had been in mathematics teacher education for between 0 and 10 years. Since this case focuses on language of instruction, it was crucial to determine the official medium of instruction in the participants' universities; English was the LoLT in 80% of the institutions, while English and Afrikaans were used as LoLTs in 20% of the institutions. It is noteworthy that 50% of the teacher educator participants are white, 30% black and 20% Indian. Thus, a cumulative of 70% of the teacher educators possess native fluency in English.

The information from the teacher educators is analysed thematically based on the relatively small number of the teacher educators sampled. Hence the main focus in this section is establishing common themes emanating from the teacher educators' responses.

FINDINGS

The research question was: **To what extent do South African university teacher education departments prepare IP mathematics teachers' in using English as LoLT?** and the key finding is that there is no sufficient preparation of mathematics teachers with adequate skills to use English as language of instruction in their practice. The section below presents, analyses and discusses data collected from 10 teacher educator questionnaires, followed by data gathered from 2 teacher educator interviews.

QUESTIONNAIRES: TEACHER EDUCATORS

IP Mathematics qualifications offered at universities

For an overview of the teacher education programmes offered at the different institutions where the sampled teacher educators are based, it was crucial to establish the IP mathematics education qualifications offered by their institutions. Responses in this regard are indicated in Table D.2.

Table D.2: IP Mathematics qualifications offered at universities

	Question: What in-service IP mathematics qualification(s) does your institution offer?		
	Duration (years)	**Name of Degree / Diploma / Certificate**	**Inclusion of methodology modules Yes / No**
TE1	4	BEd Intermediate Phase Mathematics	✗
	2	PGCE	✗
TE2	4	BEd Intermediate Phase Mathematics	✓
TE3	2	ACT	✓
TE4	2	ACT	✗
	2	BEd Honours	No, some students research on it in their projects
TE5	None		
TE6	3	Bed	✓
TE7	2	ACE	✓
TE8	3	NPDE	✓
	2	ACE	✓
	2	ACT	✓
	2	BEd Honours	✗
TE9	None		
TE10	Not for IP courses, but for Masters and Honours courses		

Key: None

Pedagogical Implications: IP Mathematics qualifications offered at universities

All the institutions surveyed indicated that they offered some pre-service qualifications in IP mathematics education; however, only 70% of these institutions offered these qualifications to in-service teachers. A number of reasons could explain this, such as capacity whereby the in-service courses are offered at other levels such as the FET and SP bands, but not to the IP level. It was also noted that in some institutions the in-service courses were offered by specialised units of the university, running parallel to the pre-service qualifications. The majority of the institutions offering IP courses had some module focusing on methodology; however, at one institution methodology was an elective module which some students pursued in their research projects. Clarity on these modules and their contents would have been sought through a content analysis of the module outlines of the different universities, but that falls beyond the scope of this case.

IP mathematics modules focusing on English as medium of instruction

The data deemed crucial were whether the institutions offered any short courses focusing on language, specifically English, as a medium of instruction in IP mathematics during or after initial teacher education. Responses in this regard are illustrated in Table D.3.

Table D.3: Prevalence of IP mathematics modules focusing on English as medium of instruction

	Question: What other in-service short courses in methodology of teaching IP mathematics does your institution offer?			
	Years	**Months**	**Name of Diploma / Certificate**	**Inclusion of English LoLT modules Yes / No**
TE1			None	
TE2			None	
TE3			None	
TE4			None	
TE5			None.	
TE6			None	
TE7		20 days	Maths Primary Project Course	Integrated content and methodology
TE8			None	
TE9			None	
TE10			Expressing Mathematics Language and Communication in Mathematics	

Key: None

Pedagogical Implications: IP mathematics modules focusing on English as medium of instruction

Out of the 10 participating teacher educators sampled from the 26 South African universities, only 2 respondents revealed that their institutions offer modules sensitising teachers on how to teach specialised mathematics vocabulary to IP mathematics pre-service teachers during initial teacher education. One institution that offered a comprehensive module, *'Language and Communication in Mathematics'*, offered it only at Master's level, a level which most of the teachers would not have attained by the time they are deployed to teach in schools. None of the institutions offered short courses to in-service teachers. One institution offered some in-service programmes focusing on IP mathematics methodology in general, but these were not accredited. This information strongly tallies with the information provided by IP teacher participants in the sense that the departments of teacher education in which they had their initial teacher education did not offer specialised modules focusing on the use of the English language as a medium of instruction for IP mathematics.

As of 2019, one of the sampled teacher education institutions has introduced a 'Literacy for IP/SP/FET teachers' courses in its Bachelor of Education degree programme. This is a commendable exercise, and the courses need to be constantly monitored and revamped to ensure that they address the peculiar needs of multilingual under-resourced settings. Furthermore, for consistency purposes, similar courses need to be rolled out across all university departments of teacher education; and instead of being generalistic, the courses need to be specific for each teaching subject and phase level such as literacy for IP mathematics teaching or literacy for FET geography teaching.

Teacher educator strategies for teaching specialised mathematical vocabulary

In was necessary to establish the teacher educators' views on how they teach or present specialised mathematical vocabulary in their lectures. Establishing this information would determine whether the teacher educators modelled 'good practice' to their students, so that the pre-service teachers may adapt these strategies for use in their own classrooms. Teacher educator responses on strategies used to teach specialised mathematical vocabulary are compiled in Table D.4.

Table D.4: Teacher educator strategies for teaching specialised mathematical vocabulary

	Question: How do you teach specialised mathematical vocabulary in your lectures?
TE1	a) I 'integrate' the vocabulary by emphasising the many meanings that mathematical symbols symbolise. b) I thus emphasise many representations of a single mathematical symbol or word. These representations are numerical, geometrical, visual and 'in words' i.e. semantic.
TE2	a) Building experiences and descriptions in which the specialised vocabulary is used. b) Use gestures and manipulatives. c) Use maths dictionary.
TE3	a) Key words displayed on the classroom wall. b) Discussion in groups of meaning of keywords. c) Games and activities to familiarise learners with keywords and phrases, Vocabulary jotter. d) Both languages used in teaching material until the learners can cope in English alone.
TE4	a) The teaching of specialised mathematical vocabulary is infused into the teaching of mathematics content.
TE5	

	Question: How do you teach specialised mathematical vocabulary in your lectures?
TE6	a) Mathematical vocabulary is part of one of the modules called curriculum studies/mathematics Education. During this time, students have to create a list of mathematics vocabulary words together with meanings and submit this as a short assignment. Terms like 'volume' which are shared with everyday English; sound alike terms within maths and English, examples of mathematical terms that are related, but have distinct meanings which are confused by learners are submitted by each student. b) Students are sensitised to a number of readings on language from local and international journals and part of this aspect includes the language of mathematics. c) Students also compile a mathematics dictionary for the class that they teach, not only Intermediate Phase students but Foundation Phase, Senior Phase and FET students as well. The dictionary comprises words and the correct meaning of the words in Mathematics. d) Students also make a set of flashcards with the mathematical vocabulary words written on them which then gets displayed on their classroom walls. All specialised mathematical vocabulary is identified, highlighted and explained during lectures. e) Students also have to submit an assignment on the language of mathematics and have to critically discuss mathematical vocabulary as part of their assignment..
TE7	a) Emphasis on teacher talk, use of representations and explanations is central to the 20-day course. b) Feedback on written assignments includes attention to mathematical precision.
TE8	a) I believe that teacher-students need to "know" the basics and fundamental principles of the mathematics they are going to teach. The teacher-students should feel capable and empowered to do their thing in the math class. The way I go about it is to emphasize the correct use of mathematical terms at all times. While I am teaching content, I would give examples of typical mistakes learners and math users make with respect to the language of math. I also encourage the students to compile their own math vocabulary-list as they are teaching and working through the course.
TE9	
TE10	a) Talk about the mathematics register: 'What is it about the mathematics register that makes it special?' b) Discuss the informal use of mathematics. c) Show the students how the lack of register can impede understanding. d) Show the students how different language forms can be interchanged.

Key: [shaded] None

Pedagogical Implications: Teacher educator strategies for teaching specialised mathematical vocabulary

80% of the teacher educators shared valuable strategies and the rest left the section incomplete. A number of strategies are cited by the teacher educators on how to teach specialised mathematics vocabulary in IP classrooms. These strategies include integration of vocabulary into content teaching and learning, using mathematics dictionaries and freely available software, keeping word walls and promoting discussions on the formal and informal use of mathematics register. The key point therefore is further researching these strategies for their feasibility in under-resourced settings and the actual filtering down of the strategies to IP mathematics teachers' everyday practice.

Resources recommended by teacher educators for addressing language-based mathematical problems

For detailed information, teacher educators were requested to share the resources they recommended for use in addressing the language-based mathematical problems. The responses are compiled in Table D.5.

Table D.5: Teaching and learning resources recommended by teacher educators for addressing language-based mathematical problems

	Question: What teaching resources do you recommend for use by in-service teachers to address language based mathematical problems?					
	a	**b**	**c**	**d**	**e**	**Other**
TE1	✓	✓	✓	✓	✓	
TE2	✓	✓	✓	✓	✓	
TE3	✓	✓	✓	✓	✓	
TE4	✗	✓	✓	✗	✓	
TE5	✓	✓	✓	✗	✓	▪ Physical resources such as egg boxes, beads, Cuisenaire rods and Lego
TE6	✗	✓	✓	✓	✓	▪ There are many errors on the GDE workbooks ▪ Textbooks written in isiXhosa for teachers might help. They will help teachers to code switch precisely and consistently.
TE7	✗	✗	✗	✗	✗	▪ We use the Beckmann text on Mathematics for primary teachers, not specifically for language but for integrated attention to explanations and representations of mathematical ideas.
TE8	✓	✓	✓	✓	✓	▪ I advise teachers to use everything they can get hold of. Extra to the list above are the many mathematics interactives that they can engage in for free on the internet. These give so many ideas and support for the struggling teacher.
TE9	✗	✗	✗	✗	✗	▪ We design our own resources, give them freely, and put them on the website – including homework books, fraction charts, dice games and booklets.
TE10	✗	✓	✓	✗	✓	▪ The GDE workbooks have a lot of errors; therefore, I do not recommend them. ▪ Multilingual dictionaries would be a useful tool if they were standardised.
Totals: % where applicable	50%	80%	80%	50%	80%	

Key: a = DBE Workbooks • b = Textbooks • c = Mathematics dictionaries • d = Computer based activities • e = Word walls / charts

[shaded] None

Pedagogical Implications: Resources recommended by teacher educators for addressing language-based mathematical problems

Textbooks, mathematics dictionaries and word wall/charts were recommended by 80% of the teacher educators for use in addressing language-based IP mathematical problems, while GDE workbooks and computer-based activities were recommended by 50% of the teacher educators. Recommended textbooks which have been evaluated by the DBE are listed on the 'national catalogue' and schools order the books for use in their classrooms. As such, the textbooks used in different schools may vary; however, the common factor is that the DBE's assessment of the textbooks includes checking whether the books comply with the CAPS requirements. One of the teacher educators suggested the use of isiXhosa textbooks for assisting teachers with precise and consistent code switching:

> *'Textbooks written in isiXhosa for teachers might help. They will help teachers to code switch precisely and consistently.'*

Interestingly, this statement suggests that the textbooks alone may be insufficient without formal instruction on how precisely and consistently code switching should be done during initial teacher education. In addition, the fact that teaching and learning material available in indigenous languages is still being developed is a significant factor. The use of mathematics dictionaries for addressing language-based mathematical problems is re-emerging as an issue in this section, after it was suggested by teacher participants in their responses on a similar question:

> *'Multilingual dictionaries would be a useful tool if they were standardised.'*

Therefore, it may be prudent to invest in mathematics dictionaries as suggested by the teacher educators. Word walls and charts were also a popular suggestion from the teacher participants; therefore, there may be a need to increase teachers' awareness of their usefulness as well as engage policy makers and publishers in standardising the dictionaries. Word walls and charts can be created in such a way that they convey more meaning to the learners beyond just being a compilation of new words or an endless list of words that end up watering down the intended meaning of the target word. Only 50% of the teacher participants recommended the use of DBE workbooks and computer-based activities. The decline in interest in these two teaching and learning resources as compared to the above-mentioned more popular resources is explained below. Two teacher educators indicated that the DBE workbooks had many errors in them:

> *'The DBE workbooks have a lot of errors; therefore, I do not recommend them'*

Determining the accuracy of this fact is a suggestion for further research; however, as a relatively newly introduced resource used to mitigate the shortage of textbooks in public schools, DBE workbooks may need to be revised regularly, so that they meet the intended requirement of being an additional teaching and learning resource. Constant revision of the workbooks will provide a balanced and adequate curriculum coverage for learners in under-resouced schools where the DBE workbooks are the only resource available for learners:

> *'These workbooks have been developed for the children of South Africa... as part of the Department of Basic Education's range of interventions aimed at improving the performance of South African learners in the first six grades... They provide every learner with worksheets to practise the language and numeracy skills they have been taught in class...'*
>
> DBE (2010)

Regarding the use of computer-based activities, two teacher educators indicated that they make these available on their website; however, the accessibility of these resources to teachers in under-resourced schools without computers and internet connectivity may be limited. Therefore, a lot of work may still need to be done to ensure that computer-based activities are easily accessible to teachers for use in IP classroom to address language-based mathematical problems. Other suggestions included the use of tangible objects such as Lego, beads, egg boxes and dice to help learners better understand abstract mathematical concepts.

Challenges experienced by teacher educators when lecturing and observing IP mathematics teachers using English as LoLT

To gain more insight into the use of English as LoLT from the teacher educators' perspective, the next question intended to establish the challenges experienced by the teacher educators when lecturing and observing IP mathematics teachers using English as a language of instruction. Responses from the teacher educators are compiled in Table D.6.

Table D.6: Challenges experienced by teacher educators when lecturing and observing IP mathematics teachers using English as LoLT

	Question: What major challenges do you experience when lecturing and observing IP mathematics teachers using English as LoLT?
TE1	a) Many or most of them struggle with 'engaging' the learners in terms of the many meanings attached to symbols and the required mathematical actions that need to be performed.
TE2	a) Intermediate Phase teachers tend to over-rely on explaining mathematical concepts, and procedures in the 1st language e.g. isiXhosa which is not the official language of instruction and assessment. This is a disservice to the learners as they are assessed in English. b) Often objectivity is lost when discussing the use of English as the LoLT as the issue of language is often politicised, hence the debate on language tends to lose objectivity.
TE3	a) Communication is difficult if the pre-service teachers do not have a good grasp of reading and listening to English. b) Strong regional accents also cause problems with communication. c) Mathematical concepts are very difficult to understand if you do not understand the language.
TE4	a) In the Eastern Cape (former Transkei) where communities in rural areas use one monolingual, teachers grapple to teach in English and find themselves switching to isiXhosa and ultimately using mother tongue. b) Over years, teachers' proficiency in English drops and eventually they struggle to express themselves in English. c) Lack of using appropriate learning and teaching devices. d) Lack of basic resources in schools (libraries).
TE5	a) knowledge of vocabulary b) knowing how to explain things c) knowing when to code switch
TE6	a) The biggest challenge is teachers being unable to reason and understand the mathematics they teach. They then translate these incorrectly. b) Some teachers struggle to use the correct mathematical words/vocabulary when they teach and very often code switch to isiXhosa which very often does not give a clear description of the mathematical term that is being emphasised. c) We have observed a few teachers (especially in the rural areas) who teach in mother tongue only during the whole lesson. We have stressed that teachers must teach in English and do minimum code switching.

	Question: What major challenges do you experience when lecturing and observing IP mathematics teachers using English as LoLT?
TE7	a) Frequent disruptions to mathematical coherence in teacher talk – this is sometimes related to mathematical problem solving, sometimes to limited explanatory repertoires, leading to repetition rather than elaboration of ideas, imprecise language, and slippage in the ways mathematical ideas are named and connected.
TE8	a) I honestly believe that most of the students who teach mathematics in the IP as well as in the other phases, do not understand the math concepts themselves. Therefore, they struggle with the language of math and also with teaching the math in English, which in most cases is not their mother tongue.
TE9	a) Learners not being able to speak English fluently and thus not being able to participate in the learning conversation.
TE10	a) Teachers who: ▫ are not aware how language can be used to impart mathematical understanding ▫ battle with communication in the prescribed LoLT. ▫ have the knowledge but have no language to explain the concepts. ▫ use code switching inappropriately, thus use code switching not to help the learner but to help themselves to continue with the utterance.

Pedagogical Implications: Challenges experienced by teacher educators when lecturing and observing IP mathematics teachers using English as LoLT

One of the challenges raised by the teacher educators in the above responses is the lack of reading material in English. In addition, it emerges that teachers are not sure of how and when to code switch:

> *'...knowing when to code switch'*
>
> *'We have observed a few teachers (especially in the rural areas) who teach in mother tongue only during the whole lesson. We have stressed that teachers must teach in English and do minimum code switching.'*

Given the vast amounts of research in mathematics learning in multilingual classrooms and the large number of studies devoted to the effectiveness of code-switching strategies, the issue of code switching should not be causing so much concern. Could it be that this literature is not available to IP mathematics teachers? Or could it be that research findings do not filter down to primary school mathematics pre- and in-service teachers?

The teacher educator responses above highlight teacher language proficiency in the LoLT as one of the most significant challenges observed among IP mathematics teachers. Without labelling teachers as ineffective, can the investigation extrapolate that the teachers have limited access to good practice during the initial teacher education? Since all the teachers who participated are qualified by different universities as able to teach, could it be the South African education system is embroiled in a chain reaction where teacher educators fail teachers who, in turn, fail the learners and the entire education system?

Improving teaching and learning of IP mathematics using English as LoLT

Having established the challenges experienced by the teacher educators, it was imperative to establish different ways in which the teacher educators thought could improve the teaching and learning of IP mathematics using English as LoLT. Responses obtained are compiled in Table D.7.

Table D.7: Strategies of improving teaching and learning of IP mathematics using English as LoLT

	Question: Suggest other strategies that could improve the teaching of IP mathematics using English as LoLT
TE1	a) I suggest that special emphasis be given to the multiple meanings of mathematical symbols. The challenge is then to work with learners who may believe that such multiple meanings can be 'confusing'. b) I often hear them complaining about 'confusing' and 'confusion'. It is challenging for the BEd (pre-service) teachers to see that experiencing and reflecting on confusion as part of learning something. c) The multiple meanings of mathematical symbols should be listed in a three-column table, for example: The first column can have the actual symbol, the middle column the English and the last column the mother-tongue equivalent and an appropriate example. This example must be one that is *rich* in meaning and one that the teacher knows the learners struggle with. Through formative assessment the teacher knows which symbols and words are confusing and difficult for the learners.
TE2	a) Address the resistance teachers have towards teaching using English. b) Empower the teachers in teaching maths in English. Some of the teachers are not proficient in English, hence are not confident in teaching maths in English. c) Address the politicization of the use of English as a medium of instruction.
TE3	a) Have subtitles in the local language. b) Have fun activities using English keywords and phrases to improve level of understanding. c) Sit learners who can read and speak some English with those who have poor English.
TE4	a) Engaging in research projects in schools b) Reviving subject associations in local schools c) Collaborative programmes among schools and d) Programmes that promote learner participation such as competitions.
TE5	a) An emphasis on the issue during ITE, and finding ways for teachers in classrooms to address the issue, such as those given below: ▫ making new vocabulary explicit . ▫ discussion of mathematics vocabulary to help learners remember and understand words (e.g. where does this word come from?) ▫ playing word games with learners. ▫ using the home language as appropriate.
TE6	a) It is imperative to connect the teaching of First Additional Language (FAL) to Mathematics and work with the language teachers in the schools to get their learners to understand what they are reading, for example. Reading is key to understanding any mathematics that is being taught. b) Especially in Grade 4 it is important to have all word problems written in both isiXhosa and English so that learners can choose which language they feel more comfortable to answer the questions as they will be able to comprehend what they are reading. As the year progresses, less mother tongue should be used so that towards the end of the year, the word problems are in English only. From Grade 5 teachers should teach in English only and do minimum code switching. c) Teaching resources in mother tongue for teachers to use as reference might be useful.

	Question: Suggest other strategies that could improve the teaching of IP mathematics using English as LoLT
TE7	a) Learning mathematics through a focus on representing and explaining has helped us to help teachers learn maths in ways that are helpful for teaching.
TE8	a) Pre-service teachers need to become English-literate and they need to master the math concepts they are going to teach. They actually need to master the math from at least one phase beyond the phase they are going to teach.
TE9	a) English should be developed stronger in earlier grades and there should be ELL support for Grade 4 learners to support their EL development.
TE10	a) Awareness is important, teachers need to: ▫ be conscious of the pedagogy of mathematics and learn the appropriate pedagogic strategies. ▫ use simple language to explain mathematical concepts. ▫ speak slowly to allow the ELL's brain to capture and process the mathematical concepts.

Pedagogical Implications: Improving teaching and learning of IP mathematics using English as LoLT

The majority of the teacher educators' responses implied that teachers were not confident in using English as a language of instruction, and called for upskilling the teachers in this regard:

> *'... teachers need to become English-literate...'*
>
> *'Empower the teachers in teaching maths in English, some of the teachers are not proficient in English hence are not confident in teaching maths in English.'*

This observation resonates with suggestions raised by in-service mathematics teachers in a research where IP mathematics teachers (n=55) were requested to suggest ways of improving IP mathematics teaching using English as LoLT. Themes emerging from the respondents are summarised in Figure D.1.

Figure D.1: Ways of improving IP mathematics teaching using English as LoLT

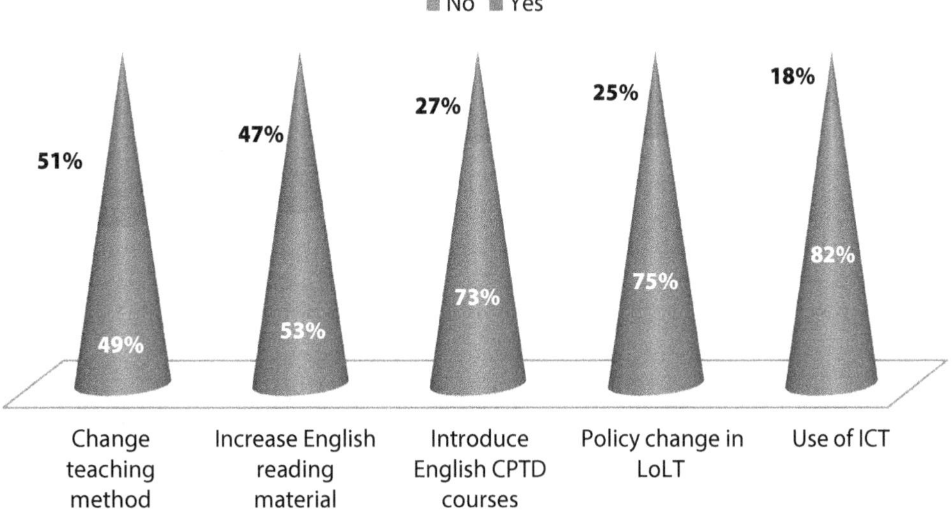

27% of the teachers suggested that the DBE needs to invest in additional CPTD courses to upskill teachers in understanding the most effective ways of using English as LoLT in IP mathematics classrooms. The following changes were suggested in support of this notion:

- *To be trained on how to use English as LoLT*
- *DBE must provide teachers with bursaries to further their studies and become competent*
- *Teachers to take part in language competency competitions*
- *Improve teachers' English language skills*
- *Provision of professional development courses for teachers*

Tshuma (2017)

By virtue of being more experienced and having access to scientific research output, the teacher educators' suggestions were more focused than those made by the teachers. Suggestions made by the teacher educators indicated that the upskilling of teachers needs to commence during initial teacher education (ITE) programmes and continue as part of teacher professional development while the teachers are in the classrooms, thus linking and providing a bridge between the two programmes instead of treating them as separate entities:

> *'... an emphasis on the issue during ITE, and finding ways for teachers in classrooms to address the issue.'*

The teacher educators' suggestions also raised the issue of awareness as an important factor. Thus, if teachers are made aware of the various ways they can tackle the problem, they become more effective and address even seemingly unimportant factors like the use of simple language for ELLS and speaking slowly when addressing ELLs:

- *use simple language to explain mathematical concepts'*
- *speak slowly to allow the ELL's brain to capture and process the mathematical concepts'*

To enable a smooth transition from FP to IP level teacher educators suggested the provision of English language support to Grade 4 learners, the entry level of IP. Making use of interactive, engaging and fun language-based activities was also suggested. Inasmuch as all these suggestions are pivotal in supporting the transition, they focus on the learners rather than the teachers. Therefore, by extrapolation, the IP teachers need to be supported during and after initial teacher education to enable them to handle the linguistic needs of their learners.

One of the most important factors suggested by the teacher educators which is not directly linked to the teaching and learning environment, but directly linked to policy makers, is the separation of education and politics:

> *'Address the politicisation of the use of English as a medium of instruction.'*

This case deems this to be an important factor as the consequences of politicising the use of English as LoLT affects the learners, and especially those in under-resourced schools with very limited choices regarding which schools they can possibly attend. As such, this case is intended to inform curriculum changes regarding the production of linguistically equipped IP mathematics teachers who would be able to better serve whichever school they are deployed in and the learners there.

The sections above have provided information gathered through the teacher educator questionnaires while the section below provides information gathered through the teacher educator interviews.

Interviews: Teacher Educators

The teacher educator interview included questions that sought to cross-check the data collected from the questionnaires. Following the design of the current case, the teacher educator interview should have been administered to 1 (10%) of the sampled teacher educators; however, for the purposes of gaining a deeper insight into the teacher educators' perspectives regarding University Education Departments' guidance on using English as language of instruction, the interview was administered to 2 (20%) of the sampled IP mathematics teacher educators after the administration of the questionnaire. Table D.8 provides the profile of the teacher educator interviewees.

Table D.8: Interviewee Profile: Teacher Educators

P	G	AGE	ETM	Q
TE1	Male	4	3½	▪ BA ▪ BEd Hons ▪ MSc ▪ PhD ▪ 20+ Peer-reviewed journal papers ▪ 35+ Conference presentations ▪ <5 Book chapters
TE2	Male	4	>10	▪ MEd ▪ 30+ Conference presentations ▪ <5 Book chapters

Key:
P – Participant
G – Gender
AGE – Age Group {**1:** 25 years and below; **2:** 26 – 29 years; **3:** 30 – 34 years; **4:** 35 years and above}
ETM – Experience Lecturing Mathematics Education Courses
Q – Qualifications

As illustrated in Table D.8, both participants are male, and although the first teacher educator has only been lecturing IP mathematics courses for three and a half years, the interviewee has a quite extensive research record which includes journal papers, conference presentations and book chapters. The interviewee's research interests lie in language and mathematics education in multilingual settings. On the other hand, the second teacher educator respondent has vast experience of over 10 years in lecturing IP CPTD mathematics courses.

The official language of learning and teaching at both the interviewees' teacher educator institutions is English. All teacher education institutions use English as a language of instruction; in addition, a few of these institutions use both English and Afrikaans as languages of instruction. Although the first teacher educator interviewee has lectured IP mathematics for less than five years, the participant can be objectively classified as experienced based on the information provided earlier highlighting the interviewee's profile. The interviewees' relatively vast research output in the field of language and mathematics in multilingual settings position the data presented as contemporary and valuable. The second teacher educator interviewee has been lecturing IP mathematics for more than 10 years and is currently directing the institution's in-service mathematics teacher education programme, which primarily provides professional development of mathematics teachers.

The Teacher Educator Interview Schedule was used to gather specific data on the following aspects: the IP mathematics teacher educator qualifications offered by their institution; whether the qualifications include a module on using English as a language of instruction; strategies used by the teacher educator to teach specialised mathematics vocabulary; teaching and learning resources recommended by the teacher educator for use in addressing language-based mathematical problems; and any major challenges usually experienced by the teacher educator while observing IP mathematics teachers using English as a language of instruction.

IP mathematics teacher education qualifications offered

Table D.9: IP mathematics teacher education qualifications offered

	Question 3: What in-service (IP) mathematics teacher education qualification(s) does your institution offer?
TE1	▪ The institution has a special unit that focuses on in-service teacher educator qualifications varying from short courses, Diplomas, Bachelor of Education, Postgraduate Certificates and Postgraduate Diplomas in Education, Master's in Education to PhD.
TE2	▪ Bachelor of Education (In-service)

Pedagogical Implications: IP mathematics teacher education qualifications offered

As illustrated in Table D.9, the first interviewee's institution provides a variety of mathematics teacher education qualifications for both in-service and pre-service teacher education and the respondent lectures modules in both streams. The second teacher educator interviewee's institution also provides both in-service and pre-service teacher education qualifications and the respondent specifically lectures on the in-service programme.

The data confirms information provided in earlier sections of this case which indicates that teacher educator institutions usually separate the provision of pre-service and in-service qualifications. Among other purposes, this separation serves to ensure the smooth administration of the different programmes as well as to cater for special needs of each unit. However, it is of paramount importance to coordinate the separate provisions so that practices in the two units inform and influence each other because as soon as a teacher graduates from pre-service training, they immediately become a possible candidate for in-service teacher education. The literature also revealed that the focus on early childhood development as well as school-leaving learners belonging to the FP and FET bands respectively has driven some teacher education institutions to cater only for these bands at the expense of the IP and SP phases. This case views the education system holistically and contends that it should be treated as such in order to combat the creation of artificial but detrimental gaps within the system.

Modules focusing on using English as LoLT in teacher education qualifications

Table D.10: Inclusion of modules focusing on using English as LoLT in teacher education qualifications

	Question 4: Do any of your IP mathematics teacher education qualifications include a module focusing on using English as a Language of Learning and Teaching?
TE1	▪ Yes, there are 2 modules focusing on the use of English as a medium of instruction; however, this module is offered as an elective course at Masters' level.
TE2	▪ Not exactly. It is only a small component of LoLT that is embedded in a section of another module that focuses on Mathematics and language.

Pedagogical Implications: Modules focusing on using English as LoLT in teacher education qualifications

The data presented in Table D.10 indicates that in the first respondent's teacher educator institution there are two modules focusing on the use of English as a language of instruction; in the second respondent's teacher educator institution there is no specific module focusing on English as a language of instruction. This confirms data gathered in the questionnaires. The second respondent further affirms that:

> '*... it is* ***only a small component of LOLT*** (my emphasis) *that is embedded in a section of another module...*'

It is worth emphasising, however, that of the 10 sampled teacher educator institutions, the first respondent's teacher educator institution is the only one providing comprehensive modules (*at Masters level*) focusing on using English as LoLT. This is despite the fact that the new Curriculum Policy of the DHE: Minimum Requirements for Teacher Education Qualifications (MRTEQ)[1] prescribes that one of the competencies of a newly qualified teacher is to "know how to communicate effectively in general, as well as in relation to their subject(s), in order to mediate learning" (MRTEQ, 2015: 20). Thus, as established in the teacher educator

1 The MRTEQ provides "a basis for the construction of core curricula for Initial Teacher Education (ITE), as well as for Continuing Professional Development (CPD) Programmes that accredited institutions must use in order to develop programmes leading to teacher education qualifications" [p 8]. The policy is based on the 2011 MRTEQ policy, with revisions in place in order to align it with the Higher Education Qualifications Sub-Framework (HEQSF).

questionnaires, it can be inferred that the majority of teacher educator institutions where the participant teacher educators are based do not provide modules focusing on the systematic and consistent use of the prescribed language of instruction in mathematics. It is also worth noting that this practice is not peculiar to South Africa alone, as Chitera's (2016) investigation (focusing on the use of mother-tongue instruction in mathematics) conducted in Malawi revealed similar findings. Thus, for the purposes of improving the quality of mathematics instruction in schools, all teacher educator institutions need to initiate and place even greater emphasis on the languages that will be used to convey the mathematical concepts taught. To reap better results, the focus on languages of instruction should not be left until advanced studies at Masters Level, as indicated in the above data, but should be compulsory in undergraduate initial teacher education programmes for the benefit of under-resourced schools and communities which cannot attract better qualified teachers.

Teacher Educator Strategies for teaching specialised mathematical vocabulary

Table D.11: Strategies for teaching specialised mathematical vocabulary used by teacher educators

	Question 5: How do you teach specialised mathematical vocabulary in your lectures?
TE1	▪ I talk about the mathematics register. ▪ I also talk about the aspects of mathematics register that make it peculiar. ▪ I make pre-service teachers aware of the informal uses of mathematical terms. ▪ I show the students how the lack of understanding of the mathematics register may impact negatively on the general understanding of mathematical concepts. ▪ I make students aware of how different aspects of a language may be interchanged.
TE2	▪ I identify such vocabulary with the students and then discuss these terms on whole class basis. ▪ They are also given tasks in which they are to identify various types of vocabulary used in the teaching and learning of mathematics. ▪ Students are required to come up with a glossary of terms that are used predominantly at their level/phase.

Pedagogical Implications: Teacher educator strategies for teaching specialised mathematical vocabulary

The strategies for teaching specialised mathematical vocabulary suggested by the two teacher educator respondents as presented in Table D.11 are very useful. The second teacher educator respondent encourages pre-service teachers '*to come up with a glossary of terms*'; however, Hinkel (2007) indicates that compiling a list of words is not good enough, so this strategy needs to be expanded further to make the compiled words more meaningful to the English-language learners in an IP mathematics class:

> … a large passive vocabulary does not necessarily result in a better active vocabulary. Active vocabulary is one a language user can easily and readily draw on for productive language use whereas passive vocabulary is somehow distant and slightly hidden.
>
> Hinkel (2007: 6)

This could mean that, inasmuch as these strategies are useful theoretically, for example, '*talking about mathematics register*', not many of them are sustained in everyday classroom practice, particularly in under-resourced settings. This may mean that the way these strategies are presented should change so that they become more practical, relevant and adaptable in the classrooms. Based on the example given above, teacher educators need to model good practice on how to cultivate an active mathematical vocabulary, instead of a passive vocabulary which is accumulated through the current strategies in use.

Recommended resources for addressing language-based mathematical problems

Table D.12: Teaching and learning resources recommended by teacher educators to address language-based mathematical problems

	Question 6: What teaching and learning resources do you recommend for use by in-service teachers to address language-based mathematical problems?
TE1	▪ Textbooks ▪ Computer-based activities ▪ Teachers may use multilingual dictionaries with caution because these are not standardised ▪ Word walls and charts
TE2	▪ I have a few textbooks that were developed here in South Africa that I use when teaching this section. ▪ I also have textbooks form other places outside South Africa that were developed to assist mathematics teaching of English Second Language (ESL) learners. These are mainly from countries like Australia, America, Malaysia and others. ▪ There are journal articles that focus on this area. I also use them as reading material for my students.

Pedagogical implications: Recommended resources for addressing language-based mathematical problems

As shown in Table D.12, the first teacher educator interviewee's responses focus on the teaching and learning resources that the teachers can possibly use within a classroom context (practice), while the second teacher educator interviewee's responses focus on the teaching and learning resources that are suitable for teacher education purposes (instruction). This disparity could indicate that there is no standardised module that guides teacher educators across different teacher education institutions on how to equip teachers with practical and relevant skills to address language-based mathematical problems and hence different institutions approach the task differently. Thus, inasmuch as the resources suggested by both the teacher educator respondents are very useful in teacher education instruction and practice, they need to be coordinated. If teachers read widely on different strategies used globally to assist ELLs in addressing language-based mathematical problems, they will be better equipped to adapt these strategies to assist learners in their classrooms.

Four different classroom-based teaching and learning resources are presented in Table D.12 as recommendations for addressing language-based mathematical problems. Textbooks, dictionaries and charts should be accessible in schools. Since multilingual dictionaries are 'yet to be standardised' teacher education programmes should provide information on how to use these resources, so that the 'use with caution' advice is left not only to the discretion of the teacher. Similarly, the use of word walls and charts, which seem to be a more accessible option among the four provided, should be interactive in nature for the purposes of benefiting the learners.

Considering that this case focuses on under-resourced schools, computer-based strategies may not be practically applicable. However, they cannot be ruled out as electrification of rural areas in South Africa is an ongoing process and, at some point in the future, these schools will eventually be in a position to use computers. Some educational partners are also in the process of providing computers and internet access to remote schools in South Africa, therefore teachers should be empowered with the skills of using computer-based activities to address language-based mathematical problems. Thus, teaching and learning strategies need to go beyond identification to encompass their application.

Challenges experienced by teacher educators when lecturing/observing IP mathematics teachers using English as LoLT

Table D.13: Major challenges experienced by teacher educators when lecturing/observing IP mathematics teachers using English as LoLT

	Question 7: What major challenges do you experience when lecturing/observing mathematics intermediate phase teachers using English as a language of learning and teaching?
TE1	▪ Teachers who are not aware of how language can impact the understanding of mathematics. ▪ Teachers who battle with communicating in English. ▪ Teachers who have the mathematical knowledge but have no language skills to explain the concepts. ▪ Teachers who use code switching inappropriately, i.e. teachers who do not code switch to help the learners but to help themselves to continue with the sentence.
TE2	▪ Challenges are complex. ▪ The teachers' English-language competence is not up to standard. ▪ Teachers are struggling with English as a language for their own purposes. This, coupled with their alarmingly low content knowledge, makes the problem complex. ▪ I work with in-service teachers who already have at least a diploma or certificate in education. At university, there are assumptions that such teachers have the necessary basic competency in English language yet this is not always the case. ▪ The other challenge is the way in which teachers are made/deployed to teach mathematics at IP. From the conversations I have had with most of my students (in-service teachers), they are not specialists in the subject, i.e. mathematics. They are either asked to volunteer because they have at least Standard Grade maths or they take up just to help out in their schools.

Pedagogical Implications: Challenges experienced by teacher educators when lecturing/observing IP mathematics teachers using English as LoLT

The challenges presented by both the teacher educator interviewees in Table D.13 confirm information collected through the teacher educator questionnaire. In general, the information presented confirms that IP mathematics teachers' linguistic skills need improvement:

- *'Teachers who battle with communicating in English'*
- *'The teachers' English-language competence is not up to standard'*
- *'Teachers are struggling with English as a language for their own purposes'*

The two respondents both confirm that the IP mathematics teachers' language competence in English is significantly low. The second respondent highlights that, although the qualified IP mathematics teachers' English language competence is significantly low, teacher educator institutions offer CPTD courses assuming that the teachers' basic competence in English is sufficient, yet this is not the case. The reiteration is that teacher educator institutions cannot continue working on this assumption as competence in the language of instruction is fundamental for meaningful learning to take place. Thus, if the English-language competence of IP mathematics teachers, which this case regards as the primary challenge, is addressed adequately; the other secondary challenges such as when and how to code switch or translanguage, teacher deployment and mathematical content knowledge would be manageable.

CONCLUSION

The data presented was gathered from teacher educator questionnaires administered to 10 sampled teacher educators and interviews, administered to 2 (20%) of the sampled teacher educators. Of greater importance in the data presented is the notion of awareness in how language impacts on the understanding of mathematical concepts which further confirms a gap in teacher education. This awareness should not only be created during initial teacher education, but promoted and constantly revived through in-service programmes. In addition, the view that some teachers might have mathematical knowledge without the supporting language skills to convey the knowledge widens the gap and presents a missed opportunity to merge content knowledge and linguistic proficiency.

Data from the teacher educator interviews confirmed that although the majority of teacher educator institutions use English as the official language of instruction, there is not sufficient consideration given to teachers who are non-native-language speakers of English in terms of equipping them with adequate skills to use English as language of instruction in their practice. Although teacher education institutions are aware of the linguistic challenges faced by the IP mathematics teachers in their daily practice, there is a gap between this awareness and what needs to be done to remedy the situation. The researcher further reiterates that it is the mandate of university departments of teacher education to narrow these gaps.

CHAPTER 10

SO WHAT? LESSONS LEARNT

INTRODUCTION

So what if IP mathematics teachers' language competency and content knowledge regarding word problems is below what is expected of a grade 7 learner?

If this is the status quo, the quality of mathematics education in South Africa is undoubtedly a cause for serious concern. This is, however, no revelation as several national and international assessments on education (TIMSS, 2013; PISA) have repeatedly narrated the popular South African mathematics story. The pertinent question therefore is: ***What needs to be done to redress the situation?*** Answering this question requires a better understanding of why mathematics is both put on a pedestal while at the same time it is a high-risk subject failed by a large number of learners in schools.

Mathematics is important for success in school, not just for some learners but for each and every learner (National Council of Teachers of Mathematics [NCTM], 2000). However, learning Mathematics is not an easy task for many learners, especially those who also need to overcome the language barrier. In contemporary South Africa, where the citizenry is enjoying cultural and linguistic diversity, poor performance in mathematics (specifically for English Language Learners [ELLs]) is evident across different levels of the education system. Although research on how the brain simultaneously processes language and mathematics information is still underway (Chen & Yeping, 2010), recent research suggests that a learner's processing of numbers in the brain is shaped by different language processes as well as other cultural factors such as mathematics teaching and learning. By extrapolation, the processing of numbers that takes place in a mathematics learner's brain is similar to the processing of numbers that takes place in a mathematics student teacher's brain. A better understanding of the learner's language and some other cultural experiences are therefore very important for mathematics teachers and teacher educators when considering and developing interventions for improving mathematics teaching and learning.

Language used to define mathematical concepts and express mathematical ideas plays a key role in mathematics teaching and learning. Learners with different language backgrounds may think about mathematics differently, and this may also influence their mathematics performance. In this compilation, learners whose first language is not English are referred to as ELLs. A number of studies have interrogated ELLs' mathematics performance (Clarkson, 1988; Adler, 2001; Setati, 2005; Sepeng, 2015) and the relevant results indicate that there is an achievement gap between ELLs and non-ELLs (Chipman, 1988, Mestre, 1988). These results suggest that learners' language competencies impact their mathematical performance over a number of different contexts, including word problems. Language in mathematics education researchers concur that language proficiency (or competency), is one of the crucial factors determining (and influencing) ELLs' mathematics performance (Clarkson, 2007). Vast as this research may be, there is a dearth of similar research interrogating the impact of teacher language competencies on mathematics instruction.

In order to reduce the mathematics performance gaps between ELLs and non-ELLs, researchers have studied the possible relationships between language proficiency and cognition (Cummins, 1979, 1981) and explained the relationship in terms of the common underlying proficiencies (CUP) and other language processing theories such as bilingualism and Second-language Acquisition. These language processing theories are detailed in this compilation in order to illuminate the impact of competency in the language of instruction on mathematics instruction. This impact is established through the following section: Language as a Resource in Mathematics Teaching and Learning, Mathematics Teaching and Learning in a Second Language, Language Competency for Mathematics Instruction, Perspectives on Language of Learning and Teaching Mathematics, Multilingual Based Bilingual Education (MTBBE) for Mathematics and Mathematics Teacher Education for Multilingual Settings.

This chapter summarises the whole and provides concise findings as well as the recommendations based on the different data-gathering techniques and sources used. It provides a summative discussion on the relationships between teachers' English-language competency and IP mathematics instruction. It also provides a concise discussion of teachers' linguistic practices when teaching mathematics in under-resourced settings. These findings, conclusions and recommendations are discussed against the backdrop of the conceptual framing and the literature reviewed. The results and findings from semi-structured teacher and teacher educator interviews as well as those from informed non-participant classroom observations corroborate the results and findings from the teacher questionnaires, teacher educator questionnaires, and the English-language competency and Mathematics word problem assessments. These results and findings shed more light on the situation regarding teacher language competency in English, the prescribed language for teaching, and IP mathematics instruction. They also illuminate the conditions in which linguistically disadvantaged IP mathematics teachers find themselves.

Interrogating how (***if at all***) language competency in the LoLT relates to content delivery by IP mathematics teachers sought to focus the readers' attention on critical concerns that need to be addressed regarding mathematics education in under-resourced multilingual settings. This text concludes that teacher language competency in the language of instruction does ***indeed*** relate to IP mathematics content delivery.

Since research is important for signposting and informing the way ahead, the section below makes some recommendations based on the findings measured against the set objectives. These findings are primarily based on the case studies included in this book, with very small sample sizes (55 Intermediate Phase mathematics teachers and 10 Intermediate Phase mathematics teacher educators). These case studies are valuable in their own right. However, their limitation is in extrapolating the findings to a broader population in the absence of bigger national longitudinal studies. Inasmuch as the suggested recommendations need to be viewed as based on case studies, these case studies may be used as a platform to conduct longitudinal studies as a national initiative to improve the quality of mathematics education in South Africa.

RECOMMENDATIONS MEASURED AGAINST THE OBJECTIVES

i. The extent to which proficiency in the LoLT relates to IP mathematics teachers' content delivery

Proficiency in the language of instruction is directly related to mathematics content delivery. Teachers with a better mastery of the language of instruction are in a better position to explain new terms to their learners and to create language learning opportunities within mathematics content delivery. On the other hand, teachers with a poor mastery of the language of instruction are not confident enough to model good linguistic practices; they are unable to effectively assist learners to perform basic reading and writing skills or to guide learners to explain mathematical concepts in their own words – skills through which the mathematical content is often assessed.

ii. Aspects of IP mathematics teacher education focusing on the use of English as LoLT

Teacher educator institutions provide very little guidance towards coherent and systematic use of code switching/translanguaging in the classroom to enhance opportunities to learn. Only 1 out of the 10 sampled teacher educator institutions provided modules that focus on the use of English as language of instruction. Unfortunately, the modules are provided as electives at Master's Level; this is a level of education which the majority of teachers based in under-resourced schools would not have attained by the time they are deployed to schools. Chapter 2 of the text: ***Language as a Resource in Mathematics Teaching and Learning*** contributes towards addressing this need. The chapter serves as a platform for further developing the prerequisite linguistic skills needed by IP mathematics teachers. Data from teacher questionnaires revealed that a few teachers in their initial teacher education qualifications covered some aspects related to the use of English for communication purposes; the data confirm that there is inconsistency in the guidance provided by teacher education systems on the use of English as a language of instruction.

iii. The extent to which IP mathematics teachers in the ECDoE schools are proficient in English, the prescribed LoLT

With a maximum of 60% and a minimum of 11% obtained in the teacher English language competency assessment, teachers revealed a lower proficiency in English than what is expected of a Grade 7 learner. Similarly, with a maximum of 70% and a minimum of 16% teachers revealed a lower performance in mathematics word problems than is expected of a Grade 7 learner. In addition, statistical and descriptive analyses indicated a positive correlation between English language competency and mathematics word problem assessments. Even though English is a foreign language to the majority of the teachers who use it as language of instruction in South Africa, the mere fact that the current LiEP requires teachers to use English as the medium of instruction calls for high levels of proficiency in that language. The MRTEQ documents, which stipulate the minimum requirements for teacher education qualifications, requires IP teachers to at least have Additional Language proficiency in English (MRTEQ, 2015: 24). However, the answers provided by the teachers in the English language proficiency assessment as well as in the Mathematics word problems revealed grammatical and idiomatic errors and it is consequently inferred that Additional Language proficiency in the medium of instruction is not good enough for an IP mathematics teacher.

iv. Teaching methods currently in use in IP classrooms that may overcome language barriers in the delivery of mathematics

As a coping strategy IP mathematics teachers over-rely on explaining mathematics concepts and procedures in the first or home language (isiXhosa), which is not the official language of instruction/assessment. Learners may not know English, but they are not passive either. When a teacher over-relies on explaining concepts in the mother tongue, learners become passive because they know the teacher is going to translate and so they simply wait for the teacher's translation instead of paying more attention to the acquisition of the mathematical register in English. Teachers' classrooms are not text-rich environments and the lessons are teacher-centred; the classroom environment does not have enough linguistic features to support the use English as a language of instruction. Thus, very little is being done to overcome the language barrier in under-resourced classrooms. Evidently, the over-reliance on isiXhosa is a disservice to the learners, but it is the teacher educator institutions that are actually performing a disservice to the nation by not serving teachers properly, who in turn are not equipped to serve the learners adequately.

v. Effectiveness of current teaching methods in overcoming language barriers in the delivery of IP mathematics

The methods currently in use are not effective in overcoming language barriers because without systematic guidance, the teachers use code switching haphazardly or prompt learners to use their mother tongue during class discussions in order to promote interaction in their classrooms. Teachers who are not competent enough in English are unable to model the required fluency in the language of instruction, and hence risk leaving learners' mistakes in English uncorrected. The haphazard use of code switching indicates that the code switching is not incorporated into the planning of the lesson.

The clearest finding arising out of this text is that competence in English is necessary for teachers to engage in high-quality mathematics instruction in English. However, it may not be sufficient. Other important components likely include: content knowledge, pedagogical content knowledge, and specific knowledge of how to teach mathematics to English learners.

vi. State decision making on Intermediate Phase mathematics teacher education curriculum design

The key findings discussed in this chapter highlight some crucial factors that policy makers, teacher education curriculum developers, teacher educators and teachers need to consider for the purposes of improving the quality of teaching and teaching primary mathematics.

As alluded to in Chapter 2 of this text mathematics education begins and proceeds in a language, and it advances or stumbles because of language, and its outcomes are often assessed in language (Durkin & Shire 1991: 3). Although this observation is applicable to most of the ordinary school curricula, the interweaving of language and mathematics is particularly intricate and this text explored some aspects of this complex interrelationship. To better serve IP multilingual learners so that mathematical concepts are linguistically accessible to them, their teachers' English-language competence needs to be improved drastically. Teacher education pedagogy needs to provide comprehensive guidance on code switching and pedagogical translanguaging, the characteristics of English-Language Learners (ELLs),and ensuring that mathematics classroom are both communicative and text-rich.

The section below provides language-related recommendations stemming directly from the comprehensive cases analysed in this text, followed by some key lessons learnt.

RECOMMENDATIONS STEMMING DIRECTLY FROM THE COMPREHENSIVE CASE STUDIES

Six recommendations stemming directly from the data collected and analysed in the comprehensive cases are listed below.

1. Improve mathematics teachers' English-language competence

Generally, mathematics teacher education in South Africa does not enforce mastery of the language of instruction. Competence in English as correlated with mathematics performance on word problems is highlighted. Thus, competence in English needs to be regarded as one of the significant predictors of mathematics performance, particularly because the country's indigenous languages have yet to be fully developed to support mathematics instruction. This recommendation needs to be interpreted in light of other settings where Indigenous African Languages (IALs) have been shown to support mathematics instruction. Hurst (2016) suggests that the dominance of English in South African education at the expense of IALs may be viewed as propagating the colonial wound. However, without politicising the debate on language in education, it is necessary to provide learners with several alternatives – alternatives that will afford learners in under-resourced communities access to better opportunities. In addition, language barriers result in gaps in content knowledge during the primary school years, which in turn usually contribute to lack of the intellectual growth necessary for high school learning (National Education Evaluation and Development Unit [NEEDU], 2013). Any education system is only as strong as its teachers; therefore, promoting language competence during and after initial teacher education may significantly improve on the quality of mathematics education in South Africa.

Teachers may not possibly know all the content, but they can be skilled in how to learn new content through language. Evidence from the teachers' English language competency and mathematics word problem assessments revealed that the teachers' mastery of English is significantly low. It is the duty of *all* public university education departments as well as other private teacher education institutions to develop and improve IP mathematics teachers' English language competence.

Adesina (2017) notes that duty is to people, not to establishments, and therefore in order to improve people's lives, there is a need to invest in brainpower, that is the *grey matter infrastructure*. This book thus contends that one of the ways of improving the quality of IP mathematics education should include investing in the *linguistic infrastructure* of the teachers.

2. Improve mathematics teachers' (pedagogical) content knowledge regarding word problems

Research focusing on learners' comprehension of mathematical word problems can be viewed from three different but related perspectives: the mathematics teacher's perspective, the language teacher's perspective, or a pedagogical perspective. Mathematics teachers emphasise the mathematics processes and procedures of the learners; thus, learners must possess certain understandings to succeed in mathematics, and that the correlation between mathematics computational skills and the reading of mathematics word problems should be positive and high. Certain reading skills are also important for success in solving mathematics word problems. Vocabulary development and literal interpretation of the problem are very important, while clarity in textbooks and other teaching and learning resources should be emphasised. Three factors interfering with reading and solving mathematics problems are absence of a diagram, presence of irrelevant information, and incorrect order of numerical information. A significant percentage of the variance observed on mathematics assessment is determined by reading comprehension strategies. This suggests that the mathematics teacher's perspective must be expanded to include something in addition to mathematics concepts, generalisations, and number facts. Therefore, the three perspectives need to form part of both pre- and in-service mathematics teacher education.

Instead of just focusing on suggested strategies for solving mathematical word problems, it is crucial that teachers also understand factors affecting word problem solving such as mathematical complexity, language proficiency, linguistic complexity, semantic factors, numerical complexity, solution strategies, information relevance and lexical consistency. Mathematics teachers also need to be aware of the interrelation of these factors. This understanding will broaden the teachers' proficiency in solving word problems, a competency which can be transferred to learners once mastered.

3. Provide guidance on the use of code switching and pedagogical translanguaging

Chitera (2016) notes that mathematics teacher education in developing countries provides minimal or no guidance towards coherent and systematic use of translanguaging and code switching in the classroom during or after initial teacher education. Pedagogical translanguaging and code switching take place when an individual uses more than one language or varieties of a language in conversation and these practices can be used by the teacher as resources for explaining mathematical content (Baker, 2011). Although Chitera focuses on indigenous language instruction, this book highlights the need not to assume that enough guidance is provided when a different lingua franca like English is used as a medium of instruction. Such a practice compromises the quality of content delivery, as pedagogically disadvantaged teachers may in turn perform a disservice to the majority of learners in public schools who are multilingual – in other words, those who are also ELLs.

Van der Walt (2016) deems language as a resource that can be used to access information; however, evidence from the classroom observations conducted in Case C confirm that teachers were not sure when and how to code switch in a manner that benefits the learner. The teachers were not aware of practices such as translanguaging, which can guide them in how to use their African indigenous languages together with English for the purposes of effectively transferring mathematics content to the learners. This book seeks to highlight the importance of code switching effectively, not as a substitute or coping mechanism for incompetence, but bearing in mind that assessment will be done in the prescribed LoLT (Tshuma, 2016). The book also seeks to highlight the positive aspects of pedagogical translanguaging so that teachers may draw on them in navigating mathematics teaching and learning in multilingual settings. It does not have to be code switching, code mixing or translanguaging, but any well-researched pedagogy that can impove the teaching and learning of mathematics in multilingual under-resourced settings needs to be incorporated into the initial teacher edication curriculum.

4. Provide knowledge about English-language learners

Current South African mathematics pedagogy does not foreground knowledge about ELLs, which reflects a gap in both practice and research. In mathematics the challenge faced by ELLs is threefold in that they have to acquire the new language of learning, as well as learning mathematics and its register (Barwell et al., 2002; Bohlmann, 2001; Setati & Adler, 2000). The mathematics register is the special vocabulary used in mathematics, as well as the phrases and methods of arguing within a given mathematics situation. The mathematics register allows learners to communicate their mathematical findings; provides analytical, descriptive and problem-solving skills; and improves their ability to listen to, question, discuss, read and record mathematical concepts.

Evidence from the classroom observations reveals that current teaching methods used by the teachers do not take cognisance of the distinguishing characteristics of ELLS. In the observed classrooms where teachers conducted most parts of the lesson in mother tongue, the teachers did not provide learners with sufficient opportunities to communicate their mathematical understanding in English. Even though using the mother tongue may be comfortable for both the teacher and the learners, the practice does not provide learners with opportunities to read, understand and interpret texts in the prescribed LoLT, the very skill which is ***always*** assessed through examinations.

5. Make mathematics classrooms text-rich

A classroom is an environment that must promote and reinforce learning. One of the linguistic features that influences mathematics teaching and learning includes making a classroom a text-rich environment. This can be done by creating opportunities to read in the classroom environment, opportunities which in under-resourced communities may be difficult to create (understandably so) outside the classroom. Some classrooms observed in Case C had very few or no charts or word walls displayed in the classrooms. Creating a text-rich classroom may encourage learners to practise reading the text as well as provide opportunities for engaging with texts outside of the conventional teaching and learning time.

To bypass a shortage of resources and to reduce the financial strain of buying ready-made charts, teachers may encourage learners to make word walls, an activity that may reduce the teachers' workload and simultaneously enhance learning opportunities for learners. Furthermore, instead of displaying ready-made charts that only serve as aesthetic artefacts rarely referred to during instruction, the teacher may celebrate the learners' work by displaying it on the walls to make the classroom text-rich.

6. Make IP mathematics classrooms communicative

With the adoption of a constructivist philosophy, mathematics education researchers now advocate more active learning for learners and a more facilitative role for teachers. A key component of most new instructional strategies is that learners are expected to discuss mathematics with fellow learners and their teachers to communicate real meaning. This new emphasis upon mathematical communication is a challenge for both teachers and learners in traditional classrooms. The language used in mathematics teaching and learning is both a means of communication and an instrument of thought (Esty, 1992).

Communication is one of the processes that need to be developed with the teaching of mathematical content and mathematical thinking: the ability to communicate findings and provide explanations is considered as an important outcome of education (CDD, 2006). Mathematics provides a powerful means of communication which can be used to present information in figures, tables, charts, graphs and symbols and the process of communication is to be developed at the same time as the teaching of mathematics content together with other skills such as mathematical skills and problem solving. Among the main aims stated in the mathematics curriculum is to develop learners' ability to interpret and communicate mathematical ideas and provide learners with the opportunity to develop the ability to communicate mathematical ideas clearly and work with others and value their contribution (DBE, 2011). In the classroom observation, communication was dominated by teachers, and in the few cases that students had an opportunity to respond, chorus answer and chanting were used. Teacher-dominated communication or chorus answers do not cultivate active learners among learners nor does it promote discussion of mathematical ideas among peers or between learners and teachers.

Teachers should provide opportunities for learners who would otherwise not share their views to communicate meaningfully in mathematics lessons. Teachers play a big role in encouraging and determining the success of communication in any class. By using language as a resource and exploiting semiotic resources for meaning making in mathematics classroom, teachers should encourage learners to take part in mathematical discussion, or explain their mathematical thinking processes, or compare alternative solutions with peers. However, in order to do this, pre- and in-service teachers need sufficient preparation and relevant continuous professional development from the teacher education institutions.

This researcher contends that the first four recommendations above are largely dependent on the macro bodies of the country's education system, i.e. the policy makers, curriculum developers and teacher education institutions, while the last two recommendations are dependent on the initiative of teachers, who are the foot soldiers of the education system. However, all these stakeholders need to work together for the improvement of the quality of mathematics education in South Africa.

The section below extracts lessons that empower IP mathematics teachers to effectively overcome language barriers in content delivery.

KEY LESSONS LEARNT

Three key lessons stemming indirectly from the literature and comprehensive cases analysed in this book are listed below. These lessons are presented as questions emanating from the researcher's evaluation of the current body of knowledge against the backdrop of other studies conducted in the field of language in mathematics education in multilingual settings.

i. Declaration without Implementation?

Findings from this book imply that teachers' competence in English needs to be improved. The researcher contends that this improvement needs to take cognisance of the following factors.

- Maybe there is no need to choose between English or an African Language to be used as the LoLT, but both are possible and desirable. Taking this direction would need an awareness of the concept of balanced instruction, which embraces the strongest components of both the foreign and indigenous languages as possible and desirable languages of instruction. If this route is taken, then teacher education programmes need to be aware of the concept of balanced instruction and teachers need to be well trained during and after initial teacher education, so that they are better prepared to function effectively in multilingual IP mathematics classrooms where the learners' home language is perceived as one of the teaching and learning resources.
- As a teacher educator, the researcher contends that whatever choice is adopted regarding the language of instruction – be it English, an indigenous language or a balance between the two – the choice must be supported by its implementation from the initial teacher education curriculum through to CPTD programmes.

ii. Perceived versus Intended Needs?

Since teachers may not know that they need these linguistic skills for improved instruction at IP level, it is therefore the duty of policy makers to conduct needs analyses that will inform the teacher education curriculum on appropriate measures for bridging this gap, which may be threatening the quality of mathematics instruction. Policy makers know what the intended outcomes are, yet teachers may not know that they have a need, because they are not aware of the acceptable language proficiency levels expected of them. Therefore the way forward may involve policy makers reaching out to the teachers through the teacher educators (and the teacher education institutions) by redesigning the mathematics teacher education curriculum to include language proficiency modules. Teachers cannot still be trained the way they were trained decades ago; the mathematics teacher education curriculum needs to be constantly reviewed in order to keep up with best practices.

iii. Deficiency versus Alternative Strategies?

Despite the data collected, the findings and the interpretations, the researcher would nevertheless like to render all due respect to the teachers who participated in the Cases A-D and who could be taken as a representative sample of a larger population of teachers who are based in under-resourced communities. Data from the language competency assessments identified a deficiency in linguistic skills among IP teachers. This deficiency is understood in the light of the context the teachers find themselves in and, most importantly, with the intention of providing alternative strategies to remedy the situation. The work that is done by the teachers is viewed with utmost respect and treated with dignity, and, being a teacher educator, the researcher would like to see alternative strategies being put in place to develop the teachers' linguistic competencies for the purposes of improving their practice.

Being linguistically equipped and proficient in the prescribed LoLT is perceived as every teacher's wish. However, this is a task that teachers perform or overcome on their own. When the teachers were trained, which could be on average 10 to 15 years ago, the curriculum might not have registered the need to equip the teachers with linguistic strategies to improve IP mathematics instruction. Moreover, the chances are that the mature teachers currently teaching primary school mathematics within the education system were trained before the advent of Foundation Phase (FP), Intermediate Phase (IP), Senior Phase (SP) and Further Education and Training (FET) band divisions. However, as the current practices in teacher educator institutions seem to be catching up (for example, one institution is providing a module in this regard), there is a need to match the intended needs of the curriculum with the current practices in teacher education institutions.

This can be achieved firstly by ensuring that (*all*) teacher education institutions provide a unified pedagogy, one that involves a module on using language as a resource in IP mathematics instruction. Such a module can be offered to both the pre-service and the in-service IP mathematics teachers. Secondly, for those teachers who might have received aspects or small portions of the information, they may well benefit from refresher courses focusing on this important aspect which significantly impacts on mathematics content delivery. Thirdly, inasmuch as prospective engineers and geologists usually undertake extended degree programmes and receive bridging courses in mathematics so that they can cope with their studies, a similar approach may be adopted for teacher education degrees. Research reveals that students enrolled for education programmes usually have low scores for mathematics, science and languages, and so do not qualify for their intended courses of choice. Therefore, great emphasis should be placed on rigorously preparing these students so that they do not continue the vicious circle of low-quality education. And lastly, it may be necessary to upgrade the entry qualifications for teacher education degrees in developing countries over a period of time, so that the teaching profession regains the prestige it once had.

The systemic concerns regarding poor mathematics performance *naturally* lead to interrogating the nature and quality of mathematics teaching and the knowledge base that the teaching draws from, and there is increasing critique too of the role of mathematics teacher education (Venkat in Webb & Roberts, 2017). Although Venkat specifically focuses on pre-service mathematics teacher education, Olivier, (2015) raises similar concerns regarding in-service mathematics teacher education:

> The sad reality in South Africa is that universities and other formal education service providers have not created a Mathematics in-service training program that has been rigorously evaluated and proven to effectively raise teacher content knowledge. Actually, no generic professional skills development program for in-service Mathematics teachers has ever existed in South Africa. This, despite many attempts by government and other academic stakeholders to draft and implement grandiose professional development programs. For example, very little if anything of what is stated in the Integrated Strategic Planning Framework for Teacher Education and Development in South Africa 2011–2025 (DBE, 2010) has been realised in practice.
>
> Olivier (2015)

Therefore, South African mathematics teacher education (both pre-and in-service) may take a leaf from the following:

1. Without literally *'getting rid of the dead wood and growing new timber'* teacher education institutions can refurbish their dead wood (in phases) by reskilling the mature in-service teachers who are products of the Bantu Education System, through Continuous Professional Teacher Development (CPTD) programmes specifically targeted to keep them abreast with the required levels of teacher (pedagogical) content knowledge. CPTD in South Africa needs to be revamped so that it moves away from daylong and weeklong workshops into more structured interventions with a duration of at least 3 – 5 years long, beginning with the Foundation and Intermediate Phases, the lower levels of the education system.

   At the same time, teacher education institutions can also ensure that the new timber they are currently growing, (pre-service teachers) fully understands the country's need to improve the quality of (*mathematics*) education and is competent enough to effectively tackle the prevailing challenges.

2. One of the biggest challenges facing university teacher education departments is the narrow career progression structure. Good lecturers are promoted to administrators (management positions) or researchers in pursuit of becoming research-intensive institutions of higher education while undergraduate teaching is relegated to graduate students. There is a need to promote

high-achieving lecturers into curriculum development or innovation sectors. There should be parallel promotion routes within universities where exceptional lecturers do not get uprooted out of lecturing or curriculum development roles and promoted to management.

3. Instead of working in silos (*and attempting to solve cross-cutting challenges individually*), university teacher education departments need to collaborate and work together in solving challenges on content development, assessment procedures, innovative lecturing methods or analysis of student teacher results. They should form links with other universities to collectively solve problems (across the ordinary school curriculum, including STEM subjects) for the greater good of the country's quality of education.
4. University teacher education departments' key role should be producing the required numbers of competent, ***not just qualified***, but competent teachers. The departments need to be able to categorically stipulate how this would be achieved and explain how the departments of teacher education will:
   a. be transformed in alignment with the production of competent teachers – competent in mathematics procedures and also proficient in the LoLT;
   b. ensure that they have the best lecturers (attract, recruit, develop and retain); and
   c. ensure that they improve their lecturing methods.

Teacher educators' responsibility should be ensuring that the teachers produced by university departments of teacher education are of good quality and the university itself keeps up with the dynamic changes in education, taking into consideration the dynamic environments to which teachers are deployed.

RELEVANCE OF THE BOOK

Considering the fact that "language is not everything in education, but without language everything is nothing in education" (Wolf in Alidou et al., 2006), the book addresses a highly significant issue. The explicit contribution made by this book is to supplement the current body of knowledge regarding language in mathematics education in general, and specifically on teacher language competencies in the prescribed language they are to teach in. As such, findings have implications for the Department of Higher Education and Training (DHET) curriculum policies.

Inasmuch as the current teacher education policy is under review by the DHET, this chapter contends that this review should ensure that teacher education courses pay particular attention to the inclusion of modules focusing on using English as a language of instruction. The current review of the primary teacher education curriculum should not only focus on initial teacher education, but also include in-service teacher programmes, in order to accommodate practising teachers who did not get the opportunity to be upskilled in the language of instruction during their initial teacher education preparation.

The book makes a contribution to current knowledge on one of the factors that need attention in IP mathematics instruction. Language plays a pivotal part in accessing knowledge and the book highlights how language plays a critical role in IP mathematics instruction, specifically if the language of instruction is foreign to both the teachers and the learners. Numerous investigations have been conducted regarding language use in multilingual classrooms; however, the majority of these investigations focus on the learners with the assumption that the teachers' language competence is up to standard. Looking at the flip side of the coin entails paying significantly more attention to teachers' language proficiency. This book has confirmed that the teachers' competence in the language of instruction is not always up to standard, with specific reference to teachers deployed to under-resourced schools that cannot attract a better workforce.

Thus the purpose of this book was not to make generalisations about the prevailing relationships, but to fill a gap in the literature by examining the linguistic competencies of IP mathematics teachers in the Eastern Cape Province's under-resourced schools. Although the empirical data presented in this book is from one province which is rather too small to base national recommendations on, the book nevertheless opens up further discussion and serves as a platform for large-scale longitudinal research. The point of this book is to make a contribution and present a different focal lens to the existing literature by presenting a case that sheds light on an often overlooked aspect – the language competencies of the IP mathematics teachers in under-resourced schools.

GAPS TO BE PLUGGED

The book provides a platform for future in-depth studies on a number of language-related matters that impede IP mathematics instruction.

There is a need to investigate how much time in-service IP mathematics teachers spend on word problem solving instruction in a given school calendar week, term or year. Furthermore, analysing the instructional time spent on mathematical word problems in under-resourced multilingual settings compared to well-resourced settings may provide comparative analyses. Other offsets of such an investigation may include the differential effects of instruction on low-, average- and high-performing learners or the effects of instruction on learners at different levels of the Intermediate Phase – grade 4 (entry level), grade 5, grade 6 and Grade 7 (exit level) – to establish whether different strategies improve learner performance in mathematical word problem solving.

Word problem research should take into account the foundational categories, properties and findings of both numerical cognition and linguistics when it examines which word problems are difficult for which groups and why. Not only the main effects of numerical and linguistic complexity should be studied, but also their interaction and strategic collaboration between linguists and numerical cognition researchers would be desirable.

There is a possibility for related studies on the relationships between language competency and mathematics instruction, such as studies focusing on other LoLTs, like Afrikaans or the South African Sign Language (SASL), which is used as an LoLT in schools for the hearing-impaired. This idea emanated from one of the national conferences for mathematics teachers. However, the idea was not followed up here as it fell beyond the scope of this book. Nonetheless, a seed was sewn.

In March 2014 and subsequently in June 2016 learners at Bartimea School for the Deaf and Blind in Thaba'Nchu, Free State Province, protested and vandalised school property. A wide spectrum of reasons was cited for the cause of the protests, including sub-standard education. According to the learners, the teachers were not well trained in South African Sign Language (SASL), which is the language of instruction for the deaf section in the school (*Bloemfontein Courant*, 2016). Although the use of SASL as a language of instruction is beyond the scope of this book, the area needs further research for the benefit of improving the quality of instruction using the particular language across the curriculum. Even though hearing-impaired learners constitute a relatively small fraction of the learners within the ordinary education system in the country, they still need to be provided for linguistically.

Similarly, the majority of IP mathematics teachers who undergo initial teacher education in South African institutions should be linguistically empowered as a contribution to improving the quality of education in the Eastern Cape Province and in the country as a whole. Teacher education institutions should not miss the opportunity of joining hands between proficiency in the language of instruction and mathematics content knowledge. Metaphorically, this is like two people walking alongside each other and smiling without shaking hands – and that is a missed opportunity. **In reality, the missed opportunity between language of instruction and mathematics content knowledge is not only a matter of kindness – it is an essential one for the sake of effective education**.

CONCLUSION

The various chapters have explored the relation between language competency and IP mathematics instruction, but why does this matter? The importance of language for the purposes of teaching, learning, understanding and communication of mathematics cannot be left unattended, and the findings from the cases herein confirm the existence of an intricate relationship between language competency and mathematics content delivery.

One of the factors directly linked to the substandard quality of mathematics education in the Eastern Cape Province's multilingual classrooms is the teachers' inadequate mastery of the language of instruction. Having established the fact that teacher education institutions do not insist on mastery in the LoLT during or after initial teacher training, there is a gap in the delivery of mathematics content in the

IP years; this book intends to fill this gap. The value of the book is its empirically founded attempt to broaden understanding of the relationship between teacher language competence and IP mathematics instruction. The knowledge will inform further research on appropriate interventions to facilitate effective delivery of mathematics content in the prescribed LoLT in the IP years.

The history and politics of South Africa is marred by apartheid policies which had some particularly negative political interventions. In the light of the propagation of the Bantu Education Act of 1952, and the authoritarian philosophy of Fundamental Pedagogies whose sole purpose was to stunt critical reflection and produce compliant and docile teachers (Webb & Roberts, 2017), black South Africans may tend to feel negatively about any practices that seem to enshrine English, despite the established fact that English is undoubtedly the language of opportunity. As such, teacher education curricula, especially for the primary school level, which has a significant impact on the formative years of learners, needs to support and promote the use of English beyond the currently required Additional Language proficiency level. Whether English (or any African Indigenous Language) is chosen as language of instruction, IP teachers need to be upskilled in using the chosen language beyond basic proficiency levels to include the use of that language as a resource in mathematics instruction. This means that language would be viewed through the educational lenses that are appropriate, instead of being viewed through political lenses.

In a nutshell, this book contends that the ECDoE is quite a long way from reaching the goals of quality mathematics education, and teachers cannot achieve this goal alone. There is an urgent need to disrupt initial teacher education and CPTD programmes in South Africa, and this disruption needs to acknowledge that CPTD cannot *and should not* replace quality initial teacher education. In order to replace all the negatives identified here with positives, teachers need the support of their teacher education institutions as well as other educational development stakeholders. Adesina (2017) sums it up best by stating that **"If you want development for a year, grow a grain; for ten years, grow a tree; for a lifetime, grow people".** Developing the linguistic infrastructure of IP teachers will go a long way towards improving the quality of mathematics instruction in under-resourced communities and raising the quality of South African mathematics education in international rankings as well as in the international community at large.

GLOSSARY

As the terms listed below are used repeatedly in this book, they are operationally against the backdrop of the text's context.

African languages: the primary languages spoken by Africans in South Africa. These primary languages are formally called Bantu languages (Adler, 2001; Essien, 2013).

Bilingual: one who is proficient in two languages (Adler, 2001; Essien, 2013; Sibanda, 2014).

Bilingual classroom: a classroom that includes one or more learners who have varying levels of language proficiency across two languages.

Code switching: a trait of bilingualism and a form of translanguaging in which there is "juxtaposition within the same speech exchange of passages of speech belonging to two different grammatical systems of subsystems". It refers to the alternating use of two or more languages in the same utterance or conversation (Gumperz, 1982: 59).

Code mixing: the embedding of various linguistic units such as affixes, words, phrases and clauses from a cooperative activity, where the participants must reconcile what they hear with what they understand (Bokamba, 1989).

Content delivery: the transmission of attitudes, behaviour, knowledge and skills (Bernstein, 1990). In the text, the term 'content delivery' generally refers to how mathematical content is relayed from the teacher to the learner. In some cases, *mathematics* **instruction** is used instead of *mathematics* content delivery and the terms are used interchangeably.

Didactician: teacher educators, including university education lecturers and personnel whose core function is to instruct teachers who may not necessarily be based in university departments of teacher education.

Dual-language education: In South Africa dual-language programmes tend to share the following characteristics: (a) languages are separated during instruction, and (b) both English and non-English speakers are present in nearly balanced numbers (Baker, 2011).

English-language learner (ELL): a learner who "has sufficient difficulty in the use of English to prevent that individual from learning successfully in classrooms in which the language of instruction is English" (Kindler, 2002: 9); a learner whose first language/mother tongue is not English/a learner who comes from a place where English is not spoken at all or where limited English is spoken. In some cases **second-language learners** (learners who are taught mathematics in a language other than their home or first language) is used instead of English-language learners; in this book the terms are used interchangeably.

First language: a language that a child acquires from birth and in which he or she is most proficient. In some cases, *mother tongue* and *home language* are used instead of first language (Adler, 2001; Essien, 2013). In this text the terms are used interchangeably.

Foreign language: any language which learners are unlikely to hear or read outside the classroom in which they are learning it, because it is not in use in the wider community (Adler, 2001; Essien, 2013).

Indigenous language: a language that originates and is native to a specified place or region and spoken uniquely by *indigenous* people of that place or region. In some cases Indigenous African Language (IAL) is used in this book for the sake of clarity.

Language of learning and teaching (LoLT): the language medium in which learning and teaching, including assessment, take place (DBE, 2010).

Language-minority learners: learners from homes where the primary language spoken is not English (Menken & Antunez, 2001).

Language competence: an individual's ability to speak, read, write, interpret and generally use a language well and efficiently. It is the quality of being adequately or well qualified to use a language for both basic communication tasks and academic purposes (DBE, 2010).

Language proficiency: an individual's level of competence to use a language for both basic communication tasks and academic purposes; proficiency also refers to the quality of great facility and competence (DBE, 2010).

Linguistic competence: an individual's implicit internalised knowledge of the rules of a language (DBE, 2010). In this book the terms 'language competency' and 'linguistic competency' are used interchangeably.

Learners: refers to school children.

Mathematics register: the meanings of the natural language used in mathematics. A mathematics register is more precise than the natural language itself, because the meanings of the terms are much narrower in scope and hence more precise. Mathematical terms give rise to an almost totally non-redundant and relatively unambiguous language.

Multilingual: a person who is competent in more than two languages. Many South Africans, whose main language is not English or Afrikaans, speak three or more languages. Generally in South Africa the majority of the population is described as multilingual rather than bilingual.

Multilingual classroom: a situation where learners bring into the class a range of home languages. The term does not imply that all learners are multilingual. The meaning in this book is that there are more than two languages used in the classroom. Thus, a multilingual class in this context refers to a class where learners (and teachers) come to class with different levels of proficiencies in two or more languages, and where mathematics is taught and learnt in a language other than the first or home language of the majority of learners. This understanding of multilingualism as encompassing bilingualism is adopted in the text.

Official languages: the eleven languages which are stipulated and recognised in the South African Language Policy, namely *English, Afrikaans, isiNdebele, isiXhosa, isiZulu, Sepedi, Sesotho, Setswana, siSwati, Tshivenda* and *Xitsonga*.

Psycholinguistics: the study of psychological states and mental activity associated with the use of language (*Psycholinguistics*, 2013).

Register: "set of meanings that is appropriate to a particular function of language, together with the words and structures which express these meanings. We can refer to a 'mathematics register', in the sense of the meanings that belong to the language of mathematics (the mathematical use of natural language, that is: not mathematics itself), and that a language must express if it is being used for mathematical purposes" (Halliday, 1975: 195).

Sociolinguistics: an orientation to the study of language that stresses the interrelationship between language and social life rather than focusing narrowly on language structure (*Sociolinguistics*, 2004).

Student teacher: a university or college student studying to become a qualified teacher. Part of their student's practical work involves teaching (at a normal school) under the supervision of qualified teacher in order to qualify for a degree in education. In this compilation, the term is also used interchangeably with pre-service teacher.

Translanguaging: "the process of making meaning, shaping experiences, understandings and knowledge through the use of two languages" (Baker, 2011: 288).

References

A

ABEDI, J. & Lord, C. (2001). The language factor in mathematics tests. *Applied Measurement in Education*, 14(3), 219-234. https://doi.org/10.1207/S15324818AME1403_2

ADESINA, A. A. (2017). *Education Financing in Africa: Adopting innovative and effective approaches to achieve the 2063 Africa agenda and the African Development* Bank's *High Fives*. Keynote speech presented at the Association for the Development of Education in Africa (ADEA) 2017 Triennale. 14 -17 March 2017. Dakar: CICAD.

ADETULA, L. O. (1990). Language factor: Does it affect children's performance on word problems? *Educational Studies in Mathematics*, 21(4): 351-365. https://doi.org/10.1007/BF00304263

ADLER, J. (2001). *Teaching Mathematics in Multilingual classrooms*. Dordrecht: Kluwer Academic Publisher.

AHMED, A., Marriot, C. & Pollitt, A. (2000). *Language, contextual and cultural constraints on examination performance*, Paper presented at International Association for Educational Assessment Conference, Jerusalem, May 2000. Available on: www.uclesred.cam.ac.uk/conferencepapers. Accessed on 22 July 2020.

AINLEY, J. & Casterns, R. (2018). *The OECD Teaching and Learning International Survey, (Talis)*. www.oecd.org; last accessed on 12 July 2020.

AKINNASO, F. (1993). Policy and Experiment in Mother Tongue Literacy in Nigeria. *International Review of Education*, 39 (4): 255-285. https://doi.org/10.1007/BF01102408

AKINDELE, D. & Letsoela, M. (2001). Code-switching in Lesotho secondary and high schools lessons and its effects on teaching and learning. *BOLESWA Educational Research Journal*, 18: 83-100.

ALIDOU, H. & Brock-Utne, B. (2005). Teaching practices – teaching in a familiar language. Namibia: *GTZ Report*.

ALIDOU, H, Boly, A., Brock-Utne, B., Diallo, Y., Heugh, K. & Wolff, H. (2006). Optimizing learning and education in Africa – the language factor: A stock taking research on mother tongue and bilingual education in sub-Saharan Africa. Paper presented at the *Association of Development in Africa (ADEA) 2006 Biennial Meeting*, Libreville, Gabon.

ALEXANDER, R. (2000). *Culture & Pedagogy: international comparisons in primary education*. Oxford: Blackwell.

AMSTRONG, T. (2009). *Multiple Intelligences in the classroom*. (3rd Edition.). Alexandra, VD: ASCD.

ANTHONY, A. R. B. (2008). Output Strategies for English-Language Learners: Theory to Practice. *The Reading Teacher*, 61(6): 472-482. https://doi.org/10.1598/RT.61.6.4

ARTHUR, J. (2001). Code switching and collusion: Classroom interaction in Botswana primary schools. In: M. Helle & M. Martin-Jones (Eds.), *Voices of Authority: Education and Linguistic Differences* (pp. 57-75). Stanford, USA: Abex Publishing.

ASHKENAZI, S., Rubinsten, O. & Henik, A. (2009). Attention, automaticity, and developmental dyscalculia. *Neuropsychology*, 23: 535-540. https://doi.org/10.1037/a0015347

B

BACA, L. & Cervantes, H. (2004). *The bilingual special education interface*. Upper Saddle River, NJ: Pearson/Merrill/Prentice Hall.

BAKER, C. (2011). *Foundations of bilingual education and bilingualism* (5th Ed.). Bristol, UK: Multilingual Matters.

BAKER, C. (2001). Foundations of Bilingual Education and Bilingualism. In: Grigorenko, M.C. *Improving CALP of low-achieving sixth grade students: a catalyst for improving proficiency scales*. Unpublished Master's Thesis.

BAKER, C. (2000). *A* Parents' *and* Teachers' *Guide to Bilingualism*. Clevedon, UK: Multilingual Matters.

BAKER, C. (1996). *Foundations of Bilingual Education and Bilingualism*. Clevedon: Multilingual Matters.

BAKER, C. & Hornberger, N. (2001). *An Introductory Reader to the Writings of Jim Cummins*, Clevedon: Multilingual Matters.

BAKER, C. & Prys Jones, S. (1998). *Encyclopaedia of Bilingualism and Bilingual Education*, Clevedon: Multilingual Matters.

BALDAUF, R. B. & Kaplan, R. B. (2005). Language-in-education policy and planning. In: E. Hinkel (Ed.), *Handbook of research in second language teaching and learning*. Mahwah, NJ: Erlbaum.

BALL, D. L., Thames, M. H. & Phelps, G. (2008). Content knowledge for teaching: What makes it special? *Journal of Teacher Education*, 59: 389-407. https://doi.org/10.1177/0022487108324554

BALLANTYRE, P. (2008). Zone of proximal development: a new approach. Available on: http://www.comnet.ca~pballan/vygotsky. Accessed 17 July 2020.

BALLSTAEDT, S. P. (2003). "Technische Kommunikation mit Bildern," in J. Hennig and M. Tjarks-Sobhani (Hgg.), *Visualisierung in Technischer Dokumentation (tekom Schriften zur Technischen Kommunikation 7)*, Lübeck, Schmidt-Römhild, 11-31.

BAMGBOSE, A. (2000). *Language and Exclusion: The Consequences of Language Policies in Africa*. London: Transaction Publishers.

BAMGBOSE, A. (2004). *Language of Instruction Policy and Practice in Africa*. Available on: http://www.unesco.org/education/languages_2004/languageinstruction_africa.pdf. Accessed 20 November 2019.

BARJESTEH, H. & Vaseghi, R. (2012). Acculturation Model for L2 Acquisition: Review and Evaluation. *Advances in Asian Social Science* (AASS), 2(4): 579-584.

BARKHUIZEN, G. P. (2002). Language-in-education policy: students' perceptions of the status and role of Xhosa and English. *Science Direct*, 30(4): 499-515. https://doi.org/10.1016/S0346-251X(02)00051-9

BARTON, B. & Neville-Barton, P. (2003). Language issues in undergraduate mathematics: A report of two studies. *New Zealand Journal of Mathematics, 32* (Supplementary Issue), 19-28.

BARWELL, R. (Ed.). (2009). *Multilingualism in mathematics classrooms – global perspectives*. Bristol: Multilingual Matters. https://doi.org/10.21832/9781847692061

BARWELL, R., Barton, B. & Setati, M. (2007). Multilingual issues in mathematics education: Introduction, *Educational Studies in Mathematics*, 64(2): 113-119. https://doi.org/10.1007/s10649-006-9065-x

BARWELL, R., Leung, C., Morgan, C. & Street, B. (2002). The Language dimension of mathematics teaching. *Mathematics Teaching*, 180, 12-15.

BEARDE, D. K. (1993). Oral Language Proficiency as a Predictor of mathematics achievement on the Woodcock-Johnson Psycho-Educational Battery-Revised. Unpublished PhD Thesis, Texas: Texas Women's University.

BENSON, C. (2002). Bilingual education in Africa: An exploration of encouraging connections between language and girls' schooling. In: Melin, M. (Ed.), *Education – A way out of poverty? Research presentations at the poverty conference 2001*. New Education Division Documents No. 12. Stockholm: Sida.

BENSON, C. (2000). The Primary Bilingual Education Experiment in Mozambique, 1993 to 1997, *International Journal of Bilingual Education and Bilingualism*, 3(3): 149-166. https://doi.org/10.1080/13670050008667704

BERENDS, I. E., & Van Lieshout E. C. D. M. (2009). The effect of illustrations in arithmetic problem-solving: effects of increased cognitive load. *Learning and Instruction*, 19: 345-353. https://doi.org/10.1016/j.learninstruc.2008.06.012

BERNHARDT, E. B. & Kamil, M. L. (1995). *Interpreting relationships between L1 and L2 reading: Consolidating the linguistic threshold and the linguistic interdependence hypotheses.* Applied Linguistics, 16: 15-34. https://doi.org/10.1093/applin/16.1.15

BERNSTEIN, B. (1990). *Class Codes and Control: The Structuring of Pedagogic Discourse*, Volume 4. London: Routledge.

BESWICK, K., Callingham, R. & Watson, J. (2012). The nature and development of middle school mathematics teachers' knowledge. *Journal of Mathematics Teacher Education*, 15(2): 131-157. https://doi.org/10.1007/s10857-011-9177-9

BIALYSTOK, E. (1982). On the relationship between knowing and using linguistic forms. *Applied Linguistics*, 3(3): 181-206. https://doi.org/10.1093/applin/3.3.181

BIALYSTOK, E. (2009). Bilingualism: The good, the bad, and the indifferent. *Bilingualism: Language and Cognition*, 12: 3-11. https://doi.org/10.1017/S1366728908003477

BIALYSTOK, E. (2001). *Bilingualism in development: Language, literacy, and cognition*. Cambridge, UK: Cambridge University Press. https://doi.org/10.1017/CBO9780511605963

BLACHOWICZ, C. & Fisher, P. (2000). Vocabulary instruction. In: M. Kamil, P. Mosenthal, P. D. Pearson, & R. Barr (Eds.), *Handbook of reading research (3rd ed.*, 503-523). Mahwah, NJ: Lawrence Erlbaum Associates.

BLOCH, G. (2009). *The toxic mix:* what's *wrong with South* Africa's *schools and how to fix it?* Cape Town: Tafelberg.

BLOCH, C. & Edwards, V. (1998). Young children's literacy in multilingual classrooms: comparing development in South Africa and the United Kingdom. *Southern African Review of Education,* 4: 11-22.

BLOEMFONTEIN Courant (2016). Bartimea teachers banned from entering school yard, by Gaeswe, R.; 16 March. Available on: www.bloemfonteincourant.co.za. Accessed 15 April 2020.

BLOMEKE, S. & Delaney, S. (2012). Assessment of Teacher Knowledge Across Countries: A review of the State of Research. *ZDM*, 44(3): 223-247. https://doi.org/10.1007/s11858-012-0429-7

BOHLMANN, C. (2001). Reading skills and mathematics. *Communications: Third Southern Hemisphere Symposium on Undergraduate Mathematics Teaching*, 5-14.

BOKAMBA, E. (1989). Are there syntactic constraints on code-mixing? *World Englishes*, 8(3), 277-292. Available on: doi:org/10.1111/j.1467-971X.1989.tb00669.x. Accessed 10 April 2020.

BOONEN, A. J. H., Van Wesel, F. Jolles, J. & Van Der Schoot, M. (2014). The role of visual representation type, spatial ability, and reading comprehension in word problem solving: an item-level analysis in elementary school children. *Int. J. Educ. Res.* 68: 15-26. https://doi.org/10.1016/j.ijer.2014.08.001

BOWIE, L., Venkat, H. & Askew, M. (2019). Pre-Service Primary Teachers' Mathematical Content Knowledge: An Exploratory Study. *African Journal of Research in Mathematics, Science and Technology Education*, 23 (3), 286-297. https://doi.org/10.1080/18117295.2019.1682777

BRISSIAUD, R., & Sander, E. (2010). Arithmetic word problem solving: a situation strategy first framework. *Dev. Sci.*, 13: 92-107. https://doi.org/10.1111/j.1467-7687.2009.00866.x

BRITISH Council, (2013). *English as a medium of instruction – a growing global phenomenon*, J. Dearden (Ed.)., London: British Council.

BROCK-UTNE, B. (2007). Language of Instruction and Student Performance: New Insights from Research in Tanzania and South Africa, *International Review of Education*, 53: 509-530. https://doi.org/10.1007/s11159-007-9065-9

BROCK-UTNE, B. (2002). *The most recent developments concerning the debate on language of instruction in Tanzania*. Institute for Educational Research. University of Oslo, Paper presented to the NETREED conference 7th-9th January 2002. Available on: http://www.uio.no/~bbrock/EduDev.html. Accessed 19 October 2019.

BROCK-UTNE, B. & Alidou, H. (2005). Active learners – learning through a language they master. Namibia, *GTZ report*.

BRODIE, K. (1989). Learning Mathematics in a Second Language. *Educational Review*, 41(1), 35-39. https://doi.org/10.1080/0013191890410105

BRODIE K. in P. Webb, & N. Roberts (2017). *The Pedagody of Mathematics in South Africa: Is there a unifying pedagogy?* Johannesburg: Mapungubwe Institute for Strategic Reflection (MISTRA) and Real African Publishers.

BROWN, H. D. (2007). *Principles of language learning and teaching*. White Plains, NY: Pearson Longman.

BROWN, R. (1973). *A first language. The Early Stages*. Cambridge, MA: Howard University Press. Available on: http://www.tip.psychology.org/language.html. Accessed 14 July 2020.

BROWNE, C. (2002). *To push or not to push: A vocabulary research question*. Aoyama Ronshu, Gakuin University Press.

BRYANT, D. P., Roberts, G., Bryant, B. R. & DiAndreth-Elkins, L. (2011). Tier 2 for K – 2 early numeracy number sense interventions for kindergarten and first-grade students with mathematics difficulties. In: Gersten & Newman-Gonchar, B. (Eds.), *Mathematics Book* (pp. 65-83). Baltimore, MD: Brookes.

BULLOCK, B. & Toribio, A. J. (2009). Reconsidering Dominican Spanish: Data from the rural Cibao. *Revista Internacional de Lingüística Iberoamericana,* 14: 49-73.

BULTÉ, B. & Housen A. (2012). "Defining and operationalising L2 complexity," in *Dimensions of L2 Performance and Proficiency – Investigating Complexity, Accuracy and Fluency in SLA,* Eds A. Housen, F. Kuiken, I. Vedder (Amsterdam: John Benjamins), 21-46. https://doi.org/10.1075/lllt.32.02bul

BURTON, M. B. (1988). A linguistic basis for student difficulties with algebra. *For the Learning of Mathematics,* 8(1), 2-7.

C

CDD, (2006). *Mathematics syllabus for lower and upper primary school*. Curriculum Department, Ministry of Education: Brunei Darussalam.

CARROLL, J. (1991). Cognitive abilities in foreign language aptitude: then and now. In: T. Parry and C. Stansfield (Eds.), 1991: *Language aptitude reconsidered*. Englewood Cliffs, New Jersey: Prentice Hall.

CARRUTHERS, E. & Worthington, M. (2006). Children's *mathematics: Making marks, making meaning*. (2nd Edition). London: Paul Chapman Publishing.

CASALE, D. & Posel, D. (2011). English language proficiency and earnings in a developing country: The case of South Africa, *The Journal of Socio-Economics*, 40: 385-393. https://doi.org/10.1016/j.socec.2011.04.009

CAZDEN, C. & Clay, M. (1992). A Vygotskian interpretation of Reading Recovery. In: Nathanson, R. 2009. *The continuum of literacy learning, Grades K-8*, Portsmouth: Heinemann.

CHAMOT, A. U. & O'Malley, J. M. (1994). The CALLA handbook: implementing the cognitive academic language learning approach. In: Brown, L. C. 2004. Context Based ESL curriculum and academic language proficiency. *The Internet TESL Journal*, 2: 1-5.

CHAPMAN, O. (2006). Classroom practices for context of mathematics word problems. *Educational Studies in Mathematics*, 62, 21-230. https://doi.org/10.1007/s10649-006-7834-1

CHEN, X. & Yeping, L. (2010). *Research in brief: Language Proficiency and Mathematics Learning, School Science and Mathematics*, 108(3), https://doi.org/10.1111/j.1949-8594.2008.tb17811.x. Accessed 15 April 2020.

CHEN, Y. R., Leung, K. & Chen, C. C. (2009). Bringing National Culture to the Table: Making a Difference with Cross-Cultural Differences and Perspectives, *Academy of Management Annals*, 3: 217-49. https://doi.org/10.5465/19416520903047244

CHEUNG, A. & Slavin, R. (2005). Effective Reading Programs for English Language Learners and Other Language-Minority Students. *Bilingual Research Journal*, 29 (2 Summer): 241-267. https://doi.org/10.1080/15235882.2005.10162835

CHEUNG, A. & Slavin, R. (2012). Learners (ELLs) in the Elementary Grades: A Synthesis of Research Effective Reading Programs for Spanish-Dominant English Language, *Review of Educational Research*, 82(4): 351-395. https://doi.org/10.3102/0034654312465472

CHICK, J. K. (1992). Language policy in education. In: R. McGregor,& A. McGregor (Eds.), McGregor's *Education Alternatives*. Pretoria: Juta.

CHIKIWA, C. (2016). An exploration of how consistently and precisely mathematics teachers code-switch in multilingual classrooms. Unpublished PhD Thesis, Grahamstown: Rhodes University.

CHILORA, H. G. (2000). *School Language Policy, Research and Practice in Malawi*. Paper presented at the Comparative and International Education Society (CIES) 2000 Conference, San Antonio, Texas, USA. 8th-12th March. Malawi/ IEQ II Project.

CHITERA, N. (2016). A journey in the teaching and learning of mathematics in home languages: Where do we stand in 2016? *Conference Proceedings of the 13th International Congress on Mathematics Education (ICME 13) 665(2776)*, Hamburg.

CHITERA, N. (2012). Language-in-Education Policies in conflict: Lessons from Malawian Mathematics Teacher Training Classrooms. *African Journal of Research in Mathematics Science Technology Education*, 16(1): 60-70. https://doi.org/10.1080/10288457.2012.10740729

CHITERA, N. (2009a). Discourse practices of Mathematics teacher educators in initial teacher training colleges in Malawi. Unpublished PhD Thesis, Johannesburg: University of the Witwatersrand.

CHITERA, N. (2009b). Code-Switching in a college mathematics classroom. *International Journal of Multilingualism*, 6(4): 426-442. https://doi.org/10.1080/14790710903184850

CHOMSKY, N. (1975). *Logical structures of linguistic theory*. London: Plenum Press. https://doi.org/10.1515/9783110867565

CLARKE, J. (1989). Students' Perceptions of Different Tertiary Learning Environment. *Higher Education Research and Development*, 17(1): 107-115. https://doi.org/10.1080/0729436980170106

CLARK, H. & Clark, E. (1977). *Psychology and language*. New York: Harcourt Brace Jovanovich. Available on: http://www.tip.psychology.org/language.html. Accessed 14 July 2020.

CLARKSON, P. C. (2007). Australian Vietnamese students learning mathematics: High ability bilinguals and their use of their languages. *Educational Studies in Mathematics*, 64(2): 191-215. https://doi.org/10.1007/s10649-006-4696-5

CLARKSON, P. C. (1980). The Newman error analysis – Some extensions. In: B. A. Foster (Ed.), *Research in mathematics education in Australia*, (1): 11-22. Hobart: Mathematics Education Research Group of Australia.

CLARKSON, P. C. (1992). *Language and Mathematics: A comparison of Bilingual and Monolingual students of Mathematics*. Oslo: Kluwer Academic Publishers. https://doi.org/10.1007/BF00302443

CLARKSON, P. C. (1991). *Bilingualism and mathematics learning*. Geelong, Vie: Deakin University Press.

CLEMENTS, M. A. (1980). Analysing children's errors on written mathematical tasks. *Educational Studies in Mathematics*, 11(1): 1-21. https://doi.org/10.1007/BF00369157

CLOUD, N. (2005). The dialogic process of capturing and building teacher practical knowledge in dual language programs. In: Tedick, D. J. (Ed.), *Second-language teacher education*. London: Lawrence Erlbaum: 273-280.

COHN, A. D. (1990). Child-Mother Attachment of Six-Year-Olds and Social Competence at School. *Child Development*, Available on: http://onlinelibrary.wiley.com/doi/10.1111/j.1467-8624.1990.tb02768. Accessed 13 July 2020. https://doi.org/10.2307/1131055

COLLIER, V. P. (1995). Acquiring a second language for school. *Directions in Language & Education National Clearing House for Bilingual Education*, 1(4): 11-21.

COLOMÉ, A., Laka, I. & Sebastiaen-Galles N. (2010). *Language effects in addition: how you say it counts*. Q. J. *Exp. Psychol.*, 63: 965-983. https://doi.org/10.1080/17470210903134377

COOK, V. (Ed.) (2003). Effects of the Second Language on the First, *Multilingual Matters*, Second Language Acquisition Series. https://doi.org/10.21832/9781853596346

CRAWFORD, M. & Witte, M. (1999). Strategies for mathematics: Teaching in context. *Educational Leadership*, November 1999.

CSIKOS, C., Kelemen, R. & Verschaffel L. (2011). Fifth-grade students' approaches to and beliefs of mathematics word problem solving: a large sample Hungarian study. *ZDM* 43: 561-571. https://doi.org/10.1007/s11858-011-0308-7

CSÍKOS, C. & Szitányi, J. (2020). Teachers' pedagogical content knowledge in teaching word problem solving strategies. *ZDM Mathematics Education* 52, 165-178. https://doi.org/10.1007/s11858-019-01115-y

CSÍKOS, C., Szitányi, J. & Kelemen, R. (2012).

The effects of using drawings in developing young children's mathematical word problem solving: A design experiment with third-grade Hungarian students. *Educational Studies in Mathematics*, 81, 47-65. https://doi.org/10.1007/s10649-011-9360-z

CUERVO, M. M. (1991). Bilingual instruction in college mathematics: Effects on performance of Hispanic students on CLAST mathematics competencies examination, *Dissertation Abstract International*, 52(12): 42-53

CUMMINS, J. (2000). *Language, power and pedagogy: Bilingual children in the crossfire.* Clevedon: Multilingual Matters. https://doi.org/10.21832/9781853596773

CUMMINS, D. D. (1991). Children's interpretations of arithmetic word problems. *Cogn. Instr.* 8, 261-289. https://doi.org/10.1207/s1532690xci0803_2

CUMMINS, J. (1981). *Bilingualism and Minority-Language Children.* Toronto, Onto: OISE Press.

CUMMINS, J. (1979). Linguistic interdependence and the development of bilingual children. *Review of Educational Research*, 49: 222-251. https://doi.org/10.3102/00346543049002222

CUMMINS, J. (1976). The influence of bilingualism on cognitive growth: A synthesis of research findings and explanatory hypotheses. *Working Papers on Bilingualism*, 9: 1-43.

CUMMING-POTVIN, W. (2009). Social Justice, Pedagogy and Multiliteracies: Developing Communities of Practice for Teacher Education. *Australian Journal of Teacher Education*, 34(3). https://doi.org/10.14221/ajte.2009v34n3.4.

CUVELIER, P., Du Plessis, T. & Teck, L. (2003). *Multilingualism, Education and Social Integration. Belgium, Europe, South Africa.* Pretoria: Van Schaik.

D

DAKROUB, H. M. (2002). The relationship between Arabic language literacy and academic achievement of Arab-American middle school students in English reading, language, and mathematics in a suburban public middle school. Available on http://digitalcommons.wayne.edu/dissertations/AAI3047545/ (10/12/2008). Accessed 19 August 2020.

DALVIT, L.., Murray, S., Mini, B., Terzoli, A. &

Zhao, X. (2005). Providing increased access to English L2 Students of Computer Science at a South African University. *Journal of US-China Education Review*, 2(9): 72-75.

DAROCZY G, Wolska M, Meurers WD and Nuerk H-C (2015) Word problems: a review of linguistic and numerical factors contributing to their difficulty. Front. Psychol. 6:348

DAWE, L. (1983). Bilingualism and mathematical reasoning in English as a second language. *Educational Studies in Mathematics,* 14(1): 325-353. https://doi.org/10.1007/BF00368233

DAWE, L. & Mulligan, J. (1997). Classroom views of language in mathematics, In B. Doig, & J. Lokan (Eds.), *Learning from Children: Mathematics from a Classroom Perspective*, Melbourne: ACER Press.

DE Corte, E. (2004). Mainstreams and

perspectives in research on learning mathematics from instruction. *Centre for Instructional Psychology and Technology,* 53(2): 279-310. https://doi.org/10.1111/j.1464-0597.2004.00172.x

DELL Inc. (2016) *STATISTICA* (data analysis software system), version 13. http://www.statsoft.com/

DEPAEPE, F., De Corte, E. & Verschaffel, L. (2010). Teachers' metacognitive and heuristic approaches to word problem solving: Analysis and impact on students' beliefs and performance. *ZDM-Mathematics Education*, 42, 205-218. https://doi.org/10.1007/s11858-009-0221-5

DEPARTMENT of Arts and Culture (DAC). (2003). *National language policy framework*. Pretoria: National Language Service.

DEPARTMENT of Basic Education (DBE). (2015). *Education Statistics in South Africa*: Pretoria. Education Management Information Systems.

DEPARTMENT of Basic Education (DBE). (2014). *Annual School Survey Report (ASS)*. Pretoria: Department of Basic Education.

DEPARTMENT of Basic Education (DBE). (2014). *Report on the Annual National Assessments*. Pretoria: Department of Basic Education.

DEPARTMENT of Basic Education (DBE). (2013). *Atlas of Education Districts in South Africa*, Pretoria: Department of Basic Education.

DEPARTMENT of Basic Education (DBE). (2011). *Curriculum Assessment Policy Statements (CAPS). Intermediate Phase Mathematics Grade 4 – 6*, Pretoria: Department of Basic Education.

DEPARTMENT of Basic Education (DBE). (2010). *The status of the Language and Learning and Teaching (LOLT) in South African public schools: A quantitative overview*. Pretoria: Department of Basic Education.

DEPARTMENT of Basic Education (DBE). (2010). Available on: http://www.education.gov.za/Curriculum/LearningandTeachingSupportMaterials(LTSM)/Workbooks.aspx Accessed 05 April 2020.

DEPARTMENT of Education (DoE). (2011). *Technical Report: Integrated Strategic Planning Framework for Teacher Education and Development in South Africa, 2011-2025*, Pretoria: Department of Basic & Department of Higher Education and Training Publication.

DEPARTMENT of Education (DoE). (1997). *Language in Education Policy*, section 3(4)(m) of the National Education Policy Act, 1996 (Act 27 of 1996). Pretoria. Department of Education.

DESCHUYTENEER, M., De Rammelaere, S. & Fias W. (2005). The addition of two-digit numbers: exploring carry versus no-carry problems. *Psychol. Sci.*, 47: 74-83.

DE Wet, C. (2002). Factors influencing the choice of English as a Language of Learning and Teaching (LoLT) - A South African Perspective. *South African Journal of Education*, 22(2): 119-124.

DOMAHS, F., Delazer, M. & Nuerk, H. C. (2006). What makes multiplication facts difficult: problem size or neighbourhood consistency? *Exp. Psychol.*, 53: 275-282. https://doi.org/10.1027/1618-3169.53.4.275

DOMAHS, F., Janssen, U., Schlesewsky, M., Ratinckx, E., Verguts, T. & Willmes, K. (2007). Neighbourhood consistency in mental arithmetic: behavioral and ERP evidence. *Behav. Brain Funct*, 3: 66. https://doi.org/10.1186/1744-9081-3-66

DOWKER, A. (2005). Linguistic influences on numeracy. In: D. V. Jones,, A. Dowker, & D. Lloyd (Eds.), *Mathematics in the Primary School* (pp. 21-35). Bangor School of Education (University of Wales): Educational Transactions.

DOWKER, A., Bala, S. & Lloyd, D. (2008). Linguistic influences on mathematical development: how important is the transparency of the counting system. *Philos. Psychol.*, 21: 523-538. https://doi.org/10.1080/09515080802285511

DICKMEYER, N. (1989). Metaphor, model, and theory in education research. *Teachers College Record*, 91(2): 151-160.

DUFFY, G. G. (1993). Teachers' progress toward becoming expert strategy teacher. *Elementary School Journal*, 94: 109-120. https://doi.org/10.1086/461754

DU Plessis, P. (2014). Problems and complexities in rural schools: Challenges of education and social development. *Mediterranean Journal of Social Sciences*, 5(20): 1109. Available on: http://www.mcser.org/journal/index.php/mjss/article/download/3842/3759. Accessed 08 February 2020.

DURKIN, K. & Shire, B. (1991). *Language in mathematical education: Research and practice*. Milton Keynes: Open University Press.

DUTCHER, N. (2004), *Language Policy and Education in Multilingual Societies: Lessons from Three Positive Models*. Bristol: Centre for Applied Linguistics.

DUVAL, R. (2006). A cognitive analysis of problems of comprehension in a learning of mathematics. *Educational Studies in Mathematics*, 61: 103-131. https://doi.org/10.1007/s10649-006-0400-z

E

EASTERN Cape Department of Education (ECDoE). (2014). Map of Districts. Available on http://www.ecdoe.gov.za/ Accessed on 17 July 2020.

EASTERN Cape Department of Education (ECDoE). (2010). Language of Learning and Teaching: Xhosa instruction reaps big rewards in Eastern Cape. Available on http://www.ecdoe.gov.za/news/page/175# Accessed 06 March 2020.

EID, A. 2012. *Language learning strategy use of Saudi EFL students in an intensive English learning context.* Asian Social Science, 8(13): 115-127. https://doi.org/10.5539/ass.v8n13p115

ELLERTON, N. F. (1989). The interface between mathematics and language. *Australian Journal of Reading,* 12: 92-102.

ELLERTON, N. F., & Clarkson, P. C. (1996). Language factors in mathematics teaching and learning. In: A. J. Bishop (Ed.), *International Handbook of Mathematics Education,* 4: 987-1033. Dordrecht, The Netherlands: Kluwer Academic Publishers. https://doi.org/10.1007/978-94-009-1465-0_30

ELLIS, N. C. (2006). SLA: The Associative Cognitive CREED. In: B. van Patten, B. Williams, & A. F. Williams (Eds.), *Theories in second language acquisition: An introduction.* Mahwah New Jersey: Erlbaum.

ELLIS R. (2003). *Task-based Language Learning and Teaching*. Oxford: Oxford University Press.

ELLIS, R. (1997). Second language acquisition.

In: Woozley, I. *Second language acquisition and communication approach.* Available on: http://www.Niigatajet/documents/second-languageacquisition.pdf Accessed 04 July 2020.

ESSIEN, A. (2013). Maths still best taught in English. *Mail & Guardian*, June 02.

ESTY, W. W. (1992). Language concepts in mathematics. *Focus on Learning Problems in Mathematics*, 14(4): 31-54.

EVEN, R. & Tirosh, D. (1995). Subject-matter knowledge and knowledge about students as sources of teacher presentations of the subject matter. *Educational Studies in Mathematics,* 29: 1-20. https://doi.org/10.1007/BF01273897

F

FAFUNWA, A. B. (1989). Using national languages in education: A challenge to African education. In: *African thoughts on the prospects of education for all* (pp. 97-110). Dakar, Senegal: UNESCO Regional Office for Education in Africa.

FAST, R. (2012). Semiotics and the classroom experience. Unpublished MA Thesis. University of Athabasca, Alberta.

FERRO, S. (1983). Language influence on mathematics achievement of Cape Verdean students. Unpublished PhD Thesis, Boston University.

FISCHER, M. H. & Shaki S. (2014). Spatial associations in numerical cognition – from single digits to arithmetic. *Q. J. Exp. Psychol.*, 67: 1461-1483. https://doi.org/10.1080/17470218.2014.927515

FONSECA, B. A. & Chi, M. T. (2011). Instruction based on self-explanation. In: R. E. Mayer & P. A. Alexander (Eds.), *Handbook of research on learning and instruction* (pp. 296-321). New York: Routledge.

FREDRICKSON, N. & Cline, T. (1996). The development of a model of curriculum related assessment, In: N. Fredrickson & T. Cline (Eds.), *Curriculum Related Assessment, Cummins and Bilingual Children*, Clevedon: Multilingual Matters.

FUCHS, L. S., Powell, S. R., Seethaler, P. M., Cirino, P. T., Fletcher, J. M. & Fuchs D. (2009). Remediating number combination and word problem deficits among students with mathematics difficulties: a randomized control trial. *J. Educ. Psychol.*, 101: 561-76. https://doi.org/10.1037/a0014701

FURST, A. & Hitch, G. J. (2000). Separate roles for executive and phonological components of working memory in mental arithmetic. *Mem. Cogn.*, 28: 774-782. https://doi.org/10.3758/BF03198412

G

GABELA, R. V. (2005). *Rural and peri-urban schools in South Africa*. Paper presented at the education workshop held at Chicago States University, USA, 25 – 28 April 2005.

GARCÍA, O. (2012). Theorizing translanguaging for educators. In: C. Celic & K. Seltzer (Eds.), *Translanguaging: A CUNY-NYSIEB guide for educators* (pp. 1-6). New York, USA: The City University of New York.

GARCÍA, O. (2009). Education, multilingualism and translanguaging in the 21st century. In: A. Mohanty, M. Panda, R. Phillipson & T. Skutnabb-Kangas (Eds.), *Multilingual education for social justice: Globalising the local* (pp. 128-145). New Delhi, India: Orient Blackswan. https://doi.org/10.21832/9781847691910-011

GARCIA, A., Jimenez, L. & Hess S. (2006). Solving arithmetic word problems: an analysis of classification as a function of difficulty in children with and without arithmetic. *LD. J. Learn. Disabil.*, 39: 270-281. https://doi.org/10.1177/00222194060390030601

GARCÍA, O. & Wei, L. (2015). Translanguaging, bilingualism and bilingual education. In: W. Wright, S. Boun & O. Garcia (Eds.), *Handbook of bilingual education* (pp. 223-240). Malden, USA: John Wiley. https://doi.org/10.1002/9781118533406.ch13

GASCO, J. & Villarroel J. D. (2014). The motivation of secondary school students in mathematical word problem solving. Electron. *J. Res. Educ. Psychol.* 12: 83-106.

GASCO, J., Villarroel, J. D. & Zuazagoitia D. (2014). Different Procedures for Solving Mathematical Word Problems in High School. *Int. Educ. Stud.*, 7: 77. https://doi.org/10.5539/ies.v7n7p77

GASKINS, I. W. (2004). Word detectives. *Educational Leadership,* 61(6), 70-73.

GAWNED, S. (1990). An emerging model of the language of mathematics. J. Bickmorebrand (Eds.), *Language in mathematics* (27 – 42). Carlton, VIC: Australian Reading Association.

GAY, J. & Ryan, M. (1999). *What guides the teacher educators? The influence of standards movements on teaching.* Melbourne, Australia: Victoria University of Technology.

GENESEE, F., Paradis, J. & Crago, M. B. (2004). *Dual language development and disorders: A handbook on bilingualism and second language learning.* Baltimore, MD: Paul H. Brookes.

GERMAN Organisation for Technical Cooperation (GTZ). (2005). Optimizing learning and education in Africa – the language factor. *Report on the conference on bilingual education and the use of local languages, Namibia.*

GERSTEN, R., Baker S. K., Haager, D. & Graves, A. (2005). Exploring the role of teacher quality in predicting reading outcomes for first-grade English learners: An observational study. *Remedial & Special Education*, 38: 212-222.

GEVA, E. & Ryan, E. B. (1993). Linguistic and cognitive correlates of academic skills in first and second languages, *Language Learning*, 43(1): 5-42. https://doi.org/10.1111/j.1467-1770.1993.tb00171.x

GLOBAL Competitiveness Report for the World Economic Forum (2013/2014). World Economic Forum. (2 September 2014). Available on: http://www.3.weforum.org/docs/WEF_GlobalCompetitivenessReport_2013-14.pdf. Accessed 07 July 2020.

GODINO, J. D. & Font, V. (2010). The theory of representations as viewed from the onto-semiotic approach to mathematics education. *Mediterranean Journal for Research in Mathematics Education,* 9(1): 189-210.

GOEBEL S., Moeller K., Pixner S., Kaufmann L., Nuerk H.-C. (2014). Language affects symbolic arithmetic in children: the case of number word inversion. *J. Exp. Child Psychol.*, 119: 17-25. https://doi.org/10.1016/j.jecp.2013.10.001

GOEBEL S. M., Shaki S., Fischer M. H. (2011). The cultural number line: a review of cultural and linguistic influences on the development of number processing. *J. Cross Cult. Psychol.*, 42: 543-565. https://doi.org/10.1177/0022022111406251

GOLDIN G.A. (2020) Mathematical Representations. In: Lerman S. (Eds.) *Encyclopedia of Mathematics Education*. Springer, Cham. https://doi.org/10.1007/978-3-030-15789-0_103

GOUGH, D. (1994). Myths of multilingualism: demography and democracy. *Bua*, 9: 9-11.

GOVERNMENT Gazette Number 38487. (2015). National Qualifications Framework Act (67/2008): *Revised policy on the Minimum Requirements for Teacher Education Qualifications (MRTEQ)*, Pretoria: Republic of South Africa.

GRABNER R. H., Ansari D., Koschutnig K., Reishofer G., Ebner F., Neuper C. (2009). To retrieve or to calculate? Left angular gyrus mediates the retrieval of arithmetic facts during problem solving. *Neuropsychologia,* 47: 604-608. https://doi.org/10.1016/j.neuropsychologia.2008.10.013

GREENE, J. (1997). A meta-analysis of the Rossell and Baker review of bilingual education research. *Bilingual Research Journal*, 21(2/3): 103-122. https://doi.org/10.1080/15235882.1997.10668656

GRIFFITHS, C. (2003). Patterns of language learning strategy use, *System*, 31: 367-383. https://doi.org/10.1016/S0346-251X(03)00048-4

GROSJEAN, F. & Moser – Mercer, B. (1997). The bilingual individual Interpreting: *International Journal of Research and Practice in Interpreting*, 2(1 & 2): 163-187. https://doi.org/10.1075/intp.2.1-2.07gro

GROSJEAN, F. (1998). Studying bilinguals: Methodological and conceptual issues. *Bilingualism: Language and Cognition*, 1: 131-149. https://doi.org/10.1017/S136672899800025X

GROSSMAN, H. (1995). *Special education in a Diverse Society*. Boston: Allyn and Bacon.

GROSSMAN, P. L., Wilson, S. M. & Shlman, L. S. (1989). Teachers of substance: Subject Matter Knowledge for Teaching. In: M. C. Reynolds (Ed.), *Knowledge Base for the Beginning Teacher* pp 23-36. New York: Pergamon.

GUMPERZ, J. J. (1982). *Discourse strategies*. London: Cambridge University Press. https://doi.org/10.1017/CBO9780511611834

GUNNING, T. (2003). The role of readability in today's classrooms. *Topics in Language Disorders*, 23: 175-189. https://doi.org/10.1097/00011363-200307000-00005

GUTIERREZ, R. (2002). Enabling the practice of mathematics teachers in context: toward a new equity research agenda. *Mathematical Thinking and Learning*, 4(2&3): 145-187. https://doi.org/10.1207/S15327833MTL04023_4

H

HAGHVERDI, M., Semnani, A. S. & Seifi, M. (2012). The relationship between different kinds of students' errors and the knowledge required to solve mathematics word problems. *Bolema Bol. Educ. Matem.*, 26: 649-660. https://doi.org/10.1590/S0103-636X2012000200012

HAKUTA, K. (1990). Bilingualism and bilingualism education: a research perspective focus. In: E. J. Pretorius, What they can't read will 'hurt them': reading and academic achievement. *Innovation*, 2: 33-41.

HAKUTA, K., Butler, Y. G. & Witt, D. (2000). How long does it take English language learners to develop oral proficiency and academic proficiency? *Policy Report 2000-2001*. Palo Alto, California: Stanford University.

HALAI A. (2016) Teaching and Learning Mathematics: Insights from Classrooms in East Africa. In: A. Halai & G. Tennant (Eds.), *Mathematics Education in East Africa*. Springer Briefs in Education. Springer, Cham. https://doi.org/10.1007/978-3-319-27258-0

HALAI, A. (2009). Politics and practice of learning mathematics in multilingual classrooms: Lessons from Pakistan. In: R. Barwell (Ed.), *Multilingualism in mathematics classrooms: Global perspectives* (pp. 47-62). Bristol, UK: Multilingual Matters. https://doi.org/10.21832/9781847692061-006

HALAI, A. & Karuku, S. (2013). Implementing language-in-education policy in multilingual mathematics classrooms: Pedagogical implications. *Eurasia Journal of Mathematics, Science & Technology Education*, 9(1): 23-32. https://doi.org/10.12973/eurasia.2013.913a

HALL, G. (1996). Working paper: Overview of current situation and the role of colleges of education in the provision of (general) post-secondary education. In: *Report of the Technical Committee on Teacher Education*. Pretoria: National Commission on Higher Education.

HALLIDAY, M. A. K. (1975). Some aspects of sociolinguistics. In: E. Jacobsen (Ed.). *Interactions between linguistics and mathematical education*: Final report of the symposium sponsored by UNESCO, CEDO and ICMI, Nairobi, Kenya, September 1-11, 1974 (UNESCO Report No. ED-74/CONF.808, pp. 64-73). Paris: UNESCO.

HAMESO, S. (1997). *Ethnicity and nationalism in Africa*. New York: Nova Science Publishers.

HAN, Y. A. (1998). Chinese and English mathematics language: The relationship between linguistic clarity and mathematics performance, Unpublished PhD Thesis, Columbia University Teachers College. Dissertation Abstracts International, 59, 2405.

HART, B. & Risley, T. R. (1995). *Meaningful differences in the everyday experience of young American children*. Baltimore, MD: Brookes.

HARTSHORNE, K. (1992). *Crisis and Challenge. Black education 1910 – 1990*. Cape Town: Oxford University Press.

HEUGH, K. (2015). Epistemologies in multilingual education: Translanguaging and genre-companions in conversation with policy and practice. *Language and Education*, 29(3): 280-285. https://doi.org/10.1080/09500782.2014.994529

HEUGH, K.,Benson, C., Bogale, B. & Gebre Yohannes, M. A. (2007). *Final Report: Study on Medium of Instruction in Primary Schools in Ethiopia*. Research report commissioned by the Ministry of Education, Addis Ababa, September to December 2006. Available on: http://www.hsrc.ac.za/research/output/outputDocuments/4379_Heugh_Studyonmediumofinstruction.pdf. Accessed 30 April 2020.

HEREDIA, R. R. & Alterriba. (2001). Bilingual Language Mixing: Why Do Bilinguals Code-Switch? *Current Directions in Psychological Science*, 10(5): 164-168. https://doi.org/10.1111/1467-8721.00140

HERSHKOVITZ, S., & Nesher, P. (2003). The role of schemes in solving word problems. *The Mathematics Educator*, 7: 1-24.

HILL, H. C., Ball, D. L. & Schilling, S. G. (2008). Unpacking pedagogical content knowledge: Conceptualizing and measuring teachers' content-specific knowledge of students. *Journal for Research in Mathematics Education*, 39: 372-400.

HINES, T. M. (2013). Parity influences the difficulty of simple addition and subtraction but not multiplication problems in children. *Psychol. Rep*, 113: 1048-1065. https://doi.org/10.2466/10.11.PR0.113x16z4

HINKEL, E. (2007). Academic writing and how to grow vocabulary. Seattle: TESOL: Available on: http://www.EliHinkel.Org/TESL2012. Accessed 02 April 2020.

HOADLEY, U. (2012). What do we know about teaching and learning in South African primary schools? *Education as Change*, 16(2): 187-202. https://doi.org/10.1080/16823206.2012.745725

HOME, F. & Heinemann, G. (2003). *English in perspective*. Oxford: Oxford University Press.

HOWIE, S. J. (2003). Language and other background factors affecting secondary pupils' performance in Mathematics in South Africa, *African Journal of Research in Mathematics, Science and Technology Education*, 7(1): 1-20. https://doi.org/10.1080/10288457.2003.10740545

HOWIE, S. J. (2001). *Mathematics and Science Performance in Grade 8 in South Africa 1998/99: TIMSS-R 1999 South Africa*. Pretoria: Human Sciences Research Council.

HUGHES, C. E., Shaunessy, E. S. & Brice, A. R. (2006). Code switching among bilingual and limited English proficient students: possible indicators of giftedness. *Journal for the Education of the Gifted*, 30(1): 7-28. https://doi.org/10.1177/016235320603000102

HUNGARIAN Academy of Sciences – Committee on Mathematics Education. (2016). A tanítói/tanári kérdő*ívre* beküldött válaszok *összesítése* [*Summary of the answers to the* teachers' *questionnaire*]. https://mta.hu/data/dokumentumok/iii_osztaly/2016/tanitoi_tanari_kerdoiv_osszegzes_2016%20(1).pdf.

HURST, E. (2016). Navigating Language: Strategies, Transition and the 'Colonial wound' in South African Education. *Language and Education*, 30(3): 219-234. https://doi.org/10.1080/09500782.2015.1102274

HURST, E. & Mona, M. (2017). Translanguaging as a socially just pedagogy. *Education as Change*. SCIELO, Vol 21(2). https://doi.org/10.17159/1947-9417/2017/2015

HYMES, D. H. (1972) "On Communicative Competence" In: J.B. Pride and J. Holmes (Eds.), *Sociolinguistics. Selected Readings*. Harmondsworth: Penguin, (2): 269-293.

I

ILANY, B. & Margolin, B. (2010). Language and Mathematics: Bridging between Natural Language and Mathematical Language in Solving Problems in Mathematics, *Creative Education*, 1(3): 138-148. https://doi.org/10.4236/ce.2010.13022

IMBO I., Bulcke C. V., De Brauwer J. & Fias, W. (2014). Sixty-four or four-and-sixty? The influence of language and working memory on children's number transcoding. *Front. Psychol.*, 5: 313. https://doi.org/10.3389/fpsyg.2014.00313

IMHOFF, G. (1990). The position of U.S. English on bilingual education, in English Plus. Issues in bilingual education, by C . Cazden & C. Snow, (Eds.), (March): 48-61. *The Annals of the American Academy of Political and Social Science*.

J

JEGEDE, O. O. (2012). Roles of code switching in multilingual public primary schools in Ile-Ife, Nigeria. *American Journal of Linguistics*, 1(3): 40-46.

JET Education Services, (2015a). *Intermediate/Senior Phase Teacher Assessment, English. Initial Teacher Education Research Project.* Johannesburg: JET Education Services.

JET Education Services, (2015b). *Intermediate/Senior Phase Teacher Assessment, Mathematics. Initial Teacher Education Research Project.* Johannesburg: JET Education Services.

JET Education Services, (2010). *The South African national education evaluation system: What will make it work?* Occasional Paper 1. Johannesburg: JET Education Services.

JIMENEZ, L.. & Verschaffel L. (2014). Development of children's solutions of non-standard arithmetic word problem solvings. *Rev. Psicodidactica*, 19: 93-123. https://doi.org/10.1387/RevPsicodidact.7865

JONES, D. (2001). That compulsory African language in the spotlight again. *Business Day*, 21 February.

JORNET, A. & Roth, W. M. (2015). The joint work of connecting multiple (re)presentations in science classrooms. *Science Education*, 99: 378-403. https://doi.org/10.1002/sce.21150

JOSTA K., Khadera P., Burkea M., Bienb S. & Röslera F. (2009). Dissociating the solution processes of small, large, and zero multiplications by means of MRI. *Neuroimage*, 46: 308-318. https://doi.org/10.1016/j.neuroimage.2009.01.044

JUEL, C. & Deffes, R. (2004). Making words stick. *Educational Leadership*, 61 (6): 30-34.

K

KAJORO, P. M. (2016). Transition of the medium of instruction from English to Kiswahili in Tanzanian primary schools in A. Halai & P. Clackson (Eds.), *Teaching and Learning Mathematics in Multilingual Classrooms*. Sense Publishers, Rottendam. https://doi.org/10.1007/978-94-6300-229-5_6

KALANTZIS, M., Cope, B., Chan, E., & Dalley-Trim, L. (2016). *Literacies* (2nd Ed.). Port Melbourne, VIC, Austalia: Cambridge University Press. https://doi.org/10.1017/9781316442821

KAMBULE, T. W. (1984). The language factor in teaching mathematics. *Lengwitch: The Journal of Language Across the Curriculum*, 1984-1989.

KASULE, D. & Mapolelo, D. (2005). Teachers' strategies of teaching primary school mathematics in a second language: A case of Botswana. *International Journal of Educational Development*, 25(6): 602-617. https://doi.org/10.1016/j.ijedudev.2004.11.021

KEMPERT, S., Saalbach, H. & Hardy, I. (2011). Cognitive benefits and costs of bilingualism in elementary school students: The case of mathematical word problems. *Journal of Educational Psychology*, 103(3): 547-561. https://doi.org/10.1037/a0023619

KHISTY, L. L. (1995). Making Inequality: Issues of language and meanings in mathematics teaching with Hispanic students. In: W. G. Secanda, E. Fennema & L. B. Adajian, *New directions for equity in mathematics education* (pp. 279-297). Cambridge: Cambridge University Press.

KILLEN, R. (2007). *Teaching strategies for outcomes-based education.* Mercury Crescent, Wetton: Juta and Company Ltd.

KINDLER, A. (2002). Survey of the states' limited English proficient students and available educational programs and services: 2000-2001 summary report (NCELA Report BE021853). Available on: National Clearinghouse for English Language Acquisition http://www. ncela.gwu.edu/files/rcd/BE021853/Survey_of_ the_States.pdf. Accessed 10 April 2020.

KLEIN, E., Moeller, K., Kettl, D., Zauner, H., Wood, G. & Willmes, K. (2013). Object-based neglect in number processing. *Behav. Brain Funct.*, 95. https://doi.org/10.1186/1744-9081-9-5

KLINGLER, K. L. (2012). Mathematics strategies for teaching problem solving: the influence of teaching mathematical problem solving strategies on students' attitudes in middle school, Unpublished PhD Thesis. Florida: University of Central Florida Orlando.

KINGSDORF, S. & Krawec, J. (2014). Error analysis of mathematical word problem solving across students with and without learning disabilities. *Learn. Disabil. Res. Pract.*, 29: 66-74. https://doi.org/10.1111/ldrp.12029

KOCK, C. E. & Woods, A. S. (2004). *Global Business Language*, 9: 67-77.

KODA, K. (2007). Reading and language learning: Crosslinguistic constraints on second language reading development. In: K. Koda (Ed.), Reading and language learning (Special issue of) *Language Learning Supplement*, 57: 1-44. https://doi.org/10.1111/0023-8333.101997010-i1

KOEDINGER K. R. & Nathan M. J. (2004). The real story behind story problems: effects of representations on quantitative reasoning. *J. Learn. Sci.*, 13: 129-164. https://doi.org/10.1207/s15327809jls1302_1

KRASHEN, S. D. & Brown, C. L. (2007). What is academic language proficiency? Available on: www.Sdkrashen.Com.\articles/krashen-brownAlp.Pdf.2014. Accessed 20 August 2020.

KRASHEN, S. D. & Terrell, T. D. (1983). *The Natural Approach: language acquisition in the classroom.* Oxford: Pergamon Press.

KRUGEL, R. & Fourie, E. (2014). Concerns for the language skills of South African learners and their teachers. *International Journal of Educational Sciences, 7*(1): 219-228. https://doi.org/10.1080/09751122.2014.11890184

L

LACHMAIR, M., Dudschig, C., De La Vega, L. & Kaup B. (2014). Relating numeric cognition and language processing: do numbers and words share a common representational platform? *Acta Psychol.*, 148: 107-114. https://doi.org/10.1016/j.actpsy.2013.12.004

LAMBELL, R., Ramia, G., Nyland, C. & Michelotti, M. (2008). NGOs and International Business Research, *International Journal of Management Reviews*, 10(1): 195-229. https://doi.org/10.1111/j.1468-2370.2007.00218.x

LASAGABASTER, D. (1998). The threshold hypothesis applied to three languages in contact at school, *International Journal of Bilingual Education and Bilingualism*, 1(2): 119-134. https://doi.org/10.1080/13670059808667678

LEE, C. (2006). *Language for Learning Mathematics: Assessment for Learning in Practice.* New York: Open University Press.

LEE, K., Ng, S. F., Ng, E. L. & Lim R. Y. (2004). Working memory and literacy as predictors of performance on algebraic word problems. *J. Exp. Child Psychol.*, 89: 140-158. https://doi.org/10.1016/j.jecp.2004.07.001

LEHOTANOVÁ, B. (2013). *Matematika az alapiskolák 1. osztálya számára 2. rész.* [Mathematics for the 1st grade of the elementary school: 2nd Part]. Bratislava: AITEC Kiadó.

LEMKE, J. L. (2003). Mathematics in the middle: Measure, picture, gesture, sign, and word. In: M. Anderson, A. Sáenz-Ludlow, S. Zellweger, & V. V. Cifarelli (Eds.), *Educational perspectives on mathematics as semiosis: From thinking to interpreting to knowing* (pp. 215-234). Brooklyn, NY, and Ottawa, Ontario: Legas.

LEMMER, E. M. (1995). Selected linguistic realities in South African schools: problems and prospects. *Educare*, 24: 82-96.

LEWIS, G., Jones, B. & Baker, C. (2012). Translanguaging: Origins and development from school to street and beyond. Educational Research and Evaluation: *An International Journal on Theory and Practice,* 18(7): 641-654. https://doi.org/10.1080/13803611.2012.718488

LIM, B. S. (1998). Factors associated with Korean-American students' mathematics achievement, *Dissertation Abstract International*, 59(06): 19-55.

LIVINGSTON, K. (2016). Teacher education's role in educational change. *European Journal of Teacher Education.* Volume 39 (1): 1-4. https://doi.org/10.1080/02619768.2016.1135531

LYONS, J. (1981). *Language, Meaning & Context.* Philadelphia: Fontana Paperbacks.

M

MACDONALD, C. A. (1990). *Crossing the Threshold into Standard Three*: Pretoria: Human Sciences Research Council.

MACSWAN, J. (2000). The architecture of the bilingual language faculty: Evidence from code switching. *Bilingualism: Language and Cognition*, 3(1): 37-54. https://doi.org/10.1017/S1366728900000122

MAHADHIR, M. & Then, C. O. (2007). Code-switching in the English language classrooms in Kuching secondary schools. Sarawak Museum Journal, *Social Sciences and Humanities*, Special Issue 7(850): 197-219.

MAHOOTIAN, S. (2006). Code switching and mixing. In: K. Brown (Ed.), *Encyclopaedia of Language and Linguistics,* (2nd Edition) (pp. 511-527). Oxford: Elsevier. https://doi.org/10.1016/B0-08-044854-2/01507-8

MANYIKE, T. V. (2007). The acquisition of English academic language proficiency among grade 7 learners in South African Schools. Unpublished PhD Thesis, University of South Africa, Pretoria.

MARINOVA-TODD, S. H. (2003). Know your grammar: what the knowledge of syntax and morphology in an L2 reveals about the critical period for Second/Foreign language acquisition. In: M. Mayo & M. Lecumberri (Eds.), *Age and the acquisition of English as a foreign language.* Clevedon: Multilingual Matters Ltd., pp 59-73. https://doi.org/10.21832/9781853596407-004

MARO, R. A. (1994). The Effect of Learning Mathematics in a Second Language on Reasoning Ability (Tanzania). Unpublished MEd Thesis, University of New Brunswick (Canada).

MARTIN, T. S. (2007). *Mathematics today: Improving practice, improving student learning. (2nd Edition).* Reston, VA: NCTM.

MARTIN, S. A. & Bassok, M. (2005). Effects of semantic cues on mathematical modelling: evidence from word-problem solving and equation construction tasks. *Mem. Cognit.*, 33: 471-478. https://doi.org/10.3758/BF03193064

MARTINIELLO M. (2008). Language and the performance of English-Language Learners in math word problems. *Harv. Educ. Rev.*, 78: 333-368. https://doi.org/10.17763/haer.78.2.70783570r1111t32

MATI, X. (2004). *Using code switching as a strategy for bilingual education in the classroom.* Paper presented at the 21st Annual AEAA Conference, 25-28 August. Pretoria: HSRC. Available on: http://www.hsrc.ac.za/en/research-data/view/1615. Accessed 20 August 2020.

MAXWELL, J. A. (1998). Designing a qualitative study. In: L. Bickman & D. J. Rog (Eds.), *Handbook of applied social research methods* (pp. 69-100). Thousand Oaks, CA: Sage.

MAY, S., Hill, R. & Tiakiwai, S. (2004). Bilingual/Immersion Education: Indicators of Good Practice. Final Report to the Ministry of Education, New Zealand: Wilf Malcolm Institute of Educational Research, School of Education, University of Waikato.

MAYO, M. P. G. & Soler, E. A. (2002). Introduction to the role of interaction in instructed language learning. *Guest editorial, International Journal of Educational Research*, 37: 233-236. https://doi.org/10.1016/S0883-0355(03)00002-8

MBUDE-SHALE, N. (2013). Exploring The Correlation Between Language Medium And Academic Achievement: A Comparative Study Of The Language Of Learning And Teaching (LoLT) And Mathematics Results In The 2010 Grade 12 National Senior Certificate Examinations in the Eastern Cape. Unpublished Master's Thesis, Grahamstown: Rhodes University.

McKelvey, B. (2003). From Fields to Science: Can Organisation Studies Make the Transition? In: R. Westwood & S. Clegg (Eds.), *Debating Organization: Point-Counterpoint in Organization Studies*. Malden: Blackwell, pp. 47-73.

Mekonnen, A. G. (2005). Socio-cultural and educational implications of using mother tongues as languages of instruction in Ethiopia. Unpublished Master's Thesis, Norway: University of Oslo.

Menken, K. & Antunez, B. (2001). *An overview of the preparation and certification of teachers working with limited English proficient students*. Washington, DC: The National Clearinghouse for English Language Acquisition.

Metila, R. A. (2009). Decoding the switch: The functions of code switching in the classroom. *Education Quarterly*, 67(1): 44-61.

Mikhailov, F. T. (2006). Problems of the method of cultural-historical psychology. *Journal of Russian and East European Psychology*, 44(1): 21-54. https://doi.org/10.2753/RPO1061-0405440103

Ministry of Education. (2002). Language policy for higher education. Available on http://education.pwv.gov.zalDoE_siteslHigher_Educationlrevised language policy 5 November2oo2.doc. Accessed 20 August 2017.

Mitchell, R. & Myles, F. (2004). *Second language learning theories*. London: Arnold.

Moeller, K., Klei, E., Kucian, K. & Willmes, K. (2014). Numerical development – from cognitive functions to neural underpinnings. *Front. Psychol.*, 5: 1047. https://doi.org/10.3389/fpsyg.2014.01047

Mohanty, A. K. (1994). *Bilingualism in a Multilingual Society: Psychological and Pedagogical Implications,* Mysore: Central Institute for Indian Languages.

Mohlabi-Tlaka, N. H. (2016). The contribution of a text-based approach to English education for communicative competence. Unpublished PhD Thesis, Pretoria: University of Pretoria.

Moloi, F., Morobe, N. & Urwick, J. (2008). Free but inaccessible primary education: A critique of the pedagogy of English and mathematics in Lesotho. *International Journal of Educational Development*, 28(5): 612-621. https://doi.org/10.1016/j.ijedudev.2007.12.003

Morgan, J. & Rinvolucri, M. (2004). *Vocabulary (2nd Edition.),* Oxford: Oxford University Press.

Moschkovich, J. N. (2002). A situated and sociocultural perspective on bilingual mathematics learners. *Mathematical Thinking and Learning*, 4(2&3): 189-212. https://doi.org/10.1207/S15327833MTL04023_5

Moschkovich, J. N. (1999). Supporting the participation of English language learners in mathematical discussions. *For the Learning of Mathematics*, 19(1): 11-19.

Mouton, J. (2001). How to Succeed in Your Master's and Doctoral Study. Hatfield. Pretoria: Van Schaik.

Mufanechiya, T. (2011). The use of English language as a medium of instruction in the Zimbabwean Junior Primary Schools. *NJLC* 4(2): 115-127.

Murray, H. (2012). *Problems with Word Problems in Mathematics.* Learning and Teaching Mathematics, 13: 55-58.

Myslìn, M. & Levy, R. (2014). Code switching and predictability of meaning in discourse. San Diego 9500: Department of Linguistics.

N

NASTASI, B. K. & Clements, D. H. (1990). Metacomponential functioning in young children. *Intelligence*, 14: 109-125. https://doi.org/10.1016/0160-2896(90)90017-N

NATION, I. S. P. (2003). Effective ways of building vocabulary knowledge. *ESL Magazine*, 6(4): 14-15.

NATIONAL Council for Curriculum and Assessment (NCCA). (2006). *English as an additional language in Irish primary schools: guidelines for teachers*. Dublin: Government Stationery Office. Available on http://www.ncca.ie/uploadedfiles/publications/EALangIPS(1).pdf. Accessed 07 July 2020.

NATIONAL Council of Teachers of Mathematics (NCTM). (2000). *Principles and Standards for School Mathematics*. Reston, VA: National Council of Teachers of Mathematics.

NATIONAL Education Evaluation and Development Unit (NEEDU). (2013). *National Report 2013: Teaching and Learning in Rural Primary Schools*. Pretoria: National Education Evaluation and Development Unit.

NATHANSON, R. (2008). A school-based, balanced approach to early reading instruction for English additional language learners in grades one to four. Unpublished PhD Thesis, Stellenbosch: Stellenbosch University.

NAVSARIA, I., Pascoe, M. & Kathard, H. (2010). Department of Health and Rehabilitation Sciences, University of Cape Town.

NCANYWA, T. (2014). The state of the Eastern Cape schools in a period almost the second decade of democracy. Economic Research Southern Africa (ERSA). Available on https://econrsa.org/system/files/publications/working_papers/working_paper_486.pdf. Accessed 17 July 2020.

NEL, N. & Swanepoel, E. (2010). Do the language errors of ESL teachers affect their learners? *Per Linguam*, 26(1): 47-60. https://doi.org/10.5785/26-1-13

NELSON Mandela Foundation (2004). *How far have we come? Reflections from the first 10 years of basic education transformation*. Johannesburg: Nelson Mandela Foundation.

NESHER, P., Greene, J. G. & Riley, M. S. (1982). The development of semantic categories for addition and subtraction. *Educational Studies in Mathematics*, 13: 373-394. https://doi.org/10.1007/BF00366618

NEMÉNYI, C. & Szendrei, E. (1997). *Szöveges feladatok* [Word problems]. Budapest: Budapesti Tanítóképző Főiskola.

NEWBY, T. J. & Stepich, D. A. (1987). Learning Abstract Concepts: The Use of Analogies as a Mediational Strategy. *Journal of Instructional Development*, 10(2): 20-26. https://doi.org/10.1007/BF02905788

NEWMAN, M. A. (1983). *Strategies for diagnosis and remediation*. Sydney: Harcourt, Brace Jovanovich.

NEWMAN, M. A. (1977). An analysis of sixth-grade pupils' errors on written mathematical tasks. *Victorian Institute for Educational Research Bulletin*, 39: 31-43.

NÍ Ríordáin, M. (2013). *A comparison of Irish and English language features and the potential impact of differences*. Paper presented at the Eighth Congress of European Research in Mathematics.

NUERK, H. C., Moeller, K. Klein, E., Willmes, K. & Fischer, M. H. (2011). Extending the mental number line – a review of multi-digit number processing. *J. Psychol.*, 219: 3-22. https://doi.org/10.1027/2151-2604/a000041

O

OECD. (2010). *PISA 2009 Results. What Students Know and Can Do. Student Performance in Reading, Mathematics and Science*, Vol. 1. Paris: OECD Publishing.

OLIVIER, W. A. (2015). Reflection on the Implementation of CAPS Mathematics in the Classroom. www.samf.ac.za, Accessed 09 April 2020.

OLIVIER, O. & Olivier, J. (2013). The influence of affective variables on the acquisition of academic literacy. *Per Linguam*, 29(2): 56-71. https://doi.org/10.5785/29-2-525

O'MALLEY, J. M., Chamot, A. U., Stewner-Manzanares, G., Küpper, L. & Russo, R. P. (1985). Learning strategies used by beginning and intermediate ESL *students*. In: E. Macaro, *Teaching and learning a second language*. New York: The Bath Press. https://doi.org/10.1111/j.1467-1770.1985.tb01013.x

O'NEILL, D. K. (2013). Preschool children's narratives and performance on the Peabody individualized achievement test – revised: evidence of a relation between early narrative and later mathematical ability. *First Lang.*, 2(4): 149-183. https://doi.org/10.1177/0142723704043529

OOSTERMEIJER, M., Boonen, A. J. H. & Jolles, J. (2014). The relation between children's constructive play activities, spatial ability, and mathematical word problem-solving performance: a mediation analysis in sixth-grade students. *Front. Psychol.*, 5: 782 https://doi.org/10.3389/fpsyg.2014.00782

ORRANTIA, J., Rodriquez, L. & Vicente, S. (2010). Automatic activation of addition facts in arithmetic word problems. *Q. J. Exp. Psychol.*, 63: 310-319. https://doi.org/10.1080/17470210902903020

OROSCO, M. J. & Hoover, J. J. (2009). Characteristics of second language acquisition, cultural diversity and learning/behaviour disabilities. In: J. J. Hoover, *Differentiating Learning Differences from Disabilities: Meeting Diverse Needs through Multi-Tiered Response to Intervention*. New York: Prentice Hall.

OVANDO, C., Collier, V. & Combs, M. (2003). *Bilingual and ESL Classrooms: Teaching Multicultural Contexts*. (3rd Edition.). Boston: McGraw-Hill.

P

PALM, T. (2006). *Word problems as simulations of real-world situations: A proposed framework for the Learning of Mathematics*, 26: 42-47.

PAPE, S. J. (2003). Compare word problems: consistency hypothesis revisited. *Contemp. Educ. Psychol.*, 28: 396-421. https://doi.org/10.1016/S0361-476X(02)00046-2

PARADES, E. (2010). *Language learning strategy use by Colombian adult English language learners: a phenomenological study*. In: A. Eid, 2012. Language learning strategy use of Saudi EFL students in an intensive English learning context. *Asian Social Science*, 8 (13): 115-127. https://doi.org/10.5539/ass.v8n13p115

PASSOLUNGHI, M. C. & Siegel L. S. (2001). Short-term memory, working memory and inhibitory control in children with difficulties in arithmetic problem solving. *J. Exp. Child Psychol.*, 80: 44-57. https://doi.org/10.1006/jecp.2000.2626

PARKER, B. (2013). Institutional landscapes and curriculum mindscapes: An overview of teacher education policy in South Africa from 1990 to 2011. In: K. Lewin, M. Samuel & Y. Sayed (Eds.), *Changing patterns of teacher education in South Africa*. London: Heinemann.

PATRINOS, H. & Velez, E. (1996). *Costs and benefits of bilingual education in Guatemala: a partial analysis.* World Bank: Human Capital Development Working Paper No. 74.

PAVÓN, V. & Cabezuelo, R. (2019). Analysing mathematical word problem solving with secondary education CLIL students: A pilot study. *Latin American Journal of Content & Language Integrated Learning,* 12(1): 18-45. https://doi.org/10.5294/laclil.2019.12.1.2

PENNYCOOK, A. (1998). *English and the Discourses of Colonialism.* London and New York.

PEREGOY, S. F. & Boyle, O. F. (2005). *Reading, writing, and learning in ESL: A resource book for K-12 teachers* (4th ed.). Boston: Pearson Allyn and Bacon.

PETERS, J. (2011). A mathematics teacher's beliefs and knowledge of grade 7 learners' problem-solving strategies during a problem-solving task. Unpublished PhD Thesis. University of Johannesburg. Available on: http://ujdigispace.uj.ac.za/handle/10210/8050. Accessed 24 March 2020.

PHONAPICHAT, P., Wongwanich, S. & Sujiva, S. (2014). An analysis of elementary school students' difficulties in mathematical problem solving. *Social and Behavioural Sciences,* 116: 3169-3174. Available on: http://www.sciencedirect.com/science/article/pii/S1877042814007459. Accessed 08 February 2020. https://doi.org/10.1016/j.sbspro.2014.01.728

PIMM, D. (1987). *Speaking Mathematically.* London: Routledge & Kegan Paul.

PIPER, B. & Miksic, E. (2011). The Early Grade Reading Assessment: Applications and Interventions to Improve Early Grade Literacy gap. *Mother Tongue and Reading: Using Early Grade Reading Assessments to Investigate Language – of – Instruction Policy in East Africa.* RTI Press. https://doi.org/10.3768/rtipress.2011.bk.0007.1109.5

PIXNER, S., Kaufmann, L., Moeller, K., Hermanova, V. & Nuerk H. C. (2011). Whorf reloaded: language effects on non-verbal number processing in 1st grade – a trilingual study. *J. Exp. Child Psychol.,* 108: 371-382. https://doi.org/10.1016/j.jecp.2010.09.002

PIXNER, S., Zuber, J., Hermanova, V., Kaufmann, L. & Nuerk, H. C. (2011). One language, two number-word systems and many problems: numerical cognition in the Czech language. *Res. Dev. Disabil.,* 32: 2683-2689. https://doi.org/10.1016/j.ridd.2011.06.004

PLÜDDEMANN, P., Xola, M. & Mahlahela-Thusi, B. (2000). Problems and possibilities in multilingual classrooms in the Western Cape. *PRAESA occasional papers No. 2.* Cape Town: PRAESA.

PLÜDDEMANN, P. (2010). *Home-language based bilingual education: Towards a learner-centred language typology of primary schools in South Africa,* PRAESA Occasional Papers No. 32: Cape Town; University of Cape Town, PRAESA.

PRESMEG, N. (2016). *Semiotics in Mathematics Education,* ICME-13 Topical Surveys. https://doi.org/10.1007/978-3-319-31370-2_1

PRETORIUS, E. J. (2014). Supporting transition or playing catch-up in Grade 4? Implications for standards in education and training. *Perspectives in Education,* 2014: 32(1).

PRETORIUS, E. J. (2002). Reading and applied linguistics – a deafening silence? *South African Linguistics and Applied Language Studies,* 20: 91-103. https://doi.org/10.2989/16073610209486300

PRETORIUS, E. J. & Mampuru, D. M. (2007). Playing football without a ball: language, reading and academic performance in a high- poverty school. *Journal of Research in Reading,* 30(1): 38-58. https://doi.org/10.1111/j.1467-9817.2006.00333.x

PROBYN, M. J. (2015). Pedagogical translanguaging: bridging discourses in South African science classrooms. *Language and Education,* 29(3): 218-234. https://doi.org/10.1080/09500782.2014.994525

PROBYN, M. J. (2009). Smuggling the vernacular into the classroom: Conflicts and tensions in classroom code switching in township/rural schools in South Africa. *International Journal of Bilingual Education and Bilingualism*, 12(2): 123-136. https://doi.org/10.1080/13670050802153137

PROBYN, M. J. (2001). Teachers' voices: Teachers' reflections on learning and teaching through the medium of English as an additional language in South Africa. *International Journal of Bilingual Education and Bilingualism*, 4(4): 249-266. https://doi.org/10.1080/13670050108667731

POLLAK, H. O. (1969). How can we teach applications of mathematics? *Educational Studies in Mathematics*, 2: 393-404. https://doi.org/10.1007/BF00303471

PÓLYA, G. (1945). *How to solve it – A new aspect of mathematical method.* Princeton: Princeton University Press. https://doi.org/10.1515/9781400828678

POWELL, S. R. & Fuchs, L. S. (2014). Does early algebraic reasoning differ as a function of students' difficulty with calculations versus word problems? *Learn. Disabil. Res. Pract.* 29: 106-116. https://doi.org/10.1111/ldrp.12037

PSYCHOLINGUISTICS. (2013). In: *Columbia electronic encyclopedia* (6th Edition). Available on: http://www.credoreference.com.libproxy.txstate.edu/entry/edinburghds/psycholinguistics – psycholinguistics. Accessed 10 April 2020.

R

RADFORD, L. (2001). *On the relevance of Semiotics in Mathematics Education*. Paper presented to the Discussion Group on Semiotics and Mathematics Education at the 25th PME International Conference, The Netherlands, University of Utrecht, July 12-17.

RADFORD, L., Schubring, G. & Seeger, F. (Eds.), (2008). *Semiotics in Mathematics Education: Epistemology, History, Classroom, and Culture*, 19-38. https://doi.org/10.1163/9789087905972

RADUAN, I. H. (2010). Error analysis and the corresponding cognitive activities committed by year five primary students in solving mathematical word problems. *Procedia Soc. Behav. Sci.*, 2: 3836-3838. https://doi.org/10.1016/j.sbspro.2010.03.600

RAKGOKONG, L. (1994). Language and the construction of meaning associated with division in primary mathematics. Paper presented at the 2nd Annual Meeting of the Southern African Association for Research in Mathematics and Science Education.

RAMÍREZ, J., Pasta, D., Yuen, S, Ramey, D. & Billings, D. (1991). Executive Summary: Longitudinal study of structured English immersion strategy, early-exit and late-exit bilingual education programs for language-minority children. *(Vol. II)*. (Prepared for U.S. Department of Education). San Mateo, CA: Aguirre International.

RAMPHELE, M. (2008). Laying ghosts to rests: dilemmas of the transformation in South Africa. In: M. Le Cordeur, 2010. The struggling reader: identifying and addressing reading problems successfully at an early stage. *A Journal for Language Learning*, 26(2): 77-89. https://doi.org/10.5785/26-2-23

RABORN, D. T. (1995). Mathematics for students with learning disabilities from language minority backgrounds: Recommendations for teaching. *New York State Association for Bilingual Education Journal*, 10: 25-33.

REAGAN, T. G. (1985). Language planning in South African education. A conceptual overview. *Journal for Language Teaching*, 19: 71-87.

REDDY, V. (2012). Towards Equity and Excellence: Highlights from TIMSS 2011. http://www.hsrc.ac.za/uploads/pageContent/2929/TIMSSHighlights2012Dec7final.pdf Accessed 7 October 2019.

REDDY, V. (2006). *Mathematics and Science Achievement at South African Schools in TIMSS 2003. Cape Town*: Human Sciences Research Council Press.

REYNDERS, A. (2014). Obstacles that hamper learners from successfully translating mathematical word problems into number sentences, Unpublished Master's Thesis. Bloemfontein: University of the Free State.

RICHARDS, J. C. & Rodgers, T. S. (2001). *Approaches and methods in language teaching. (2nd Ed.)*. New York. Cambridge University Press. https://doi.org/10.1017/CBO9780511667305

RICHES, C. & Genesee, F. (2006*)*. Literacy: Cross-linguistic and cross-modal issues. In: F. Genesee, K. Lindholm-Leary, W. Saunders, & D. Christian, D. *Educating English language learners: A synthesis of empirical findings*, pp. 64-108. NY: Cambridge University Press. https://doi.org/10.1017/CBO9780511499913.004

RISKU, H. & Pircher, R. (2008). Visual Aspects of Intercultural Technical Communication: A Cognitive Scientific and Semiotic Point of View. *Meta,* 53 (1): 154-166. https://doi.org/10.7202/017980ar

ROBERTS, N. (2017). School Maths: What is our story? *Daily Maverick*, 8 January 2017. Available at https://www.dailymaverick.co.za/article/2017-01-08-school-maths-what-is-our-story/#.WLOejSN97sE. Accessed 28 March 2020.

ROETTGER, T. B. & Domahs, F. (2015). Grammatical number elicits SNARC and MARC effects as a function of task demands. *Q. J. Exp. Psychol*. https://doi.org/10.1080/17470218.2014.979843

ROLLNICK, M. (2000). Current issues and perspectives on second language learning of science. *Studies in Science Education*, 35: 93-121. https://doi.org/10.1080/03057260008560156

ROSSELL, C. & Baker, K. (1996). The Education Effectiveness of Bilingual Education, *Research in the Teaching of English*, 30(1): 7-74.

ROTHWELL, J. (2010). *In the company of others on introduction to communication. (3rd Edition.)*. New York: Oxford University Press.

RUBIN, J. (1980). Study of cognitive processes in second language learning. *Applied Linguistics*, 11: 117-- 131. In: C. Griffiths, 2004. Some influences on the syllable structure of interlanguage phonology. *International Review of Applied Linguistics*, 4: 143-163. https://doi.org/10.1093/applin/2.2.117

S

SARNECKA, B. W. (2013). On the relation between grammatical number and cardinal numbers in development. *Front. Psychol.*, 5:1132 https://doi.org/10.3389/fpsyg.2014.01132

SAVILLE-TROIKE, M. (1988). Private speech: Evidence for second language learning strategies during the "silent period." *Journal of Child Language*, 15: 567-90. https://doi.org/10.1017/S0305000900012575

SCHLEPPEGRELL, M. J. (2007). The Linguistic Challenges of Mathematics Teaching and Learning: A Research Review, *Reading & Writing Quarterly*, 23: 2, 139-159. https://doi.org/10.1080/10573560601158461

SCHLEY, D. R. & Fujita K. (2014*)*. Seeing the math in the story: on how abstraction promotes performance on mathematical word problems. *Soc. Psychol. Personal. Sci.*, 5: 953-961. https://doi.org/10.1177/1948550614539519

SCHUMACHER, R. F. & Fuchs, L. S. (2012). Does understanding relational terminology mediate effects of intervention on compare word problems? *Journal of Experimental Child Psychology,* (4): 607-628. https://doi.org/10.1016/j.jecp.2011.12.001

SCHUNK, D. H. (2005). Self-regulated learning: The educational legacy of Paul R. Pintrich. *Educational Psychologist*, 40: 85-94. Available on https://libres.uncg.edu/ir/uncg/f/D_Schunk_Self_2005.pdf. Accessed 25 March 2020. https://doi.org/10.1207/s15326985ep4002_3

SECADA, W. & Cruz, Y., (2000). Teaching mathematics for understanding to bilingual students. http://eric-web.tc.columbia.edu/ncbe/immigration/mathematics.htm Accessed 22 September 2016.

SEPENG, P. (2015). Home Language and the Language of Learning and Teaching in Mathematics Classrooms. *International Journal of Educational Sciences*, 8(3): 655-664. https://doi.org/10.1080/09751122.2015.11890286

SEPENG, P. (2014a). Use of common-sense knowledge and reality in mathematics word problem solving. *African Journal of Research in Mathematics, Science and Technology Education,* 18(1): 14-24. https://doi.org/10.1080/10288457.2014.890808

SEPENG, P. (2014b). Learners' voices on the issues of language use in learning mathematics. *International Journal of Educational Sciences*, 6(2): 207-215. https://doi.org/10.1080/09751122.2014.11890133

SEPENG, P. (2013). Issues of language and mathematics: contexts and sense-making in word problem-solving. *Special Issue of Mediterranean Journal of Social Sciences*, 4(13): 51-62. https://doi.org/10.5901/mjss.2013.v4n13p51

SETATI, M. (2008). Access to mathematics versus access to the language of power: The struggle in multilingual mathematics classrooms. *South African Journal of Education*, 28: 103-116.

SETATI, M. (2005). Mathematics education and language: Policy, research and practice in multilingual South Africa. In: R. Vithal, J. Adler & C. Keitel, *Math Education Research in South Africa: Perspectives, practices and possibilities* (pp. 73-109). Pretoria: HSRC.

SETATI, M. (2003). Language use in a multilingual mathematics classroom in South Africa: A different perspective. In: N. A. Plateman, B. J. Dougherty & J. Zillox, *Proceedings of the 2003 joint meeting of PME & PMENA (Vol. 4).* Honolulu: University of Hawaii.

SETATI, M. (2002). Researching mathematics education and language in multilingual South Africa. *The Mathematics Educator,* 12(2): 6-12.

SETATI, M. (1999). Innovative language practices in the classroom. In: N. Taylor & P. Vinjevold (Eds.), *Getting Learning Right*. Johannesburg: JET Education Services.

SETATI, M. (1998). Code-switching in a senior primary class of second language learners. *For the Learning of Mathematics*, 18(1): 34.

SETATI, M. & Adler, J. (2001). Between languages and discourses: Code-switching practices in primary mathematics classrooms in South Africa. *Educational Studies in Mathematics,* 43: 243-269. https://doi.org/10.1023/A:1011996002062

SETATI, M., Adler, J., Reed, Y., & Bapoo, A. (2001). Incomplete journeys: Code-switching and other language practices in mathematics, science and English language classrooms in South Africa. *Language and Education,* 16: 128-149. https://doi.org/10.1080/09500780208666824

SHAFTEL, J., Belton-Kocher, E., Glasnapp, D. & Poggio, J. (2006). The impact of language characteristics in mathematics test items on the performance of English language learners and students with disabilities. *Educ. Assess.*, 11: 105-126. https://doi.org/10.1207/s15326977ea1102_2

SHAKI, S., Fischer, M. H. & Petrusic, W. M. (2009). Reading habits for both words and numbers contribute to the SNARC effect. *Psychon. Bull. Rev.*, 16: 328-331. https://doi.org/10.3758/PBR.16.2.328

SHIUNDU, J. O. & Mohammed, A. (1996). Issues in Social Studies Teacher Education in Africa, Kenya. *Journal of Education*, 6 (1): 1-17.

SHULMAN, L. S. (1986). Those who understand: Knowledge growth in teaching. *Educational Researcher*, 15: 4-14. https://doi.org/10.3102/0013189X015002004

SHULTZ, T. R., & Horibe, F. (1974). Development of the appreciation of verbal jokes. *Developmental Psychobiology*, 10: 13-20. https://doi.org/10.1037/h0035549

SIBANDA, J. (2014). Investigating the English Vocabulary Needs, Exposure, and Knowledge of isiXhosa Speaking Learners for Transition from Learning to Read in the Foundation Phase to Reading to Learn in the Intermediate Phase: A Case Study. Unpublished PhD Thesis, Grahamstown: Rhodes University.

SIDLE, S. D. (2009). Building a Commited Global Workforce: Does what Employees Want Depend on Culture? *Academy of Management Perspectives*, 23 (1): 79 -80. https://doi.org/10.5465/amp.2009.37008007

SIGLEY, R. & Wilkinson, L. C. (2015). Ariel's cycles of problem solving: An adolescent acquires the mathematics register. *Journal of Mathematical Behavior*, 40: 75-87. https://doi.org/10.1016/j.jmathb.2015.03.001

SKEHAN, P. (1989). *Individual Differences in Second-Language Learning*. London: Edward Arnold.

SLAVIN, R. E., Madden, N., Calderon, M., Chamberlain, A. & Hennessy, M. (2011). Reading and Language Outcomes of a Multiyear Randomized Evaluation of Transitional Bilingual Education, *Educational Evaluation and Policy Analysis*, 33(1): 47-58. https://doi.org/10.3102/0162373711398127

SOCIOLINGUISTICS (Sociolinguistic). (2004). In: *A Dictionary of Sociolinguistics*. Available on: http://www.credoreference.com.libproxy.txstate.edu/entry/edinburghds/sociolinguistics – sociolinguistic. Accessed 10 April 2020.

SOMYO, S. (2015). Eastern Cape 2015/16 Budget Vote. Available on: http://www.gov.za/speeches/mec-sakhumzi-somyo-eastern-cape-201516-budget-vote-6-mar-2015-0000. Accessed 07 July 2015.

SOUDIEN, C. (2008). *Misrecognition: policy in pursuit of justice?* Keynote address to the EASA/OVSA Conference, January 2008, 1-11. Langebaan: EASA/OVSA.

SOUTH African University Vice Chancellors Association (SAUVCA). (2003). Summary Report. September 2003. *The FET* schools' *policy: The National Curriculum Statement and FET (General) exit qualification*. Pretoria: SAUVUCA.

SPAULL, N. (2016). *Reading in the Foundation Phase: Current developments and remaining gaps*. Presentation at University of Johannesburg. 28 April 2016.

SPAULL, N. (2013). *Report for ECD, South* Africa's *Education Crisis: The quality of education in South Africa, 1994 – 2011*. Centre for Development and Enterprise.

SPAULL, N. (2013). Teachers can't teach what they don't know: On one key factor behind the academic bankruptcy of many SA schools, Politicsweb, http://www.politicsweb.co.za/politicsweb/view/politicsweb/en/page71619?oid=399970&sn=Detail (Accessed 09 April 2020).

SPOLSKY, B. (1989). *Conditions for second language Learning*. Oxford: Oxford University Press.

STACEY, K. (2012). "The international assessment of mathematical literacy: PISA 2012 framework and items," in *Proceedings of the 12th International Congress on Mathematical Education*, Seoul.

STARKEY, A. (1998). Interim report on the English language project. In: Stephen, D.F., Welman, J. C. & Jordan, W. J., 2004. English proficiency as an indicator of academic performance at a tertiary institution. *South African Journal of Human Resource Management*, 2(3): 48-52. https://doi.org/10.4102/sajhrm.v2i3.48

STATISTICS South Africa. (2012). *Census 2011: Census in Brief.* Pretoria: Statistics South Africa.

STEPHEN, D. F., Welman, J. C. & Jordaan, W. J. (2004). English proficiency as an indicator of academic performance at a tertiary institution. *South African Journal of Human Resource Management*, 2(3): 48-62. https://doi.org/10.4102/sajhrm.v2i3.48

SUNDAY Times News, (2015). Big boost for learning of African languages by Govender, P. 17 July. Available on: www.timeslive.co.za. Accessed 10 October 2019.

SWAIN, M. (2005). The output hypothesis: Theory and research. In: E. Hinkel (Ed.), *Handbook of research in second language teaching and learning* (pp. 471-483). Mahwah, NJ: Erlbaum.

SWANSON, H. L. (2004). Working memory and phonological processing as predictors of children's mathematical problem solving at different ages. *Mem. Cognit.*, 32: 648-661. https://doi.org/10.3758/BF03195856

SWANSON, H. L., Lussier, C. & Orosco, M. (2013). Effects of cognitive strategy interventions and cognitive moderators on word problem solving in children at risk for problem solving difficulties. *Learn. Disabil. Res. Pract.*, 28: 170-183. https://doi.org/10.1111/ldrp.12019

SWETZ, F. J. (2012). *Mathematical expeditions: Exploring word problems across the ages*. Baltimore, MD: John Hopkins University Press.

T

TAOLE, J. K. (1981). A study of effect on pupils' achievement of studying a selected secondary school mathematics topic in the vernacular. *Dissertation Abstract International*, 42(05), 2009. (UMI No. 8122985).

TARONE, E. (1981). Language learning strategies: Theory and research. In: C. Griffiths, 2004. Some influences on the syllable structure of interlanguage phonology. *International Review of Applied Linguistics*, 4: 143-163.

TAYLOR, N. (2015). *Initial Teacher Education Research Project: An examination of aspects of initial teacher education curricula at five higher education institutions. Summary Report*. Johannesburg: JET Education Services.

TAYLOR, S. & Coetzee, M. (2013). *Estimating the impact of language of instruction in South African primary schools. A fixed effects approach*: Stellenbosch Economic Working Papers 21/13: Stellenbosch. Stellenbosch University.

TAYLOR, N. & Moyane, J. (2004). Khanyisa Education Support Programme: Baseline study Part 1: Communities, Schools and Classrooms, Memorandum (April 2005): 38-41.

TAYLOR, N., & Taylor, S. (2013). Teacher knowledge and professional habitus. In: N. Taylor, S. van der Berg, & T. Mabogoane, *What makes schools effective? Report of the National Schools Effectiveness Study* (pp. 202-232). Cape Town: Pearson Education, South Africa.

TERAO, A., Koedinger, K. R.. Sohn, M. H., Qin, Y., Anderson, J. R. & Carter, C. S. (2004). "An fMRI study of the interplay of symbolic and visuo-spatial systems in mathematical reasoning." in *Proceedings of the 26th Annual Conference of the Cognitive Science Society,* Chicago, 1327-1332.

THEVENOT, C., Devidal, M., Barrouillet, P. & Fayol, M. (2007). Why does placing the question before an arithmetic word problem improve performance? A situation model account. *Q. J. Exp. Psychol.*, 60: 43-56. https://doi.org/10.1080/17470210600587927

THOMAS, W. P. & Collier, V. (2002). *A national study of school effectiveness for language minority* students' *long term academic achievement.* Available on ERIC database. (ED475048).

THOMASON, S. G. & Kaufman, T. (1988). *Language contact, creolization and genetic linguistics.* California: University of California Press.

THOMPSON, D. R., Kersain, G., Richards, J. C., Hunsader, P. D. & Rubenstein, R. N. (2012). *Mathematical literacy: Helping students make meaning in the middle grades.* Portsmouth, NH: Heinemann.

TOLLEFSON, J. (1991). *Planning Language, Planning Inequality: Language Policy in the Community*, London: Longman.

TOWNSEND, J. & Turner, M. (2000). Dyslexia in practice: a guide for teachers. In: M. Le Cordeur, *The struggling reader: identifying and addressing reading problems successfully at an early stage. Per Linguam*, 26(2): 77-89. https://doi.org/10.1007/978-1-4615-4169-1

TRAORE, S. (2001). *La Pedagogie convergente: Son experimentation au Mali et son impact sur le systeme Educatif.* Geneve: UNESCO Bureau International d'Education.

TRASK, R. L. (1995). *Language: the basics.* London: Routledge.

TREFFERS-DALLER, J. (2007). *Handbook of Pragmatics.* Chicago: John Benjamins Publishing Company.

TRENDS in Mathematics and Science Studies (TIMSS). (2013). Assessment Frameworks. International Association for the Evaluation of Educational Achievement (IEA). TIMSS & PIRLS International Study Centre, Lynch School of Education, Boston College, Chestnut Hill, MA.

TREWBY, R. & Fitchat, S. (2001). *Language development in Southern Africa: Making the right choices.* Windhoek, Namibia, Gamsberg: Macmillan Press.

TSHABALALA, L. (2012). Exploring Language Issues in Multilingual Classrooms. *Learning and Teaching Mathematics, A Journal of AMESA*, No. 13: 22-25.

TSHUMA, L. (2017). Multiple levels and aspects of language competency in English and Intermediate Phase mathematics teachers: An analysis of the case of the Eastern Cape Province. Unpublished PhD Thesis. Stellenbosch: University of Stellenbosch.

TSHUMA, L. (2016). Relationship between language competency and Intermediate Phase Mathematics Instruction. *Proceedings of the 13th International Congress on Mathematics Education (ICME 13) 337(1047).* Hamburg.

U

UNITED Nations Development Programme (UNDP). (2004). *Human Development Report 2004: Cultural Liberty in* Today's *Diverse World.* New York: Oxford University Press, Clarendon Press.

UNITED Nations Educational, Scientific and Cultural Organisation (UNESCO), (2016). Policy Paper 24: If you don't understand, how can you learn? *Global Education Monitoring Report.*

UYS, M, J.,Van Der Walt, C., Van Den Berg, R. & Botha, S. (2007). English medium of instruction: a situation analysis. *South African Journal of Education*, 27(1): 69-82.

V

VACCA, J. A., Vacca, R. T., Gove, M. K., Burkey, L. C., Lenhart, L. A. & McKeon, C. A. (2008). *Reading and learning to read.* New York: Allyn & Bacon.

VAN der Berg, S., Burger, C., Burger, R., Vos, M. D., Randt, R. D., Gustafsson, M., Shepherd, D., Spaull, N., Taylor, S., Van Broekhuizen, H. & Von Fintel, D. (2011). *Low quality education as a poverty trap,* Stellenbosch: University of Stellenbosch, Department of Economics. Research report for the PSPPD project for Presidency. https://doi.org/10.2139/ssrn.2973766

VAN de Walle, J. A. (2007). *Elementary and middle school mathematics (6th Edition).* Boston: Pearson.

VAN der Schoot, M., Bakker, A. H., Horsley, T. M. & Van Lieshout E. C. D. M. (2009). *The consistency effect depends on markedness in less successful but not successful problem solvers: an eye movement study in primary school children. Contemp. Educ. Psychol.*, 34: 58-66. https://doi.org/10.1016/j.cedpsych.2008.07.002

VAN der Walt, C. (2016). Your university future is English – or is it? *Stellenbosch University Faculty of Education Research Bulletin February 2016*, 37-38.

VAN Dooren, W., Verschaffel, L. & Onghena P. (2002). The impact of pre-service teachers' content knowledge on their evaluations of students' strategies for solving arithmetic and algebra word problems. *J. Res. Math. Educ.*, 33: 319-351. https://doi.org/10.2307/4149957

VAN Dyk, T. J. (2005). Towards providing effective academic literacy intervention. *Per Linguam* 21(2): 38-51. https://doi.org/10.5785/21-2-75

VAN Lier, L. (1996). *Interaction in the curriculum: Awareness, autonomy & authenticity.* London: Addison Wesley Longman.

VAN Patten, B. (2003). *From input to output:* A teacher's *guide to second language acquisition.* Boston: McGraw-Hill.

VAN Patten, B. & Williams, J. (2007). *Theories in second language acquisition: an introduction.* Mahwah, New Jersey: Lawrence Erlbaum Associates.

VAN Rinsveld, A., Brunner, M., Landerl, K., Schiltz, C., & Ugen, S. (2015). The relation between language and arithmetic in bilinguals: Insights from different stages of language acquisition. *Frontiers in Psychology*, 6(265): 1-15. https://doi.org/10.3389/fpsyg.2015.00265

VAN Staden, S. & Bosker, R. (2014). Factors that can affect South African reading literacy achievement: Evidence from prePIRLS 2011. *South African Journal of Education*, 34 (3): 131-137. https://doi.org/10.15700/201409161059

VERGNAUD, G. (2009). The theory of conceptual fields. *Hum. Dev.*, 5: 83-94. https://doi.org/10.1159/000202727

VERGUTS, T. & Fias, W. (2005). Interacting neighbors: a connectionist model of retrieval in single-digit multiplication. *Mem. Cogn.*, 33: 1-16. https://doi.org/10.3758/BF03195293

VARUGHESE, N. A. & Glencross, M. (1996). Mathematical language among first year university students. Paper presented at the Fourth Annual Meeting of the Association for Research in Mathematics, Science and Technology Education, Pietersburg, South Africa.

VERSCHAFFEL, L., & De Corte, E. (1990). "Do non-semantic factors also influence the solution process of addition and subtraction word problems?" In: H. Mandl, E. De Corte, N. Bennett & N. H. H. F. Friedrich (Eds.), *Learning and Instruction. European Research in an International Context*. Oxford: Pergamon, 415-429.

VERSCHAFFEL, L., De Corte, E. & Pauwels A. (1992). Solving compare problems: an eye movement test of Lewis and Mayer's consistency hypothesis. *J. Educ. Psychol.*, 84: 85-94. https://doi.org/10.1037/0022-0663.84.1.85

VERSCHAFFEL, L., Greer, B. & De Corte, E. (2000). *Making Sense of Word Problems*. Netherlands: Swets and Zeitlinger.

VERSCHAFFEL, L., Van Dooren, W., Greer, B. & Mukhopadhyay, S. (2010). Reconceptualising word problems as exercises in mathematical modelling. *Journal für Mathematik-Didaktik*, 31: 9-29. https://doi.org/10.1007/s13138-010-0007-x

VICENTE S., Orrantia, J. & Verschaffel, L. (2007). Influence of situational and conceptual rewording on word problem solving. *Br. J. Educ. Psychol.*, 77: 829-848. https://doi.org/10.1348/000709907X178200

VIRGINIA Department of Education. (2006). *English strategies for teaching limited English proficiency (LEP) students: A supplemental resource guide to the K-12 English standards of learning enhanced scope and sequence.* Richmond, Virginia.

VITHAL, R. (1992). The construct of ethnomathematics and Implications for curriculum thinking in South Africa. Unpublished MPhil Thesis, Cambridge: University of Cambridge.

VON Tetzchner, S. (2015). The semiotics of aided language development. *Cognitive Development* 36: 180-190. https://doi.org/10.1016/j.cogdev.2015.09.009

VORSTER, H. (2008). Investigating a scaffold to code switching as a strategy in multilingual classrooms. *Pythagoras*, 67 (June): 33-41. https://doi.org/10.4102/pythagoras.v0i67.72

VORSTER, C., Mayet, A. & Taylor, S. (2013). *Creating Effective Schools. The language of teaching and learning in South African schools.* Cape Town: Pearson.

VYGOTSKY, L. S. (1978). *Mind in society: The development of higher psychological process.* Cambridge, MA: Harvard University Press.

W

WALTER, S. & Benson, C. (2012). Language policy and medium of instruction in formal education. In: B. Spolsky (Ed.), *The Cambridge Handbook of Language Policy*, 278-300. Cambridge: Cambridge University Press. https://doi.org/10.1017/CBO9780511979026.017

WALTER, S. & Chuo, G. (2012). *The Kom Experimental Mother Tongue Education Pilot Project.*

WALTON, J. R. (2002). Global-mindedness and communication competency: teachers in multicultural classrooms. *Journal of Research on Minority Affairs*, April, 19: 50-81.

WATSON, I. (1980). Investigating errors of beginning mathematicians. *Educational Studies in Mathematics*, 11(3): 319-329. https://doi.org/10.1007/BF00697743

WEBB, P. & Roberts, N. (2017). (Eds.). *The pedagogy of Mathematics in South Africa: Is there a unifying pedagogy?* Johannesburg: Mapungubwe Institute for Strategic Reflection (MISTRA) and Real African Publishers.

WEIDEMAN, A. & Rensburg, C. (2000). Language proficiency current strategies, future, tendencies. *Journal for Language Teaching*, 36 (1&2): 152-164. https://doi.org/10.4314/jlt.v36i1-2.6010

WESSELS, D. C. J. (1990). The identification and the origins of a number of basic concepts in the teaching of mathematics: Implications for concept teaching. *Reader on the Role of Language in the Teaching and Learning of Mathematics.* Pretoria: UNISA Publishing Services.

WESTBY, C. (1994). Communication refinement in school age and adolescence: In: D. Van Rooyen & H. Jordaan. An aspect of language for academic purposes in secondary education: complex sentence comprehension by learners in an integrated Gauteng school. *South African Journal of Education*, 29 (2): 271-287. https://doi.org/10.15700/saje.v29n2a260

WILKINS, J. L., Baroody A. J. & Tiilikainen S. (2001). Kindergartners' understanding of additive commutativity within the context of word problems. *J. Exp. Child Psychol.*, 79: 23-36. https://doi.org/10.1006/jecp.2000.2580

WILLIS, S. (1998). *Which numeracy?* Unicorn, 24(2): 32-42.

WILLIS, G. B. & Fuson, K. C. (1988). Teaching children to use schematic drawings to solve addition and subtraction problems. *Journal of Educational Psychology*, 80, 192-201. https://doi.org/10.1037/0022-0663.80.2.192

WOOD, G., Nuerk, H.C., Moeller, K., Geppert, B., Schnitker, R. & Weber J. (2008). All for one, but not one for all. How multiple number representations are recruited in one numerical task. *Brain Res.*, 1187: 154-166. https://doi.org/10.1016/j.brainres.2007.09.094

WORLD Bank, (2005). Education Sector Strategy Update: Achieving Education For All, Broadening our Perspective, Maximizing our Effectiveness: Available on: http://siteresources.worldbank.org/EDUCATION/Resources/ESSU/Education_Sector_Strategy_Update.pdf. Accessed 07 April 2020.

Y

YEAP, B. H. & Kaur, B. (2001). *Semantic characteristics that make arithmetic word problems difficult.* Proceedings of the 24th Annual Conference of the Mathematics Education Research Group of Australasia Incorporated (MERGA) on "Numeracy and Beyond," Australia.

YUSHAU B, & Bokhari, M. A. (2005). Language and Mathematics: A Mediational Approach to Bilingual Arabs. *Electronic Journal of International Journal for Mathematics Teaching and Learning*, April 13th, www.cimt.plymouth.ac.uk.

Z

ZARNHOFER S., Braunstein V., Ebner F., Koschutnig K., Neuper C. & Ninaus M. (2013). Individual differences in solving arithmetic word problems. *Behav. Brain Funct.*. 9: 28. https://doi.org/10.1186/1744-9081-9-28

ZEVENBERGEN, R. J. (2001). Language, social class and underachievement in school mathematics. In: P. Gates (Ed.), *Issues in teaching mathematics* (pp. 38-50). London, UK: Routledge/Falmer.

ZUBER J., Pixner S., Moeller K., Nuerk H.-C. (2009). On the language-specificity of basic number processing: transcoding in a language with inversion and its relation to working memory capacity. *J. Exp. Child Psychol.*, 102: 60-77. https://doi.org/10.1016/j.jecp.2008.04.003 https://www.gooverseas.com/blog/tips-for-teaching-with-limited-classroom-resources. Last Accessed 25 March 2020. https://ealjournal.org/2016/07/26/what-is-translanguaging/ Last Accessed 13 March 2020.

www.ingramcontent.com/pod-product-compliance
Lightning Source LLC
LaVergne TN
LVHW080330110826
845155LV00024B/138

* 9 7 8 1 9 2 8 4 8 0 9 6 9 *